Functional Tactile Sensors

Woodhead Publishing Series in Electronic and Optical Materials

Functional Tactile Sensors

Materials, Devices and Integrations

Edited by

Ye Zhou
Ho-Hsiu Chou

Woodhead Publishing is an imprint of Elsevier
The Officers' Mess Business Centre, Royston Road, Duxford, CB22 4QH, United Kingdom
50 Hampshire Street, 5th Floor, Cambridge, MA 02139, United States
The Boulevard, Langford Lane, Kidlington, OX5 1GB, United Kingdom

© 2021 Elsevier Ltd. All rights reserved.

No part of this publication may be reproduced or transmitted in any form or by any means, electronic or mechanical, including photocopying, recording, or any information storage and retrieval system, without permission in writing from the publisher. Details on how to seek permission, further information about the Publisher's permissions policies and our arrangements with organizations such as the Copyright Clearance Center and the Copyright Licensing Agency, can be found at our website: www.elsevier.com/permissions.

This book and the individual contributions contained in it are protected under copyright by the Publisher (other than as may be noted herein).

Notices
Knowledge and best practice in this field are constantly changing. As new research and experience broaden our understanding, changes in research methods, professional practices, or medical treatment may become necessary.

Practitioners and researchers must always rely on their own experience and knowledge in evaluating and using any information, methods, compounds, or experiments described herein. In using such information or methods they should be mindful of their own safety and the safety of others, including parties for whom they have a professional responsibility.

To the fullest extent of the law, neither the Publisher nor the authors, contributors, or editors, assume any liability for any injury and/or damage to persons or property as a matter of products liability, negligence or otherwise, or from any use or operation of any methods, products, instructions, or ideas contained in the material herein.

Library of Congress Cataloging-in-Publication Data
A catalog record for this book is available from the Library of Congress

British Library Cataloguing-in-Publication Data
A catalogue record for this book is available from the British Library

ISBN: 978-0-12-820633-1 (print)
ISBN: 978-0-12-820912-7 (online)

For information on all Woodhead publications
visit our website at https://www.elsevier.com/books-and-journals

Publisher: Matthew Deans
Acquisitions Editor: Kayla Dos Santos
Editorial Project Manager: Amy Moone
Production Project Manager: Vignesh Tamil
Cover Designer: Christian J. Bilbow

Typeset by SPi Global, India

Contents

Contributors	ix
Preface	xiii

1   Introduction to tactile sensors — **1**
Hongye Chen and Ye Zhou
1.1 Introduction	1
1.2 The principles of tactile sensors	1
1.3 Past trends and advancements	6
References	10

2   Resistive tactile sensors — **13**
Yue Li, Lu Zheng, Xuewen Wang, and Wei Huang
2.1 Introduction	13
2.2 Principle of resistive tactile sensors	14
2.3 Pressure detection of resistive tactile sensors	19
2.4 Multi-functional tactile sensors	22
2.5 Discussion and outlooks	23
References	26

3   Tactile sensor based on capacitive structure — **31**
Ho-Hsiu Chou and Wen-Ya Lee
3.1 Introduction	31
3.2 Resistive pressure sensors	32
3.3 Capacitive pressure sensors	40
3.4 Perspective	50
References	51

4   Tactile sensors based on organic field-effect transistors — **53**
Wen-Ya Lee
4.1 Introduction	53
4.2 Operation principle of field-effect transistors	53
4.3 Tactile field-effect transistors	55
4.4 Multifunctional transistor-based pressure sensors	57
4.5 Perspective	62
References	64

5	**Conductive composite-based tactile sensor**	**67**
	Haotian Chen and Haixia Zhang	
	5.1 Introduction	**67**
	5.2 Working mechanism	**68**
	5.3 Preparation methods	**70**
	5.4 Applications	**80**
	5.5 Summary	**88**
	References	**88**
6	**Mechanoluminescent materials for tactile sensors**	**91**
	Dengfeng Peng and Sicen Qu	
	6.1 Background	**91**
	6.2 Mechanoluminescence materials for tactile sensors	**93**
	6.3 Conclusions and prospective	**107**
	Acknowledgments	**108**
	References	**108**
7	**Mechanophores in polymer mechanochemistry: Insights from single-molecule experiments and computer simulations**	**113**
	Wenjin Li	
	7.1 Introduction	**113**
	7.2 Brief theory of mechanochemistry	**115**
	7.3 Single-molecule approaches	**115**
	7.4 Computational approaches	**118**
	7.5 Covalent mechanophores	**123**
	7.6 Organometallic mechanophores	**128**
	7.7 The effect of polymer chain	**131**
	7.8 Conclusions and perspectives	**133**
	Acknowledgments	**133**
	References	**134**
8	**Perovskites for tactile sensors**	**141**
	Rohit Saraf and Vivek Maheshwari	
	8.1 Introduction	**141**
	8.2 Background on need for self-powered sensors	**142**
	8.3 Polarization effects in perovskites and their basic properties	**143**
	8.4 Design of perovskite-based light-powered tactile sensors	**150**
	8.5 Future vision and challenges	**154**
	References	**154**
9	**Electrospun nanofibers for tactile sensors**	**159**
	Yichun Ding, Obiora Onyilagha, and Zhengtao Zhu	
	9.1 Introduction	**159**

9.2	Electrospinning and electrospun nanofibers	**160**
9.3	Transduction mechanisms of tactile sensor	**164**
9.4	Tactile sensors from electrospun nanofibrous materials	**167**
9.5	Conclusions and perspective	**184**
	Acknowledgments	**186**
	References	**186**

10 Tactile sensors based on buckle structure — **197**

Yuhuan Lv, Mingti Wang, Lizhen Min, and Kai Pan

10.1	Introduction	**197**
10.2	Buckle structure in tactile sensor	**197**
10.3	Methods of buckle structures	**198**
10.4	Conductive materials for buckled tactile sensor	**206**
10.5	Overview	**214**
	References	**215**

11 Tactile sensors based on ionic liquids — **219**

Yapei Wang, Naiwei Gao, and Yonglin He

11.1	Introduction of ionic liquids	**219**
11.2	Pressure sensors based on ionic liquids	**222**
11.3	Temperature sensors based on ionic liquids	**226**
11.4	Signal separation and integration of tactile sensors based on ionic liquids	**231**
11.5	Preventing the leakage of ionic liquid-based sensors	**235**
11.6	Summary and outlook of the tactile sensor based on ionic liquid	**241**
	References	**241**

12 Self-powered flexible tactile sensors — **245**

Xuan Zhang and Bin Su

12.1	Introduction	**245**
12.2	Flexible piezoelectric nanogenerators	**246**
12.3	Flexible triboelectric nanogenerators	**250**
12.4	Flexible, self-powered magnetoelectric elastomers	**255**
12.5	Conclusion and challenges	**256**
	Acknowledgements	**257**
	References	**257**

13 Self-healable tactile sensors — **263**

Jinqing Wang, Xianzhang Wu, Zhangpeng Li, and Shengrong Yang

13.1	Introduction	**263**
13.2	The structure and functional materials of self-healing tactile sensors	**264**
13.3	Self-healing mechanisms of tactile sensors	**273**

13.4 Applications of self-healing tactile sensors	**277**
13.5 Outlook and future challenges	**283**
Acknowledgment	**284**
References	**284**

Index   **291**

Contributors

Haotian Chen Laboratory for Soft Bioelectronic Interfaces, Centre for Neuroprosthetics, École Polytechnique Fédérale de Lausanne (EPFL), Geneva, Switzerland

Hongye Chen Institute for Advanced Study, Shenzhen University, Shenzhen, China

Ho-Hsiu Chou Department of Chemical Engineering, National Tsing Hua University, Hsinchu, Taiwan

Yichun Ding CAS Key Laboratory of Design and Assembly of Functional Nanostructures, Fujian Provincial Key Laboratory of Nanomaterials, Fujian Institute of Research on the Structure of Matter, Chinese Academy of Sciences, Fuzhou, China; Biomedical Engineering Program, South Dakota School of Mines & Technology, Rapid City, SD, United States

Naiwei Gao Department of Chemistry, Renmin University of China, Beijing, China

Yonglin He Department of Chemistry, Renmin University of China, Beijing, China

Wei Huang Institute of Flexible Electronics, Northwestern Polytechnical University, Xi'an, China

Wen-Ya Lee Department of Chemical Engineering & Biotechnology, National Taipei University of Technology, Taipei, Taiwan

Wenjin Li Institute for Advanced Study, Shenzhen University, Shenzhen, China

Yue Li Institute of Flexible Electronics, Northwestern Polytechnical University, Xi'an, China

Zhangpeng Li State Key Laboratory of Solid Lubrication, Lanzhou Institute of Chemical Physics, Chinese Academy of Sciences, Lanzhou; Qingdao Center of Resource Chemistry & New Materials, Qingdao, China

Yuhuan Lv Beijing Key Laboratory of Advanced Functional Polymer Composites, State Key Laboratory of Organic-Inorganic Composites, College of Materials Science and Engineering, Beijing University of Chemical Technology, Beijing, China

Vivek Maheshwari Department of Chemistry, Waterloo Institute for Nanotechnology, University of Waterloo, Waterloo, ON, Canada

Lizhen Min Beijing Key Laboratory of Advanced Functional Polymer Composites, State Key Laboratory of Organic-Inorganic Composites, College of Materials Science and Engineering, Beijing University of Chemical Technology, Beijing, China

Obiora Onyilagha Nanoscience and Nanoengineering Program, South Dakota School of Mines & Technology, Rapid City, SD, United States

Kai Pan Beijing Key Laboratory of Advanced Functional Polymer Composites, State Key Laboratory of Organic-Inorganic Composites, College of Materials Science and Engineering, Beijing University of Chemical Technology, Beijing, China

Dengfeng Peng School of Physics and Optoelectronic Engineering, Shenzhen University, Shenzhen, China

Sicen Qu Department of Physical Education, Shenzhen University, Shenzhen, China

Rohit Saraf Department of Chemistry, Waterloo Institute for Nanotechnology, University of Waterloo, Waterloo, ON, Canada

Bin Su State Key Laboratory of Material Processing and Die & Mould Technology, School of Materials Science and Engineering, Huazhong University of Science and Technology, Wuhan, Hubei, PR China

Jinqing Wang State Key Laboratory of Solid Lubrication, Lanzhou Institute of Chemical Physics, Chinese Academy of Sciences, Lanzhou; Center of Materials Science and Optoelectronics Engineering, University of Chinese Academy of Sciences, Beijing, China

Mingti Wang Beijing Key Laboratory of Advanced Functional Polymer Composites, State Key Laboratory of Organic-Inorganic Composites, College of Materials Science and Engineering, Beijing University of Chemical Technology, Beijing, China

Xuewen Wang Institute of Flexible Electronics, Northwestern Polytechnical University, Xi'an, China

Yapei Wang Department of Chemistry, Renmin University of China, Beijing, China

Xianzhang Wu State Key Laboratory of Solid Lubrication, Lanzhou Institute of Chemical Physics, Chinese Academy of Sciences, Lanzhou; Center of Materials Science and Optoelectronics Engineering, University of Chinese Academy of Sciences, Beijing, China

Shengrong Yang State Key Laboratory of Solid Lubrication, Lanzhou Institute of Chemical Physics, Chinese Academy of Sciences, Lanzhou; Center of Materials Science and Optoelectronics Engineering, University of Chinese Academy of Sciences, Beijing, China

Haixia Zhang National Key Laboratory of Nano/Micro Fabrication Technology, Institute of Microelectronics, Peking University, Beijing, China

Xuan Zhang State Key Laboratory of Material Processing and Die & Mould Technology, School of Materials Science and Engineering, Huazhong University of Science and Technology, Wuhan, Hubei, PR China; Department of Chemical Engineering, ARC Hub for Computational Particle Technology, Monash University, Clayton, VIC, Australia

Lu Zheng Institute of Flexible Electronics, Northwestern Polytechnical University, Xi'an, China

Ye Zhou Institute for Advanced Study, Shenzhen University, Shenzhen, China

Zhengtao Zhu Biomedical Engineering Program; Nanoscience and Nanoengineering Program; Department of Chemistry, Biology, and Health Sciences, South Dakota School of Mines & Technology, Rapid City, SD, United States

Preface

Tactile sensor acts as a source of information regarding physical contact with the external world. This technology has obtained considerable attentions for potential applications in healthcare monitoring, skin prosthetics, and biomimic robotics. In order to extend the practical applications of tactile sensors, a high sensitivity and wide linearity range are of significant importance in sensing the force. On the other aspect, tactile sensors with multifunctional integration are highly desirable to satisfy various application requirements. This book will cover the functional material selection, device structure design, and integration of tactile sensors.

We hope the readers can systematically learn and understand the design, materials characteristics, device operation principles, specialized device application, and mechanisms of the latest reported tactile sensors in our book. The challenge of designing and constructing an excellent tactile sensor is a difficult balancing act. New phenomena will be discovered in the future, and more novel device applications will emerge.

We would like to acknowledge all the contributed authors in this book. We want to express our gratitude to Kayla Dos Santos, Vignesh Tamilselvvan, Narmatha Mohan, Amy Moone, and Ana Claudia Abad Garcia at ELSEVIER for all the help during the book editorial process. We also want to thank all the readers for their interest in our book on tactile sensors. It is expected that our book is well-timed and instrumental for the rapid progress of academic sector of tactile sensor as well as artificial intelligence.

Ye Zhou
Ho-Hsiu Chou

Introduction to tactile sensors

Hongye Chen and Ye Zhou[*]
Institute for Advanced Study, Shenzhen University, Shenzhen, China
[*]Corresponding author: e-mail address: yezhou@szu.edu.cn

1.1 Introduction

As humans, we can feel smoothness when we touch silk; we can feel coolness in the summer when we touch ice; we can feel softness when we touch a baby's face; we can tell weight when we hold a book. Through skin, by physical contact or touch, human beings can tell texture, temperature, softness, and even weight. What if robots could do the same? For robots, the functions of tactile sensing are similar to the aforementioned feedback during exploring and interacting with objects.

This chapter provides an introduction to tactile sensors, beginning with a review of the pros and cons of tactile sensors based on different transduction techniques with a focus on flexible and stretchable physical sensors. It also discusses the evolution of sensors over the decades and their use in various applications, such as artificial skin.

1.2 The principles of tactile sensors

Tactile sensors are based on various principles, including piezoresistivity, piezoelectricity, capacitance, optoelectricity, strain gauge, and so on. In the following sections, we list the advantages and disadvantages of three typical principles and related applications, as shown in Table 1.1. Moreover, we introduce the materials relevant to these principles.

1.2.1 Capacitance type

A capacitor consists of two conductive plates with an insulating material (dielectric) sandwiched between them. The capacitance C can be expressed as $C = A\varepsilon_0\varepsilon_r/d$, where A is the area of the overlap of the two plates, ε_0 is the vacuum permittivity, ε_r is the relative static permittivity of the dielectric, and d is the thickness of the dielectric. In order to obtain high sensitivity, elastomeric dielectrics with small modulus, such as polydimethylsiloxane (PDMS) and polyurethane (PU), are ideal for the dielectric layer.

Recently, the properties of transparence and stretchability have become relatively popular in pressure sensors. Sun et al. [1] developed a transparent pressure sensor that is sandwiched between two stretchable ionic conductors (Fig. 1.1A). Because of its biocompatibility and excellent stretching ability (1%–500%), this device is promising for use in wearable or implantable devices like E-skin (Fig. 1.1B and C). To enhance

---

Functional Tactile Sensors. https://doi.org/10.1016/B978-0-12-820633-1.00011-5
© 2021 Elsevier Ltd. All rights reserved.

Table 1.1 The pros and cons of various principles of tactile sensors and their applications.

Principle	Advantage	Disadvantage	Application
Capacitance	Good frequency response High spatial resolution Large dynamic range	Susceptible to noise	Pressure sensor [1, 2] Strain sensor [3, 4]
Piezoresistivity	Less sensitive to noise	Hysteresis Low-frequency response	Pressure sensor [5–9] Strain sensor [10–12]
Piezoelectricity	High-frequency response	No measurement on static force	Pressure sensor [13, 14] Strain sensor [15, 16]

sensitivity, Mannsfeld et al. [2] presented a highly sensitive and flexible pressure sensor. They utilized biocompatible elastomer PDMS as the insulating layer, which is one of the key layers in transistor structure. Moreover, their devices had ultra-high sensitivity that surpassed that of previous works. Excellent sensitivity and fast response properties are necessary in prosthetic skin.

In a similar way, for transparent strain sensors, materials like single-walled carbon nanotubes (SWCNTs) and silver nanowires (Ag NWs) are widely used. Cai et al. [3] designed a CNT-based capacitive strain sensor that can detect strain from 100% to 300%. This sensor presented with high sensitivity and superior stability and reliability, which makes is useful for prototypical data glove and respiration monitor applications. Yao et al. [4] reported multifunctional strain sensors based on stretchable Ag NW electrodes that can detect thumb movement and knee joint movement. Multifunctional sensors have potential to be integrated with other wearable devices.

1.2.2 Piezoresistivity type

A pressure-sensitive element whose resistance varies with applied force constitutes piezoresistive sensors. The voltage-current characteristic can be expressed as $V = IR$, where V is voltage, I is current, and R is the resistance of the material. In terms of pressure sensors, it is common that combining conductive polymers with elastomers like PDMS can fabricate piezoresistive sensors. Nevertheless, this kind of composite elastomer with a planar structure has low sensitivity. Different structures have been made to overcome this obstacle. Choong et al. [5] utilized a micropyramidal PDMS array coated by a conductive elastomeric composite that consists of poly(3, 4-ethylenedioxythiophene) polystyrene sulfonate (PEDOT:PSS) and a polyurethane (PU) dispersion (Fig. 1.1D). Compared to an unstructured film, this structure achieved

Introduction to tactile sensors 3

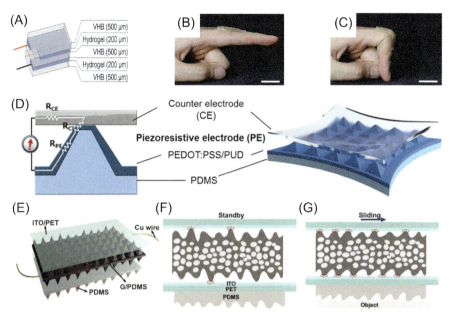

Fig. 1.1 (A) A strain senor consisting of a layer of stretchable dielectric sandwiched between two layers of a stretchable ionic conductor connected to two metallic electrodes. The device was covered with two additional layers of VHB. (B) The strain sensor was attached to a straight finger. (C) The bending of the finger stretched the strain sensor. (D) Circuit model used to derive the sensing principle of the sensor. The structure of the device. (E) The schematic diagram of the skin sensor. Panels (F) and (G) show the mechanism of the skin sensor in recognizing texture roughness.
Reproduced with permission from J.Y. Sun, C. Keplinger, G.M. Whitesides, Z. Suo, Ionic skin, Adv. Mater. 26 (45) (2014) 7608–7614; C.L. Choong, M.B. Shim, B.S. Lee, S. Jeon, D.S. Ko, T.H. Kang, J. Bae, S.H. Lee, K.E. Byun, J. Im, Y.J. Jeong, C.E. Park, J.J. Park, U.I. Chung, Highly stretchable resistive pressure sensors using a conductive elastomeric composite on a micropyramid array, Adv. Mater. 26 (21) (2014) 3451–3458; Q.-J. Sun, X.-H. Zhao, Y. Zhou, C.-C. Yeung, W. Wu, S. Venkatesh, Z.-X. Xu, J.J. Wylie, W.-J. Li, V.A.L. Roy, Fingertip-skin-inspired highly sensitive and multifunctional sensor with hierarchically structured conductive graphite/polydimethylsiloxane foams, Adv. Funct. Mater. 29 (18) (2019) 1808829.

a sensitivity of 4.88 kPa^{-1} over range of pressure from 0.37 to 5.9 kPa. Luo et al. [6] used a sponge structure consisting of a high-elastic graphene/PU nanocomposite, which showed relatively greater sensitivity (0.75–3.08 kPa^{-1}) [6]. Recently, Sun et al. [7] invented an ultrasensitive and multifunctional tactile sensor with hierarchical microstructure that showed a high sensitivity of 245 kPa^{-1} and can detect subtle wrist pulse as well as human motion. This sensor consisted of conductive graphite/PDMS foam films sandwiched by two ITO/PET films (Fig. 1.1E). Through optimizing the graphite concentration and porosity, this tactile sensor achieved high sensitivity. Moreover, combining it with a rough PDMS layer allows it to detect the roughness of an object (Fig. 1.1F and G).

However, instead of just one-time feedback, human could collect the feedback information and storage them. Bioinspired neuromorphic systems for touch sensing are popular in artificial intelligence systems such as E-skin. Hence, Zhang et al. [8] put forward the integration of a pressure sensor and a Nafion-based memristor (Fig. 1.2A). In this device, the memristor served as an artificial synapse to process information and imitate the perceptual learning process. The micropyramidal structure of Au/PDMS/PET enhanced the sensitivity (6.7×10^7 kPa^{-1} in 1–5 kPa and 3.8×10^5 kPa^{-1} in 5–50 kPa) (Fig. 1.2B and C), while the memristor had low power consumption (10–200 pJ).

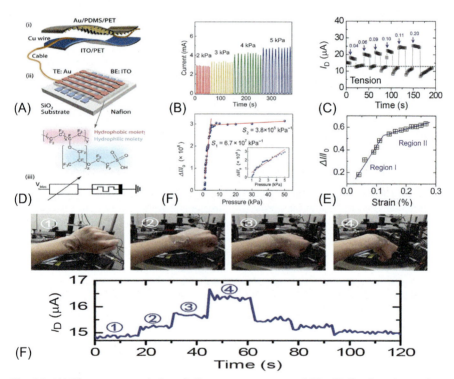

Fig. 1.2 (A) The system consisting of: (i) a pressure sensor and (ii) a Nafion-based memristor; (iii) the corresponding electrical circuit. (B) Current responses of pressure sensor under pressure pulses from 2 to 5 kPa. (C) Current variation of sensor under pressures ranging from 1 to 50 kPa. *Inset*: enlarged graph of the sensitivity responses in 1–5 kPa range. (D) Output I_D signals as a function of time obtained from a piezopotential-gated GT strain sensor under different tensile strains. (E) Sensitivity characteristics of the piezopotential-gated GT strain sensor. (F) Hand movement monitoring by a conformal piezopotential-gated GT strain sensor fabricated on a PDMS substrate.
Reproduced with permission C. Zhang, W.B. Ye, K. Zhou, H.Y. Chen, J.Q. Yang, G. Ding, X. Chen, Y. Zhou, L. Zhou, F. Li, S.T. Han, Bioinspired artificial sensory nerve based on nafion memristor, Adv. Funct. Mater. 29 (20) (2019) 1808783; Q. Sun, W. Seung, B.J. Kim, S. Seo, S.W. Kim, J.H. Cho, Active matrix electronic skin strain sensor based on piezopotential-powered graphene transistors, Adv. Mater. 27 (22) (2015) 3411–3417.

Introduction to tactile sensors

For application in robotics, electronic skin, and human motion detection, strain sensors based on piezoresistivity are also useful. Most structures applied in pressure sensors, such as sponges [6] and hierarchical microstructures [7, 9], are also suitable for strain sensors. In addition, a sandwiched structure is common in strain sensors. Fan et al. designed a sandwiched structure where PEDOT:PSS and Ag NWs were embedded into PDMS [10]. The device presented a reliable response in a cyclic stretching-releasing test and a high sensitivity of 6.5–8.0. Thanks to the excellent properties of PEDOT:PSS and Ag NWs, this strain sensor has advantages of optical transparency, light weight, and high flexibility. Graphene also has good optical transparency and electrical properties. With applied tensile strain, the hexagonal structure near the edge of the graphene film will be destroyed, which leads to the changed resistance. Bae et al. [11] reported a transparent strain sensor based on CVD graphene with transparency of 75%–80%. Attached on a hand glove, this sensor detects finger motion simultaneously. Nevertheless, these materials may break down under high strain. Therefore, highly stretchable multi-walled carbon nanotube (MWCNT) strain sensors were developed by Tadakaluru et al. [12]. These stretchable sensors not only had high stretchability (620%) and good reliability (400 cycle for 150%–500% strain range), but also exhibited multifunctionality, measuring speed, acceleration, and frequency. These sensors was applied in earthquake detection and health monitoring.

1.2.3 Piezoelectricity type

The change of resistance in piezoelectric tactile sensor depends on how large a voltage potential is generated when deforming the crystal lattice. For various materials, especially certain crystals, sensitivity depends on crystal structure.

Piezoelectric-based pressure sensors rely on the piezoelectric effect when dipoles form an internal polarization under pressure. It can transfer mechanical energy to electrical energy. Compared to other strain sensors, piezoelectric strain sensors exhibit high sensitivity and low power consumption. Usually, piezoelectric materials are lead zirconate titanate (PZT) [13] and the polymer P(VDF-TrFE) [14]. In addition, excellent sensitivity is necessary for human motion monitoring and robotics. Dagdeviren et al. introduced a skin-mounted pressure sensor that relied on PZT [13], which presented fast response time (0.1 ms) and high sensitivity (0.005 Pa). Combined with a transistor, these devices were able to measure subtle effects from blood pressure to vibration in the throat. In addition, Persano et al. [14] presented a flexible piezoelectric pressure sensor based on P(VDF-TrFE) nanofibers by using the electrospinning method. P(VDF-TrFE) presented a stable piezoelectric crystalline β-phase at room temperature. In this pressure sensor, P(VDF-TrFE) offered ultra-sensitivity for detecting small pressure (0.1 Pa), and it worked in any direction and could be mounted on skin. Moreover, Sun et al. [15] reported flexible active-matrix strain sensors based on piezopotential-gated coplanar-gate graphene transistors. Fig. 1.2D shows that a larger strain produces a higher output current. High sensitivity is shown in Fig. 1.2E; in region I, gauge factor (GF) is 389, which is relatively higher than the state-of-the-art strain sensor. This strain sensor can be attached to human skin to detect human motion (Fig. 1.2F). In addition to common materials like P(VDF-TrFE) [15], zinc oxide nanowires (ZnO NWs) are also suitable for strain sensors. Under applied tensile strain, the two ends will form piezoelectric charges, which leads to a piezoelectric potential inside the ZnO NW. Liao et al. [16] developed a flexible piezotronic strain sensor based on carbon fiber-ZnO NW. The I–V characterization showed a quick response under both static and dynamic mechanical loads.

1.3 Past trends and advancements

In this section, we discuss the evolution of tactile sensors over the decades.

1.3.1 Seed in the 1970s

In 1969, Pfeiffer et al. [17] indicated that tactile sensors had potential to become prosthetic aides, although they still needed a lot of work. As such, they created an experimental device that could detect the presence of force. However, this device could not detect the magnitude of force. In the same era, Stojiljkovic and Clot [18] combined a series of transductors to form "artificial skin" that can detect hardness.

1.3.2 Seedling in the 1980s

In the 1980s, to specify design criteria and match it as much as possible with application, Harmon [19] surveyed the industry with a set of questionnaires and interviews, and identified device and system parameters and configurations focused on robotics needs. For instance, spatial resolution between sensor points should be less than 2 mm; the range of sensor points per fingertip should be 50–200; most required force sensitivity should be at least 1 g; and stable, monotonic, and repeatable performances are desired.

Harmon [19] also discussed materials and structure. Although resistive materials were simple, cheap, heat resistant, and easily fabricated, these materials had problems of hysteresis and insufficient dynamic range. Semiconductors readily permitted preprocessing at the transducer level. However, they were fragile and particularly sensitive to temperature and noise. Piezoresistivity was compatible with silicon technology. Nonetheless, these sensors tended to be stiff, flat, and slippery. Capacitative sensing still needed more powerful materials in the 1980s. Although optical approaches were influenced by variables such as smoke, they were still considered as a possibly useful method that makes use of flexure of diaphragms or bending of waveguides.

In this decade, exploration of various transduction techniques developed rapidly. Nevertheless, the high cost of manufacturing small-scale design in electrical and mechanical devices was a major obstacle to facilitating the development of tactile sensors.

1.3.3 Evolution in the 1990s

It is difficult to find a dissenting view about the importance of tactile sensors to industry in the 1980s. However, most people were overoptimistic. In fact, technical superiority was only one dimension. Industrial managers also considered other factors, such as cost and reconfiguration. Highly structured environments in manufacturing industries made tactile sensors unnecessary. Hence, in the 1990s, researchers changed their focus to applications in hospitals and homes.

The application of tactile sensing in minimally invasive surgery (MIS) was the major focus in this era. The term MIS was first put forward by Wickham [20] in 1984 and published in 1987. MIS, also called endoscopic surgery, was important in medical history. Surgery requires a combination of vision and contact in order to master the condition of patients. Nonetheless, for example, the deep location of lung nodules was difficult to find through thoracoscopy. Hence, by using a new tactile sensor for thoracoscopy, small invisible nodules were easily detected without injuring lung tissue. Ohtsuka et al. [21] developed a tactile sensor that consisted of a sensor probe with round tip, a piezoelectric transducer, a filter, and a frequency counter. Once the sensor probe passed an object harder than the adjacent tissue, the curve of $\triangle f$ would increase dramatically. This sensor was first tested on pigs and later on in humans. In the latter, the sensor detected 10 nodules in 8 patients.

Soft materials may have more desirable characteristics for contact surfaces.

Therefore, soft polymer was attempted later. In 1995, Beebe [22] described a silicon-based piezoresistive force sensor than could be skin-like and withstand large force loadings. The packaged sensor consisted of a silicon-sensing element and Al leads sandwiched between protective skin-like polyimide layers. Finally, researchers put a Torlon dome on the sensor to distribute any applied force and protect the diaphragm. When force was applied on the dome, the sensor could measure the magnitude accurately. The device exhibited a linear response, good repeatability, and low hysteresis with flexible and durable packaging. However, it was still limited on temperature changing and spatial resolution.

Lee and Nicholls [23] predicted that tactile sensors might play an important role in three main fields: surgical applications, rehabilitation and service robotics, and agriculture and food processing. In surgery, soft tissue could only be properly identified by assessing its elasticity and softness, making tactile sensors important. MIS was still promising in the future. With the rapid development of robotics, plenty of advanced robots appeared in people's life. Good stability and safety are the key factors in this area. Thus, it is important that people and robots coexist in a "safe partner" relationship. In this aspect, tactile sensors detected force that was below the thresholds of human pain or damage between any contact of human and robots. In agriculture, high-valued products emphasized zero defects, zero contamination, and zero human error. In addition, some product processing requires an extreme environment that is not suitable for humans. These conditions require robotic handling systems with high sensitivity and dexterity.

1.3.4 Improvement in the 2000s

In the early 21st century, improvement in computation and data processing gave researchers an opportunity to pursue work unfinished in the 1970s and 1980s. Both research and commercial sectors put more emphasis on biomedical applications and tactile sensing systems for unstructured environments. As Lee and Nicholls [23] predicted, a number of attempts were made to improve tactile feedback of MIS, especially regarding force detection. Until now, most surgical operations still rely on the da Vinci Surgical System. The master console of this system allows a surgeon to accurately operate robotic arms with the index finger and thumb of both hands. However, tissue properties vary among patients. Surgeons cannot tell the applied force and tissue characteristics through the da Vinci System. Hence, King et al. [24] mounted a tactile feedback system onto the robotic end effectors, which allowed surgeons to feel the applied force at the grasper. The piezoresistive force sensors and pneumatic balloon tactile display mainly constituted the tactile system. This system was the first complete tactile feedback system known to be applied to a commercial robotic surgical system and was likely to be applied in other applications.

Although Lee and Nicholls predicted that force detection and control on robots would be a trend in 21st century, most attention about force detection and control was paid to prosthetics. Tactile sensors are needed to measure the fitness of prostheses during motion. The fit at the stump-socket interface is necessary. In the past, most fittings were unsuitable, which caused stress and limited healing of wounds. Therefore, Valdastri et al. [25] developed a novel hybrid silicon three-axial force sensor. This device could evaluate the fitting degree of the socket and optimized the shape, which dramatically reduced skin damage. Another problem of prosthetics is lack of feedback. Simple actions like walking posed a huge challenge in acquiring information from prosthetic limbs. Attempts have been made to overcome this challenge. For example, Sabolich et al. [26] invented a prosthetic limb that could provide a sense of feel.

Different from the first three decades, a major success of tactile sensing is in smart phones. Tactile sensors allow users to select and browse through content on a small screen. In terms of smarter touch interfaces and navigation interfaces, design engineers make use of tactile sensors in devices like iPods.

With the appearance of more and more successful companies such as X-sensor (Alberta, Canada), tactile sensing has become more and more important, increasing the acceptance of intelligent skin and wearable devices.

1.3.5 Advancements in the 2010s

Thanks to improved knowledge about feedback in prostheses, flexible and wearable devices have become popular. Non-invasive stimulation techniques have greatly advanced as well. Three major tactile feedback techniques are show in Table 1.2.

Due to recent improvements in soft materials and system integration technologies, wearable flexible hybrid electronics (WFHE) have been designed for human healthcare and human machine interfaces. Lim et al. considered WFHE as

Table 1.2 On-invasive stimulation techniques and their relative advantages and disadvantages.

Stimulation techniques	Mechanism	Advantages	Disadvantages	Application
Electrotactile stimulation	Stimulates afferent nerves in the skin with surface electrodes	Low power consumption Light weight Little noise	Burning pain Interference with EMG and EEG signals	[27]
Vibrotactile stimulation	Modulates vibration frequency, amplitude, duration	No interference with electric signals	The sensitivity varies with condition of subjects	[28, 29]
Mechanotactile stimulation	Provides a force/ pressure or position feedback	A natural feeling of force/ pressure	Large size Heavy High energy consumption	[30]

For in-depth discussion on these techniques, refer to Ref. [31].

mechanically soft, flexible, and stretchable electronic devices that can be tightly attached to the surface of the body [32]. Flexible and light tactile sensors play a key role in WFHE. Various types of wearable tactile sensors utilized the required properties to develop WFHE. For example, wearable temperature sensors attached to skin were able to monitor continuous temperature, allowing for detection of abnormal signals and timely medical diagnostics. Wearable electrophysiological sensors can measure bioelectric signals noninvasively through the skin. Thus, like ECG, EEG, and EMG, all these biopotentials require bioelectric signals to provide detailed information. What's more, through touching biofluids such as tears and sweat, electrochemical sensors can detect target analytes. Advances in WFHE improve yield of integrated skin-like artificial sensory systems. Although tactile sensors have plenty of applications in WFHE, there are still many aspects to be developed, like realization of multi-sensors in daily life.

1.3.6 Future direction in the 2020s

The electronics company IDTechEx [33] refutes the claim that many sensors remain at the stage of commercial evaluation or that there are relatively early commercial sales. As IDTechEx predicts, there will be 3 billion wearable sensors by 2025, 30% of which are just beginning to emerge [34]. A 5 billion sensor market will drive a 160 billion wearable technology market in 2028 [33]. Hence, numerous opportunities exist for future research. IDTechEx reported that the top fastest-growing flexible sensor type in the next decade will be a stretch and pressure sensor. In addition, Han et al. presented a state-of-the-art flexible sensor and gave the future trend according to market analysis and worthy work [35].

For flexible and stretchable tactile sensors, the biocompatibility of materials is a crucial research area, especially for invasive applications. With an aging population, improvement of human lives with prosthetic devices, increased popularity of touch-based commercial and home products, and acceptance of robotic surgery systems in hospitals, tactile sensing is an area ripe for future research and development.

References

[1] J.Y. Sun, C. Keplinger, G.M. Whitesides, Z. Suo, Ionic skin, Adv. Mater. 26 (45) (2014) 7608–7614.

[2] S.C. Mannsfeld, B.C. Tee, R.M. Stoltenberg, C.V. Chen, S. Barman, B.V. Muir, A.N. Sokolov, C. Reese, Z. Bao, Highly sensitive flexible pressure sensors with microstructured rubber dielectric layers, Nat. Mater. 9 (10) (2010) 859–864.

[3] L. Cai, L. Song, P. Luan, Q. Zhang, N. Zhang, Q. Gao, D. Zhao, X. Zhang, M. Tu, F. Yang, W. Zhou, Q. Fan, J. Luo, W. Zhou, P.M. Ajayan, S. Xie, Super-stretchable, transparent carbon nanotube-based capacitive strain sensors for human motion detection, Sci. Rep. 3 (2013) 3048.

[4] S. Yao, Y. Zhu, Wearable multifunctional sensors using printed stretchable conductors made of silver nanowires, Nanoscale 6 (4) (2014) 2345–2352.

[5] C.L. Choong, M.B. Shim, B.S. Lee, S. Jeon, D.S. Ko, T.H. Kang, J. Bae, S.H. Lee, K.E. Byun, J. Im, Y.J. Jeong, C.E. Park, J.J. Park, U.I. Chung, Highly stretchable resistive pressure sensors using a conductive elastomeric composite on a micropyramid array, Adv. Mater. 26 (21) (2014) 3451–3458.

[6] Y. Luo, Q. Xiao, B. Li, Highly compressible graphene/polyurethane sponge with linear and dynamic piezoresistive behavior, RSC Adv. 7 (56) (2017) 34939–34944.

[7] Q.-J. Sun, X.-H. Zhao, Y. Zhou, C.-C. Yeung, W. Wu, S. Venkatesh, Z.-X. Xu, J.J. Wylie, W.-J. Li, V.A.L. Roy, Fingertip-skin-inspired highly sensitive and multifunctional sensor with hierarchically structured conductive graphite/polydimethylsiloxane foams, Adv. Funct. Mater. 29 (18) (2019) 1808829.

[8] C. Zhang, W.B. Ye, K. Zhou, H.Y. Chen, J.Q. Yang, G. Ding, X. Chen, Y. Zhou, L. Zhou, F. Li, S.T. Han, Bioinspired artificial sensory nerve based on Nafion Memristor, Adv. Funct. Mater. 29 (20) (2019) 1808783.

[9] Q.J. Sun, J. Zhuang, S. Venkatesh, Y. Zhou, S.T. Han, W. Wu, K.W. Kong, W.J. Li, X. Chen, R.K.Y. Li, V.A.L. Roy, Highly sensitive and ultrastable skin sensors for biopressure and bioforce measurements based on hierarchical microstructures, ACS Appl. Mater. Interfaces 10 (4) (2018) 4086–4094.

[10] X. Fan, N. Wang, F. Yan, J. Wang, W. Song, Z. Ge, A transfer-printed, stretchable, and reliable strain sensor using PEDOT:PSS/Ag NW hybrid films embedded into elastomers, Adv. Mater. Technol. 3 (6) (2018) 1800030.

[11] S.-H. Bae, Y. Lee, B.K. Sharma, H.-J. Lee, J.-H. Kim, J.-H. Ahn, Graphene-based transparent strain sensor, Carbon 51 (2013) 236–242.

[12] S. Tadakaluru, W. Thongsuwan, P. Singjai, Stretchable and flexible high-strain sensors made using carbon nanotubes and graphite films on natural rubber, Sensors (Basel) 14 (1) (2014) 868–876.

[13] C. Dagdeviren, Y. Su, P. Joe, R. Yona, Y. Liu, Y.S. Kim, Y. Huang, A.R. Damadoran, J. Xia, L.W. Martin, Y. Huang, J.A. Rogers, Conformable amplified lead zirconate titanate sensors with enhanced piezoelectric response for cutaneous pressure monitoring, Nat. Commun. 5 (2014) 4496.

[14] L. Persano, C. Dagdeviren, Y. Su, Y. Zhang, S. Girardo, D. Pisignano, Y. Huang, J.A. Rogers, High performance piezoelectric devices based on aligned arrays of nanofibers of poly(vinylidenefluoride-co-trifluoroethylene), Nat. Commun. 4 (2013) 1633.

[15] Q. Sun, W. Seung, B.J. Kim, S. Seo, S.W. Kim, J.H. Cho, Active matrix electronic skin strain sensor based on piezopotential-powered graphene transistors, Adv. Mater. 27 (22) (2015) 3411–3417.

[16] Q. Liao, M. Mohr, X. Zhang, Z. Zhang, Y. Zhang, H.J. Fecht, Carbon fiber-ZnO nanowire hybrid structures for flexible and adaptable strain sensors, Nanoscale 5 (24) (2013) 12350–12355.

[17] E.A. Pfeiffer, C.M. Rhode, S.I. Fabric, An experimental device to provide substitute tactile sensation from the anaesthetic hand, Med. Biol. Eng. 7 (2) (1969) 191–199.

[18] Z. Stojiljkovic, J. Clot, Integrated behavior of artificial skin, IEEE Trans. Biomed. Eng. 24 (1997) 396–399.

[19] L.D. Harmon, Automated tactile sensing, Int. J. Rob. Res. 1 (2) (1982) 3–32.

[20] J.E. Wickham, The new surgery, Br. Med. J. 29 (1987) 1581–1582.

[21] T. Ohtsuka, A. Furuse, T. Kohno, J. Nakajima, K. Yagyu, S. Omata, Application of a new tactile sensor to thoracoscopic surgery: experimental and clinical study, Ann. Thorac. Surg. 60 (3) (1995) 610–614.

[22] D.J. Beebe, A.S. Hsieh, D.D. Denton, R.G. Radwin, A silicon force sensor for robotics and medicine, Sens. Actuators A Phys. 50 (1) (1995) 55–65.

[23] M. Lee, H. Nicholls, Tactile sensing for mechatronics—a state of the art survey, Mechatronics 9 (1999) 1–31.

[24] C.H. King, M.O. Culjat, M.L. Franco, C.E. Lewis, E.P. Dutson, W.S. Grundfest, J.W. Bisley, Tactile feedback induces reduced grasping force in robot-assisted surgery, IEEE Trans. Haptics 2 (2) (2009) 103–110.

[25] P. Valdastri, S. Roccella, L. Beccai, E. Cattin, A. Menciassi, M.C. Carrozza, P. Dario, Characterization of a novel hybrid silicon three-axial force sensor, Sens. Actuators A Phys. 123–124 (2005) 249–257.

[26] J. Sabolich, G. Ortega, G. Schwabe IV, System and Method for Providing a Sense of Feel in a Prosthetic or Sensory Impaired Limb, US Patent 6,500,210 (2002).

[27] G. Chai, X. Sui, S. Li, L. He, N. Lan, Characterization of evoked tactile sensation in forearm amputees with transcutaneous electrical nerve stimulation, J. Neural Eng. 12 (6) (2015) 066002.

[28] E. Rombokas, C.E. Stepp, C. Chang, M. Malhotra, Y. Matsuoka, Vibrotactile sensory substitution for electromyographic control of object manipulation, IEEE Trans. Biomed. Eng. 60 (8) (2013) 2226–2232.

[29] S. Schätzle, B. Weber, Universal Access in Human-Computer Interaction. Access to Interaction, (2015).

[30] S. Casini, M. Morvidoni, M. Bianchi, M. Catalano, G. Grioli, A. Bicchi, in: 2015 IEEE/RSJ International Conference on Intelligent Robots and Systems (IROS), (IEEE, 2015).

[31] K. Li, Y. Fang, Y. Zhou, H. Liu, Non-invasive stimulation-based tactile sensation for upper-extremity prosthesis: a review, IEEE Sens. J. 17 (9) (2017) 2625–2635.

[32] H.-R. Lim, H.S. Kim, R. Qazi, Y.-T. Kwon, J.-W. Jeong, W.-H. Yeo, Advanced soft materials, sensor integrations, and applications of wearable flexible hybrid electronics in healthcare, energy, and environment, Adv. Mater. 32 (15) (2019) 1901924.

[33] Wearable Sensors, Market Forecasts, Technologies, Players, 2018 IDTechEx, Cambridge, UK, 2018-2028.

[34] Wearable Sensors, Market Forecasts, Technologies, Players, 2016 IDTechEx, Cambridge, UK, 2016-2026.

[35] S.T. Han, H. Peng, Q. Sun, S. Venkatesh, K.S. Chung, S.C. Lau, Y. Zhou, V.A.L. Roy, An overview of the development of flexible sensors, Adv. Mater. 29 (33) (2017).

Resistive tactile sensors

2

Yue Li, Lu Zheng, Xuewen Wang[*]*, and Wei Huang*[*]
Institute of Flexible Electronics, Northwestern Polytechnical University, Xi'an, China
[*]Corresponding authors: e-mail address: iamxwwang@nwpu.edu.cn, provost@nwpu.edu.cn

2.1 Introduction

The tactile sensation of the human body comes directly from the basic sensory units of tactile cells, nerve endings, and tactile corpuscles in skin tissues, which allow us to feel mechanical stimuli like touch, pressure, and pain as well as temperature stimuli (cold or heat). Multiple dimensional perception, which collects various information from external environments and target objects, is one of the most important features for robot intelligence. A tactile sensor is a kind of sensing device used by robots to mimic human tactile functions like sensing the characteristics of objects being touched and the tactile feedback of mechanical stimuli such as impact and compression. Tactile sensors can be broadly divided into several types by their functions, such as pressure sensors [1], force-torque sensors [2–4], slip sensors [5, 6], temperature sensors, and vibration sensors [7]. Up to now, a variety of configurations for tactile sensors has been developed, including capacitive, piezoelectric, resistive, and transistor-based devices. Resistive sensors are the most widely used tactile sensors due to their simple structure, low power consumption, and high performance (e.g., high sensitivity, good stability, wide dynamic and linear range, durability) [8–10]. Recent studies have demonstrated their application in detecting force, acceleration, temperature, friction, and displacement in grasp. However, the hysteresis effect in flexible materials casts a shadow on their applications in fast and real-time detection, because it prolongs the response time of resistive tactile sensors. Temperature and light drift produce errors in detection of pressure due to high carrier concentration in piezo-resistive materials. Scientists are developing novel materials and new device structures to eliminate these barriers and fill the gap in the state of the art of intelligent tactile sensors by multifunctionality.

In this chapter, we briefly summarize tactile sensors based on resistive structure as well as their working principles, functions, and applications. We also examine the working principles of piezo-resistive effect and microstructure-enhanced piezo-resistivity in both semiconductors and conducting composite materials. Thereafter, we discuss single-function tactile sensors in detection of pressure and multifunctional hybrid tactile sensor arrays together with their performance and applications. Finally, the chapter ends with a discussion of the challenges and opportunities of tactile sensors.

Functional Tactile Sensors. https://doi.org/10.1016/B978-0-12-820633-1.00009-7
© 2021 Elsevier Ltd. All rights reserved.

2.2 Principle of resistive tactile sensors

In general, the working principle of a resistive tactile sensor is to transduce external physical information to resistive signals measured by current, voltage, and resistance. The basic principle refers to energy band, carrier dynamics, and electronic transport that tune and change under external stimulations. This section describes the piezo-resistive effect and microstructure-enhanced piezo-resistivity in both semiconductors and conducting composites, and discusses how these effects contribute to resistive tactile sensors as well as their applications.

2.2.1 Piezo-resistive effect of semiconductors

Piezo-resistive effect in semiconductors is a primary principle for designing and fabricating tactile sensors for detecting pressure and strain. It was first observed in germanium (Ge) and silicon (Si) by Smith in 1954. The energy structure and carrier concentration of semiconductors are sensitive to pressure and stress because of lattice deformation [11, 12]. The conductivity and resistivity of semiconductors can be changed under external stress. Due to the anisotropy of material structures, the piezo-resistive effect in semiconductors is anisotropic, which is largely affected by crystal orientation applied by external force. Therefore, real-time detection of the change of conductivity and resistivity provides an effective way to evaluate the force and strains applied to semiconductors. For example, N-doped Si exhibited strong electromechanical properties and good performance including high sensitivity and short response time in mechanical transducers [13, 14]. Yang et al. observed a great piezo-resistive effect in p-type Si nanowires (Fig. 2.1A and B), which exhibit a much greater piezo-resistance

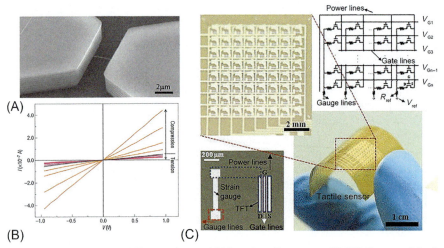

Fig. 2.1 Piezo-electricity of Si nanowires and Si-based tactile sensors. (A) SEM image of single Si nanowire bridging; (B) I–V curves of Si nanowire under compression and tension; (C) Optical images and electrical circuit of Si-based tactile sensor matrix.

coefficient $(-3550 \times 10^{-11} \, Pa^{-1})$ than that of bulk Si $(-94 \times 10^{-11} \, Pa^{-1})$ [15]. Si membrane- and nanowire-based tactile sensors have demonstrated excellent performance such as detection of scanning force with a fast speed of 100 kHz and resolution of 12.4 kPa (Fig. 2.1C) [16]. Si-based tactile sensors are promising for practical applications as most current micro-manufacturing techniques are compatible for Si. However, its intrinsic characteristics of rigidity and fragility limit its applications in bioelectronic and wearable electronic devices.

Atomic thin film of two-dimensional (2D) semiconductors such as graphene and transition metal dichalcogenides (TMDs) possess unpredictable mechanical properties and are promising for flexible electronics [17–20]. Recently, there has been much literature published on uncovering strain-induced bandgap engineering in graphene, MoS_2, WS_2, SnS, WSe_2, InSe, GaSe, and black phosphorus (Fig. 2.2A). [21–26] Their electrical properties are highly related with the mechanical stimulations. However, the flat 2D materials are limited in measurement of large deformation because of unrecoverable structural damage. Wrinkled structure and kirigami structure were designed to enhance stretchability for large tensile deformation (Fig. 2.2B and C) [27–30]. Chen et al. demonstrated a flexible tactile sensor by wrinkled graphene that displayed high sensitivity $(6.92 \, kPa^{-1})$ in detecting pressure ranging from 300 Pa to 1.5 kPa [31]. Both piezo-resistive and piezo-electric properties have been observed from monolayer molybdenum sulfide (MoS_2) [32, 33]. Park et al. reported a large-area, flexible piezo-resistive sensor array composed of a MoS_2 strain gauge with atomic thickness and an active matrix circuit (Fig. 2.2D and E) [34]. The sensor could be attached to skin for detecting pressure in the range of 1–120 kPa, which is close to the detection range of human skin. This tactile sensor array also shows a high sensitivity $(\triangle R/R_0 = 0.011 \, kPa^{-1})$ and short response time (180 ms) for multi-contact, and capability to detect the shape of the grabbed object.

2.2.2 Piezo-resistivity of conducting materials

Conducting polymers and composites belong to a new type of piezo-resistive materials that can be used in flexible tactile sensors. Piezo-resistive effect of those materials is determined by their mechanical properties and conducting channels produced by structures. Traditionally, there are two strategies to realize piezo-resistive effect in flexible materials: (1) conducting polymers such as PPy, PEDOTS, PANI and P3HT, which have tunable conductivity via the changing of molecular structure under external force [35–37] and (2) composites of polymer with conducting materials like black carbon, graphene, carbon nanotubes [38], metal nanoparticles, and nanowires.

Piezo-resistivity, conductivity, and mechanical strength are dependent on the morphologies and proportions of the conducting additive, where the density is proportional to conductivity and inversely proportional to the elasticity, flexibility, and tensile properties of flexible composites. At the original state without stress (Fig. 2.3A), the conducting materials in polymers are not in contact with each other, and there exists a polymer isolated layer between the particles of conducting materials. The transport channel of the electron is cut off and results in the material having electrical insulation or high resistivity. Upon force and strain being applied to composites

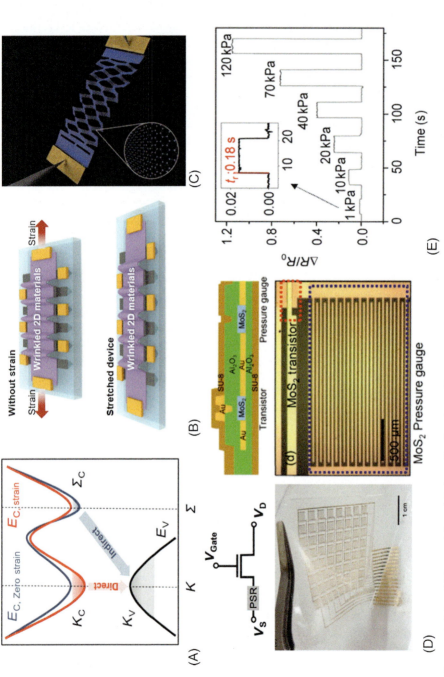

Fig. 2.2 Piezo-resistivity of 2D materials and their applications in tactile sensors. (A) Strain-induced energy band change of WSe₂; (B) wrinkled 2D materials on flexible substrate; (C) kirigami graphene-enabled stretchable device; (D and E) Structural illustration of MoS₂-based tactile sensors and their performance in detecting pressure.

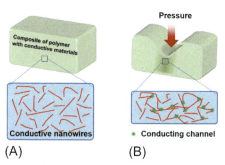

Fig. 2.3 Illustration of piezo-resistance effect of polymer composites. (A) Original state of polymer composite without pressure and strain; (B) polymer composite under external pressure.

(Fig. 2.3B), conducting particles move close to each other and connect, building electron channels that show conductivity. The magnitude of the external force affects the distribution of conductive particles in polymers, where resistance of materials is changed significantly. In order to maximize the piezo-resistive effect, the concentration of the conducting additive should be near the permeability threshold to improve resistance to rapidly changing external forces [39]. However, high concentrations of additives decreased mechanical properties such as ductility and stretchability of composites, makes the materials unable to bear great strain and pressure. Actually, the piezo-resistive effect of conducting polymers is a phenomenon that occurs at the interface of conducting additives, and could be optimized by the interface of microstructures.

2.2.3 Microstructure-enhanced piezo-resistivity

Microstructures in conducting polymers could improve the piezo-resistivity of materials because they induce more conductive channels and sensitive sites than polymers without micro/nanostructures. As shown in Fig. 2.4A, microstructures can exist both in and on materials with electrodes. Three-dimensional (3D) microstructures in polymers reduced Young's modulus of materials and made them sensitive to deformation caused by external force. For example, 3D conducting foam like graphene and its composite with polymers demonstrated high performance in flexible pressure sensors (Fig. 2.4B) [40–45]. The reversible deformation and resilience of elastic polymers give flexible sensors high sensitivity (0.26 kPa^{-1}), low detection limit (9 Pa), and excellent stability (10,000 times) [41]. The linear relationship between deformation and external force enabled large linear detection range. Bao et al. reported flexible pressure sensors by hollow-sphere-structured PPy, which present ultra-high sensitivity of 133 kPa^{-1} and detection limit as low as 1 Pa [37]. However, the mechanical response of the closed hollow sphere is non-linear, which leads to those sensors performing a non-linear response to pressure.

The interface of electrodes is important for optimizing flexible pressure sensors. A micro-patterned electrode is another way to enhance piezo-resistivity. Typically, two micro-patterned conducting layers are laminated face to face, where the interface has

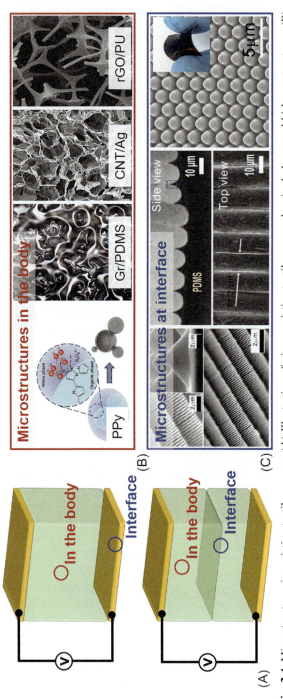

Fig. 2.4 Microstructures in resistive tactile sensors. (A) Illustration of piezo-resistive tactile sensors by single-layer and bi-layer structures; (B) microstructured PPy with hollow spheres and SEM images of 3D microstructures of graphene/PDMS, CNT/Ag foam, and rGO/PU; (C) the interface of a polymer composite and an electrode structured by micro-pyramid, silk-molded structures, and hemisphere structures.

many contact points that could improve the sensitivity of devices (Fig. 2.4C). Micro-pattern conducting film could be fabricated by lithography and molding. Patterned Si wafer, fabric silk film, sandpaper, PS spheres array, lotus leaf, and rose petals been proven to be great templates for patterning soft materials [46–49]. Chen et al. reported a flexible pressure sensor by pyramid-structure of PDMS with graphene, and demonstrated high sensitivity (5.5 kPa^{-1}) and short response time (0.2 ms) for detecting pressure less than 100 Pa [50]. The performance is largely dependent on the size and density of pyramid structures. Wang et al. employed silk fabric as a soft mold to fabricate micro-patterned PDMS with carbon nanotube film, and reported resistive tactile sensors that displayed highly sensitive to low pressure (0.6 Pa), and good reproducibility and stability (>67,000 times) [51]. The isotropy or anisotropy of microstructures makes their mechanical properties and sensors capable of detecting forces in different directions.

2.3 Pressure detection of resistive tactile sensors

Tactile sensors can be used to mimic tactile functions of the human body such as perceiving multiple external information such as pressure, prickle, roughness, and temperature. Pressure detection is one of the basic functions for tactile sensors. Detecting and mapping pressure allows for obtaining the geometry of an object for robotic perceptual reconstruction and prosthesis perceptual restoration. Up to now, much literature has been published on piezo-resistive pressure sensors because of their simple configuration and amazing performance as well as potential applications [1, 41, 50, 52–79].

Typically, performance of flexible tactile sensors in detecting pressure could be evaluated by the parameters of sensitivity, detection range, response time, power consumption, and stability under fatigue testing. Sensitivity (S) is an important parameter to indicate response between pressure and conductivity of materials, which can be calculated by dividing resistance change (dR) of the device with the change (dF) of applied force: $S = dR/dF$. Detection range is the working range where tactile sensors cannot be damaged and therefore perform well and accurately. A large detection range indicated that sensors have more extensive applications. High-frequency pressure trigging needs short response time, while hysteresis of piezo-resistive materials produced a long response time. Power consumption and stability are important for applications in wearable electronic devices. The stability of devices was mainly measured by the long-term detection of their performance under repeated mechanical impact. Table 2.1 summarizes the performance of flexible tactile sensors in detecting pressure.

The pressure detection characteristics of a flexible tactile sensor are promising for application in human-machine interaction and health care [80]. A flexible tactile sensor can be equipped to gloves or parts of the human body for detecting movement and body language for human machine interaction. Wang et al. reported highly sensitive flexible pressure sensors by micropatterned PDMS substrate and CNT film.

Table 2.1 Performance summary of flexible pressure sensors.

Materials	Detection range	Sensitivity	Response time	Stability	Reference
PDMS/AuNWs	0.013–50 kPa	>1.14 kPa^{-1}	<17 ms	>50,000	[1]
PDMS/rGO	<100 Pa	−5.5 kPa^{-1}	0.2 ms	5000	[50]
PDMS/MWNTs/RMS	0.25–260 kPa	–	–	10,000	[52]
PDMS/silver nanowires	0.015–4.5 kPa	>3.8 kPa^{-1}	<150 ms	1500	[53]
PDMS/CNT-AgNP	>0.001 kPa	80 kPa^{-1}	~100 ms	1000	[54]
Polyurethane (PU)/graphene	>0.009 kPa	> 0.03 kPa^{-1}	~4.5 s	10,000	[41]
MoS$_2$ semiconductor/graphene	1.24–240 kPa	–	–	1000	[55]
PVDF/ZnO	>0.01 kPa		<120 ms	–	[56]
CNT/graphene	~0.6–1.5 kPa	~1 kPa^{-1}	~20 ms	–	[57]
PDMS/ITO-PET/ZnO microparticles	>0.015 Pa	121 kPa^{-1}	~7 ms	>2000	[58]
PDMS/SWNTS	0–300 kPa	−3.26 kPa^{-1}	200 ms	5000	[59]
PDMS/rGO	<200 Pa	−2 kPa^{-1}	~0.15 ms	500	[60]
Microstructured PDMS/rGO	0–225 Pa	−1.17 kPa^{-1}	6 ms	–	[61]
Non-woven aramid fibers /PEI-CNT	0.0025–40 MPa	~0.05 MPa^{-1}	–	550	[62]
PDMS/pyramid ITO	0.1 Pa–10 kPa	–	120 ms	100,000	[63]
Microstructured PDMS/CNT	0.007–50 kPa	~−0.101 ± 0.005 kPa^{-1}	~10 ms	>5000	[64]
PDMS/3D graphene film	0.2–75 kPa	110 kPa^{-1}	<30 ms	10,000	[65]
PDMS/pyramidal AuNWs	<600 Pa	23 kPa^{-1}	<10 ms	10,000	[66]
MoS$_2$/graphene porous networks/Ecoflex	0.6–25.4 kPa	~0.069–6.06 kPa^{-1}	–	4000	[67]
Random distribution spinosum (RDS) microstructure PDMS/GO	0–2.6 kPa	25.1 kPa^{-1}	120 ms	3000	[68]
3D-printed PDMS/carbon nanofibers (CNFs)	0–2 kPa	−3.6 kPa^{-1}	20–50 ms	10,000	[69]

Table 2.1 Continued

Materials	Detection range	Sensitivity	Response time	Stability	Reference
PDMS/CNWs	0–10 kPa	6.64 kPa^{-1} (0–0.2 kPa) 1.26 kPa^{-1} (0.2–1 kPa)	30 ms	1000	[70]
Polyvinyl alcohol nanowires/wrinkled graphene film	0–13.6 kPa	28.34 kPa^{-1} (3–10 kPa) 4.52 kPa^{-1} (0–3 kPa)	–	6000	[71]
Polyvinyl alcohol (PVA)-polyacrylamide (PAM) hydrogel	0–6.67 kPa	0.05 kPa^{-1} (0–3.27 kPa)	150 ms	500	[72]
Multilevel microstructured (MM) PDMS/rGO	0.01–400 kPa	2.5–1051 kPa^{-1}	<150 ms	10,000	[73]
PVDF/rGO	0.0013–353 kPa	47.7 kPa^{-1}	20 ms	5000	[74]
Polypyrrole/PDMS micropyramid	≥0.0001 kPa	1907.2 kPa^{-1} (<0.1 kPa) 461.5 kPa^{-1} (<1 kPa)	50 μs	20,000	[75]
Ecoflex/SiC	>0.002 kPa	1.28 kPa^{-1}	200 ms	10,000	[76]
Ternary polymer composite of polyaniline, polyacrylic acid, phytic acid	0–5 kPa	37.6 kPa^{-1} (0–0.8 kPa) 7.1 kPa^{-1} (0.8–4.5 kPa) 1.9 kPa^{-1} (>5 kPa)	50 ms	1500	[77]
Bacterial cellulose/Ti$_3$C$_2$	0–10 kPa	12.5 kPa^{-1}	167 ms	1000	[78]
Micropatterned gold-nanowire/polyacrylamide hydrogels	0–2.8 kPa	3.71 kPa^{-1}	500 ms	10,000	[79]

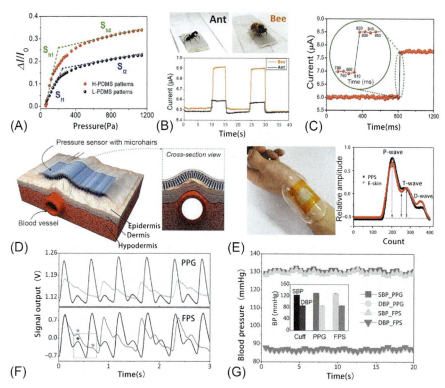

Fig. 2.5 Resistive tactile sensors in detection of pressure. (A–C) Performance of CNT-/PDMS-based flexible tactile sensors in detecting low pressure; (D) illustration of pressure sensor used in monitoring blood pressure; (E) wrist pulse monitoring by flexible tactile sensor; (F and G) comparison of flexible pressure sensor (FPS) and commercial photoplethysmogram (PPG) in monitoring blood pressure before and after sport.

The sensors showed high sensitivity to pressure of 0.6 Pa with response of 10 ms (Fig. 2.5A–C), making the flexible pressure sensor useful for detection of health-related physiological signals such as wrist pulse, heart rate, and muscle movement upon talking (Fig. 2.5D and E) [81]. It also can be attached to the wrist and neck for measuring blood pressure (Fig. 2.5F and G) [82, 83]. Therefore, wearable devices assembled with flexible tactile sensors will provide new avenues to remote medicine, and bring revolutionary techniques to traditional medicine [84, 85].

2.4 Multi-functional tactile sensors

Real touch of human skin includes simultaneous sensing of various external stimulations such as pressure, temperature, and roughness. Artificial electronic skin (e-skin) is a multi-functional tactile sensor formed by sensor arrays or a sensor matrix, which is the integration of single sensor units with different functions [4, 86–88]. Tactile sensors are gradually developing towards multi-function and integration by new device structures and sensing materials [56, 89–93]. Mu et al. proposed a fully flexible and stretchable e-skin that consists of a porous carbon nanotube/GO@PDMS thin film and a double-layer conductive structure (Fig. 2.6A) [94]. The e-skin shows high performance with sensitivity of 0.31 kPa^{-1} (0.05–3.8 kPa) and gauge factor of 2.26 for pressure and tangential force, respectively. Due to its high sensitivity, robustness, and stability, e-skin displays a variety of functions, including detection of wrist pulse, extreme slight slippage (under 20 mg), and monitoring human breathing and sound vibrations without contact (Fig. 2.6B).

Inspired by human fingertips, Park et al. reported multi-functional tactile sensors by fingerprint-like patterns and interlocked microstructures in ferroelectric films (Fig. 2.6C) [95]. The device was capable of simultaneously detecting multiple spatiotemporal tactile stimuli such as pressure, vibration, and temperature. They also demonstrated their applications in monitoring wrist pulse pressure as well as roughness and hardness of contact surface. Cao et al. reported flexible tactile sensors by micropatterned CNT /polyethylene/PDMS with interlocked structure [59]. This device was not only sensitive to low pressure (-3.26 kPa^{-1} in the pressure range of 0–300 Pa), but also capable of detecting vibrations. This flexible tactile sensor was also able to recognize the roughness of contact surfaces (Fig. 2.6D). The delicate surface texture of 15 μm could be distinguished by frequency signals of the device.

Beyond detection of pressure and temperature, Hua et al. expanded the functionality of e-skin to sense of humidity, light, magnetic field, in-plane strain, and proximity by a highly stretchable and conformable matrix network (Fig. 2.7A and B) [96]. The device showed large-area expandability (300% strain) with a tunable sensing range and 3D integration. The device was equipped to an intelligent prosthetic hand to mimic the skin of human hands, and presented the properties in simultaneous detection of spatial pressure mapping and temperatures of grasping objects. Kim et al. demonstrated an ultrathin e-skin by single crystal Si nanobelts [97]. e-skin is an integrated sensor array that consist the sense of strain, pressure, temperature, and humidity. It is also integrated with resistive heaters for nerve stimulation (Fig. 2.7C). Structural designing of the neutral mechanical plane with ultrathin thickness makes the device resist cracks under bend and strain.

2.5 Discussion and outlooks

Many resistive tactile sensors have been designed and fabricated that perform remarkably similar functions as human skin. The functionality of those artificial tactile sensors includes not only the detection of mechanical stimulations such as pressure, strain, and shear, but also the sensing of temperature, non-contact approaching, and surface roughness. Applications of tactile sensors have moved beyond prosthetic limbs and have expanded to other areas of healthcare and human-machine interfaces.

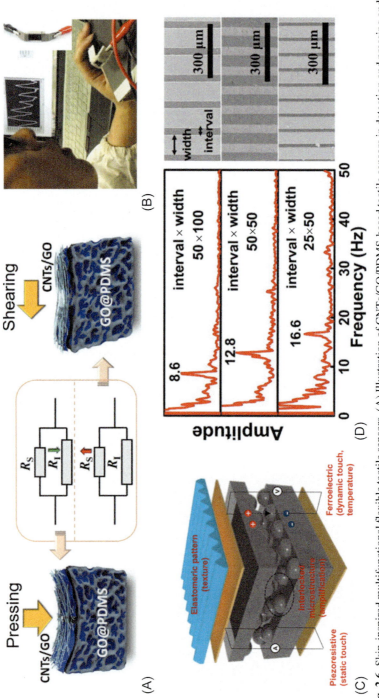

Fig. 2.6 Skin-inspired multifunctional flexible tactile sensors. (A) Illustration of CNTs/GO/PDMS-based tactile sensors in detection and pressing and shearing, and its application in monitoring (B) breath; (C) schematic illustration of fingerprint-inspired flexible tactile sensor; (D) FFT wave patterns by the flexible tactile sensor scanning on different Si micro-stripes.

Resistive tactile sensors

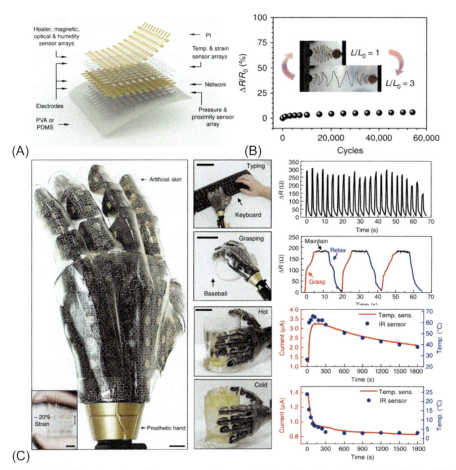

Fig. 2.7 Multifunctional tactile sensor matrixes and their applications in prosthetic limbs. (A) Schematic illustration and (B) stretching performance of highly stretchable and conformable matrix networks; (C) Photograph of prosthetic skin by integrated flexible tactile sensor arrays, and its performance in sensing typing, grasping, and hot/cold water.

These novel applications will drive intelligent and smart living technology, making our lives convenient, efficient, and comfortable. For example, a flexible tactile sensor-based health monitoring system could promote traditional medical services from the hospital to remote and wearable medicine for health care at any time and at any place. However, there are still a few questions that need to be resolved, including how to integrate multiple sensor arrays for multi-functionality with long-time stability while avoiding the crosstalk of different signals, and how to develop biocompatible and implantable devices for the interface between central neurons and tactile sensors. The functions of resistive tactile sensors will be increasingly diversified with the emergence of novel functional materials and new micro-manufacturing technologies.

References

[1] S. Gong, W. Schwalb, Y. Wang, et al., A wearable and highly sensitive pressure sensor with ultrathin gold nanowires, Nat. Commun. 5 (1) (2014) 3132.

[2] K. Noda, E. Iwase, K. Matsumoto, et al., Stretchable liquid tactile sensor for robot-joints, in: IEEE International Conference on Robotics and Automation, Anchorage, AK, 2010, 2010, pp. 4212–4217.

[3] R.a. Li, K. Zhang, L. Cai, et al., Highly stretchable ionic conducting hydrogels for str-ain/tactile sensors, Polymer 167 (2019) 154–158.

[4] K. Takei, T. Takahashi, J. Ho, et al., Nanowire active-matrix circuitry for low-voltage macroscale artificial skin, Nat. Mater. 9 (2010) 821–826.

[5] K. Noda, K. Hoshino, K. Matsumoto, et al., A shear stress sensor for tactile sensing with the piezoresistive cantilever standing in elastic material, Sensors Actuators A Phys. 127 (2) (2006) 295–301.

[6] T. Okatani, H. Takahashi, K. Noda, et al., A tactile sensor using piezoresistive beams for detection of the coefficient of static friction, Sensors 16 (5) (2016) 718.

[7] S. Ding, Y. Pan, M. Tong, et al., Tactile perception of roughness and hardness to discriminate materials by friction-induced vibration, Sensors 17 (12) (2017) 2748.

[8] D. Silvera-Tawil, D. Rye, M. Velonaki, Artificial skin and tactile sensing for socially interactive robots: a review, Robot. Auton. Syst. 63 (2015) 230–243.

[9] Y. Wan, Y. Wang, C.F. Guo, Recent progresses on flexible tactile sensors, Mater. Today Phys. 1 (2017) 61–73.

[10] T. Yang, D. Xie, Z. Li, et al., Recent advances in wearable tactile sensors: materials, sensing mechanisms, and device performance, Mater. Sci. Eng. R Rep. 115 (2017) 1–37.

[11] S. Li, J.-P. Chou, H. Zhang, et al., A study of strain-induced indirect-direct bandgap transition for silicon nanowire applications, J. Appl. Phys. 125 (8) (2018), 082520.

[12] N. Frantzis, A piezoresistive, integrated silicon pressure transducer(Piezoresistive effect and application to semiconductor silicon pressure transducers, noting diffused silicon beam and diaphragm and Wheatstone bridge), ISA Trans. 4 (1965) 344–348.

[13] J.C. Greenwood, Silicon in mechanical sensors, J. Phys. E 21 (12) (1988) 1114–1128.

[14] M. Aceves, I. Sandoval, Silicon pressure transducers: a review, Rev. Mex. Fis. 40 (4) (1994) 533–546.

[15] R. He, P. Yang, Giant piezoresistance effect in silicon nanowires, Nat. Nanotechnol. 1 (1) (2006) 42–46.

[16] M. Park, M.S. Kim, Y.K. Park, et al., Si membrane based tactile sensor with active matrix circuitry for artificial skin applications, Appl. Phys. Lett. 106 (4) (2015), 043502.

[17] H. Jiang, L. Zheng, Z. Liu, et al., Two-dimensional materials: from mechanical properties to flexible mechanical sensors, InfoMat (2019), https://doi.org/10.1002/inf2.12072.

[18] T.-H. Han, H. Kim, S.-J. Kwon, et al., Graphene-based flexible electronic devices, Mater. Sci. Eng. R Rep. 118 (2017) 1–43.

[19] J.-H. Ahn, B.H. Hong, Graphene for displays that bend, Nat. Nanotechnol. 9 (10) (2014) 737–738.

[20] K.S. Kim, Y. Zhao, H. Jang, et al., Large-scale pattern growth of graphene films for stretchable transparent electrodes, Nature 457 (7230) (2009) 706–710.

[21] T. Cao, Y. Chen, R. Sankar, et al., Ultrasensitive tunability of the direct bandgap of 2D InSe flakes via strain engineering, 2D Mater. 5 (2) (2018) 021002.

[22] J.-W. Jiang, H.S. Park, Analytic study of strain engineering of the electronic bandgap in single-layer black phosphorus, Phys. Rev. B 91 (23) (2015) 235118.

[23] S.B. Desai, G. Seol, J.S. Kang, et al., Strain-induced indirect to direct bandgap transition in multilayer WSe2, Nano Lett. 14 (8) (2014) 4592–4597.

[24] Y. Wu, H.-R. Fuh, D. Zhang, et al., Simultaneous large continuous band gap tunability and photoluminescence enhancement in GaSe nanosheets via elastic strain engineering, Nano Energy 32 (2017) 157–164.

[25] Q. Peng, Strain-induced dimensional phase change of graphene-like boron nitride monolayers, Nanotechnology 29 (40) (2018) 405201.

[26] Y. Zhang, B. Shang, L. Li, et al., Coupling effects of strain on structural transformation and bandgap engineering in SnS monolayer, RSC Adv. 7 (48) (2017) 30327–30333.

[27] Y. Wei, S. Chen, X. Yuan, et al., Multiscale wrinkled microstructures for piezoresistive fibers, Adv. Funct. Mater. 26 (28) (2016) 5078–5085.

[28] D.-Y. Khang, H. Jiang, Y. Huang, et al., A stretchable form of single-crystal silicon for high-performance electronics on rubber substrates, Science 311 (5758) (2006) 208–212.

[29] Z. Qi, D.K. Campbell, H.S. Park, Atomistic simulations of tension-induced large deformation and stretchability in graphene kirigami, Phys. Rev. B 90 (24) (2014) 245437.

[30] M.K. Blees, A.W. Barnard, P.A. Rose, et al., Graphene kirigami, Nature 524 (7564) (2015) 204–207.

[31] W. Chen, X. Gui, B. Liang, et al., Structural engineering for high sensitivity, ultrathin pressure sensors based on wrinkled graphene and anodic aluminum oxide membrane, ACS Appl. Mater. Interfaces 9 (28) (2017) 24111–24117.

[32] W. Wu, L. Wang, Y. Li, et al., Piezoelectricity of single-atomic-layer MoS2 for energy conversion and piezotronics, Nature 514 (7523) (2014) 470–474.

[33] H. Zhu, Y. Wang, J. Xiao, et al., Observation of piezoelectricity in free-standing monolayer MoS2, Nat. Nanotechnol. 10 (2) (2015) 151–155.

[34] Y.J. Park, B.K. Sharma, S.M. Shinde, et al., All MoS2-based large area, skin-attachable active-matrix tactile sensor, ACS Nano 13 (3) (2019) 3023–3030.

[35] E. Laukhina, R. Pfattner, L.R. Ferreras, et al., Ultrasensitive piezoresistive all-organic flexible thin films, Adv. Mater. 22 (9) (2010) 977–981.

[36] S. Harada, K. Kanao, Y. Yamamoto, et al., Fully printed flexible fingerprint-like three-axis tactile and slip force and temperature sensors for artificial skin, ACS Nano 8 (12) (2014) 12851–12857.

[37] L. Pan, A. Chortos, G. Yu, et al., An ultra-sensitive resistive pressure sensor based on hollow-sphere microstructure induced elasticity in conducting polymer film, Nat. Commun. 5 (1) (2014) 1–8.

[38] Alamusi, N. Hu, H. Fukunaga, et al., Piezoresistive strain sensors made from carbon nanotubes based polymer nanocomposites, Sensors 11 (11) (2011) 10691–10723.

[39] W.E. Mahmoud, A.M.Y. El-Lawindy, M.H. El Eraki, et al., Butadiene acrylonitrile rubber loaded fast extrusion furnace black as a compressive strain and pressure sensors, Sensors Actuators A Phys. 136 (1) (2007) 229–233.

[40] H. Liu, M. Dong, W. Huang, et al., Lightweight conductive graphene/thermoplastic polyurethane foams with ultrahigh compressibility for piezoresistive sensing, J. Mater. Chem. C 5 (1) (2017) 73–83.

[41] H.-B. Yao, J. Ge, C.-F. Wang, et al., A flexible and highly pressure-sensitive graphene–polyurethane sponge based on fractured microstructure design, Adv. Mater. 25 (46) (2013) 6692–6698.

[42] S. Zhao, L. Guo, J. Li, et al., Binary synergistic sensitivity strengthening of bioinspired hierarchical architectures based on fragmentized reduced graphene oxide sponge and silver nanoparticles for strain sensors and beyond, Small 13 (28) (2017) 1700944.

[43] C. Yan, J. Wang, W. Kang, et al., Highly stretchable piezoresistive graphene–nanocellulose nanopaper for strain sensors, Adv. Mater. 26 (13) (2014) 2022–2027.

[44] R. Xu, Y. Lu, C. Jiang, et al., Facile fabrication of three-dimensional graphene foam/poly (dimethylsiloxane) composites and their potential application as strain sensor, ACS Appl. Mater. Interfaces 6 (16) (2014) 13455–13460.

[45] H. Tian, Y. Shu, X. Wang, et al., A graphene-based resistive pressure sensor with record-high sensitivity in a wide pressure range, Sci. Rep. 5 (2015) 8603.

[46] S.C. Mannsfeld, B.C. Tee, R.M. Stoltenberg, et al., Highly sensitive flexible pressure sensors with microstructured rubber dielectric layers, Nat. Mater. 9 (10) (2010) 859–864.

[47] C. Pang, G.-Y. Lee, T.-i. Kim, et al., A flexible and highly sensitive strain-gauge sensor using reversible interlocking of nanofibres, Nat. Mater. 11 (9) (2012) 795–801.

[48] B. Su, S. Gong, Z. Ma, et al., Mimosa-inspired design of a flexible pressure sensor with touch sensitivity, Small 11 (16) (2015) 1886–1891.

[49] R. Guo, Y. Yu, J. Zeng, et al., Biomimicking topographic elastomeric petals (E-petals) for omnidirectional stretchable and printable electronics, Adv. Sci. 2 (3) (2015) 1400021.

[50] B. Zhu, Z. Niu, H. Wang, et al., Microstructured graphene arrays for highly sensitive flexible tactile sensors, Small 10 (18) (2014) 3625–3631.

[51] S. Kang, H. Qin, Y. Fang, et al., Preparation and electrochemical performance of yttrium-doped Li [Li0. 20Mn0. 534Ni0. 133Co0. 133] O2 as cathode material for lithium-ion batteries, Electrochim. Acta 144 (2014) 22–30.

[52] S. Jung, J.H. Kim, J. Kim, et al., Reverse-micelle-induced porous pressure-sensitive rubber for wearable human–machine interfaces, Adv. Mater. 26 (28) (2014) 4825–4830.

[53] Y. Joo, J. Byun, N. Seong, et al., Silver nanowire-embedded PDMS with a multiscale structure for a highly sensitive and robust flexible pressure sensor, Nanoscale 7 (14) (2015) 6208–6215.

[54] K. Takei, Z. Yu, M. Zheng, et al., Highly sensitive electronic whiskers based on patterned carbon nanotube and silver nanoparticle composite films, Proc. Natl. Acad. Sci. U.S.A. 111 (5) (2014) 1703.

[55] M. Park, Y.J. Park, X. Chen, et al., MoS2-based tactile sensor for electronic skin applications, Adv. Mater. 28 (13) (2016) 2556–2562.

[56] J.S. Lee, K.-Y. Shin, O.J. Cheong, et al., Highly sensitive and multifunctional tactile sensor using free-standing ZnO/PVDF thin film with graphene electrodes for pressure and temperature monitoring, Sci. Rep. 5 (2015) 7887.

[57] S. Lee, A. Reuveny, J. Reeder, et al., A transparent bending-insensitive pressure sensor, Nat. Nanotechnol. 11 (5) (2016) 472–478.

[58] B. Yin, X. Liu, H. Gao, et al., Bioinspired and bristled microparticles for ultrasensitive pressure and strain sensors, Nat. Commun. 9 (1) (2018) 1–8.

[59] Y. Cao, T. Li, Y. Gu, et al., Fingerprint-inspired flexible tactile sensor for accurately discerning surface texture, Small 14 (16) (2018) 1703902.

[60] Y. Zhu, J. Li, H. Cai, et al., Highly sensitive and skin-like pressure sensor based on asymmetric double-layered structures of reduced graphite oxide, Sensors Actuators B Chem. 255 (2018) 1262–1267.

[61] J. Zhang, L. Zhou, H. Zhang, et al., Highly sensitive flexible three-axis tactile sensors based on the interface contact resistance of microstructured graphene, Nanoscale 10 (16) (2018) 7387–7395.

[62] S.M. Doshi, E.T. Thostenson, Thin and flexible carbon nanotube-based pressure sensors with ultrawide sensing range, ACS Sens. 3 (7) (2018) 1276–1282.

[63] M. Jung, S.K. Vishwanath, J. Kim, et al., Transparent and flexible Mayan-pyramid-based pressure sensor using facile-transferred indium tin oxide for bimodal sensor applications, Sci. Rep. 9 (1) (2019) 1–11.

[64] G. Yu, J. Hu, J. Tan, et al., A wearable pressure sensor based on ultra-violet/ozone microstructured carbon nanotube/polydimethylsiloxane arrays for electronic skins, Nanotechnology 29 (11) (2018) 115502.

[65] K. Xia, C. Wang, M. Jian, et al., CVD growth of fingerprint-like patterned 3D graphene film for an ultrasensitive pressure sensor, Nano Res. 11 (2) (2018) 1124–1134.

[66] B. Zhu, Y. Ling, L.W. Yap, et al., Hierarchically structured vertical gold nanowire array-based wearable pressure sensors for wireless health monitoring, ACS Appl. Mater. Interfaces 11 (32) (2019) 29014–29021.

[67] S.J. Kim, S. Mondal, B.K. Min, et al., Highly sensitive and flexible strain–pressure sensors with cracked paddy-shaped MoS2/graphene foam/ecoflex hybrid nanostructures, ACS Appl. Mater. Interfaces 10 (42) (2018) 36377–36384.

[68] Y. Pang, K. Zhang, Z. Yang, et al., Epidermis microstructure inspired graphene pressure sensor with random distributed spinosum for high sensitivity and large linearity, ACS Nano 12 (3) (2018) 2346–2354.

[69] S. Peng, P. Blanloeuil, S. Wu, et al., Rational design of ultrasensitive pressure sensors by tailoring microscopic features, Adv. Mater. Interfaces 5 (18) (2018) 1800403.

[70] X. Zhou, Y. Zhang, J. Yang, et al., Flexible and highly sensitive pressure sensors based on microstructured carbon nanowalls electrodes, Nano 9 (4) (2019) 496.

[71] W. Liu, N. Liu, Y. Yue, et al., Piezoresistive pressure sensor based on synergistical innerconnect polyvinyl alcohol nanowires/wrinkled graphene film, Small 14 (15) (2018) 1704149.

[72] G. Ge, Y. Zhang, J. Shao, et al., Stretchable, transparent, and self-patterned hydrogel-based pressure sensor for human motions detection, Adv. Funct. Mater. 28 (32) (2018) 1802576.

[73] X. Tang, C. Wu, L. Gan, et al., Multilevel microstructured flexible pressure sensors with ultrahigh sensitivity and ultrawide pressure range for versatile electronic skins, Small 15 (10) (2019) 1804559.

[74] Y. Lee, J. Park, S. Cho, et al., Flexible ferroelectric sensors with ultrahigh pressure sensitivity and linear response over exceptionally broad pressure range, ACS Nano 12 (4) (2018) 4045–4054.

[75] H. Li, K. Wu, Z. Xu, et al., Ultrahigh-sensitivity piezoresistive pressure sensors for detection of tiny pressure, ACS Appl. Mater. Interfaces 10 (24) (2018) 20826–20834.

[76] Y. Gao, Q. Li, R. Wu, et al., Laser direct writing of ultrahigh sensitive SiC-based strain sensor arrays on elastomer toward electronic skins, Adv. Funct. Mater. 29 (2) (2019) 1806786.

[77] T. Wang, Y. Zhang, Q. Liu, et al., A self-healable, highly stretchable, and solution processable conductive polymer composite for ultrasensitive strain and pressure sensing, Adv. Funct. Mater. 28 (7) (2018) 1705551.

[78] Z. Chen, Y. Hu, H. Zhuo, et al., Compressible, elastic, and pressure-sensitive carbon aerogels derived from 2D titanium carbide nanosheets and bacterial cellulose for wearable sensors, Chem. Mater. 31 (9) (2019) 3301–3312.

[79] M.J. Yin, Y. Zhang, Z. Yin, et al., Micropatterned elastic gold-nanowire/polyacrylamide composite hydrogels for wearable pressure sensors, Adv. Mater. Technol. 3 (7) (2018) 1800051.

[80] Y. Khan, A.E. Ostfeld, C.M. Lochner, et al., Monitoring of vital signs with flexible and wearable medical devices, Adv. Mater. 28 (22) (2016) 4373–4395.

[81] X. Wang, Y. Gu, Z. Xiong, et al., Silk-molded flexible, ultrasensitive, and highly stable electronic skin for monitoring human physiological signals, Adv. Mater. 26 (9) (2014) 1336–1342.

[82] C. Pang, J.H. Koo, A. Nguyen, et al., Highly skin-conformal microhairy sensor for pulse signal amplification, Adv. Mater. 27 (4) (2015) 634–640.

[83] N. Luo, W. Dai, C. Li, et al., Flexible piezoresistive sensor patch enabling ultralow power cuffless blood pressure measurement, Adv. Funct. Mater. 26 (8) (2016) 1178–1187.

[84] Y. Li, L. Zheng, X. Wang, Flexible and wearable healthcare sensors for visual reality health-monitoring, Virtual Reality Intell. Hardw. 1 (4) (2019) 411–427.

[85] K. Takei, W. Honda, S. Harada, et al., Toward flexible and wearable human-interactive health-monitoring devices, Adv. Healthc. Mater. 4 (4) (2015) 487–500.

[86] T. Someya, T. Sekitani, S. Iba, et al., A large-area, flexible pressure sensor matrix with organic field-effect transistors for artificial skin applications, Proc. Natl. Acad. Sci. U. S.A. 101 (27) (2004) 9966–9970.

[87] T. Someya, Y. Kato, T. Sekitani, et al., Conformable, flexible, large-area networks of pressure and thermal sensors with organic transistor active matrixes, Proc. Natl. Acad. Sci. U. S.A. 102 (35) (2005) 12321–12325.

[88] M.L. Hammock, A. Chortos, B.C.K. Tee, et al., 25th anniversary article: the evolution of electronic skin (e-skin): a brief history, design considerations, and recent progress, Adv. Mater. 25 (42) (2013) 5997–6038.

[89] J. Yuji, K. Shida, A new multifunctional tactile sensing technique by selective data processing, IEEE Trans. Instrum. Meas. 49 (5) (2000) 1091–1094.

[90] H. Zhang, E. So, Hybrid resistive tactile sensing, IEEE Trans Sysyt Man Cybern B Cybern 32 (1) (2002) 57–65.

[91] A. Kimoto, N. Sugitani, S. Fujisaki, A multifunctional tactile sensor based on PVDF films for identification of materials, IEEE Sensors J. 10 (9) (2010) 1508–1513.

[92] Z. Chi, K. Shida, A new multifunctional tactile sensor for three-dimensional force measurement, Sensors Actuators A 111 (2–3) (2004) 172–179.

[93] A. Kimoto, Y. Matsue, A new multifunctional tactile sensor for detection of material hardness, IEEE Trans.Instrum. Meas. 60 (4) (2011) 1334–1339.

[94] C. Mu, Y. Song, W. Huang, et al., Flexible normal-tangential force sensor with opposite resistance responding for highly sensitive artificial skin, Adv. Funct. Mater. 28 (18) (2018) 1707503.

[95] J. Park, M. Kim, Y. Lee, et al., Fingertip skin–inspired microstructured ferroelectric skins discriminate static/dynamic pressure and temperature stimuli, Sci. Adv. 1 (9) (2015), e1500661.

[96] Q. Hua, J. Sun, H. Liu, et al., Skin-inspired highly stretchable and conformable matrix networks for multifunctional sensing, Nat. Commun. 9 (1) (2018) 244.

[97] J. Kim, M. Lee, H.J. Shim, et al., Stretchable silicon nanoribbon electronics for skin prosthesis, Nat. Commun. 5 (1) (2014) 5747.

Tactile sensor based on capacitive structure

Ho-Hsiu Chou[a,*] and Wen-Ya Lee[b,*]
[a]Department of Chemical Engineering, National Tsing Hua University, Hsinchu, Taiwan,
[b]Department of Chemical Engineering & Biotechnology, National Taipei University of Technology, Taipei, Taiwan
*Corresponding authors: e-mail address: hhchou@mx.nthu.edu.tw, wenyalee@mail.ntut.edu.tw

3.1 Introduction

Skin is an important organ for humans to sense stimuli such as temperature, tactile sense, pain, and vibration, as well as protect cells from outside threats [1]. Among these stimuli, the tactile sense plays an essential role in a person's interaction with the world. Tactile sensors in a hand can transfer the outside pressure stimuli into the signal transmitting to the brain. Tactile sensing ability helps people to distinguish a gentle touch from a handshake and to understand the difference between a smooth surface and a rough textile. For next-generation applications of robotics and prosthetic electronic skins, scientists are developing electronic tactile sensors to mimic skin-like tactile sensory capabilities [1–7].

Humans can sense a variety of pressure. For instance, the pressure of a gentle touch is usually in the range of 1–10 kPa; the pressure of a heartbeat is approximately <2 kPa [8, 9]; and the pressure of a small insect is only 0.003 kPa [1, 3]. Therefore, to measure a large variety of mechanical pressures, a highly sensitive tactile sensor is critical for the development of wearable tactile electronics. To date, several technologies have been developed for pressure sensors, including piezoelectric, piezoresistive, capacitive, and transistor-type sensors [1]. Piezoelectric sensors are mainly based on the ferroelectric polymer, polyvinylidene fluoride (PVDF) [10, 11]. The deformation of piezoelectric materials can generate voltages. This feature makes piezoelectric materials applicable for pressure sensing. However, piezoelectric devices are suitable for measuring high pressure changes (>100 kPa). They are not suitable for health monitoring or wearable electronics because of their minimal sensitivity to gentle touch (<2 kPa). Therefore, to design a tactile sensor with high sensitivity in the low-pressure range, researchers have reported two main technologies: piezoresistive and capacitive sensors [12, 13]. Through converting mechanical deformation or contact area to electrical signal changes, these sensors can be employed to measure both low- (<10 kPa) and high-pressure (10–100 kPa) regimes [7, 14].

Functional Tactile Sensors. https://doi.org/10.1016/B978-0-12-820633-1.00004-8
© 2021 Elsevier Ltd. All rights reserved.

Both piezo-resistive and capacitive sensors can couple with transistors to enhance pressure sensitivity, reduce signal crosstalk, and design a large-scale array. In this chapter, we systematically review these three types of devices and briefly introduce their benefits, current developments, and challenges.

3.2 Resistive pressure sensors

The resistive pressure sensor uses the resistance change between the conductor and the electrode to detect pressure change. In general, the resistance analysis of resistive sensors is not like the capacitive analysis of capacitive sensors, which requires expensive semiconducting instruments to carry out. The analysis of resistive sensors is only necessary to measure the voltage and current to convert the resistance, or a simple multimeter can be used for preliminary analysis. Therefore, the development of resistive sensors is faster than capacitive components. Compared with the capacitive pressure sensor, the advantage of the resistive-type pressure sensor is its greater sensitivity, which can generally reach to tens of kPa^{-1}. Pressure sensitivity is the most important parameter to evaluate the performance of pressure sensors. Pressure sensitivity S can be evaluated from the following equation:

$$S = \frac{d\left(\dfrac{\Delta y}{y_0}\right)}{dP} \tag{3.1}$$

Where y is the electrical signal obtained from the pressure sensor, which can be either a current, resistance, or capacitance type of device. For resistive pressure sensors, y can be current (I). Δy is the signal change with applied pressure, y_0 is the initial value without any applied pressure, and P is the applied pressure. From the differential results, we can calculate the S of the pressure sensors. The unit of pressure sensitivity is kPa^{-1}. Unlike the capacitive pressure sensor, which is mainly affected by the distance of the electrode, the resistive pressure sensor is affected by a few factors that can influence the changes of resistance. As such, this mechanism is much more complicated than the capacitive one. Resistive signal-sensing mechanisms can be divided into three types: piezo-resistive resistance (R_P), interface contact resistance (R_c), and electrode resistance (R_E) [12, 15], all of which can be simply expressed by the following formula:

$$R = R_P + R_c + R_E \tag{3.4}$$

Generally, the electrode resistance is extremely small $(R_E \ll R_P + R_c)$, so this influence can be ignored. Resistive pressure sensors can also be divided into two mechanisms: (a) the piezo-resistive type and (b) the contact resistance type between the rough or structured conductive material and the electrode. In the first mechanism, the applied pressure changes the energy band structure of the semiconductors or the distribution of conductive materials in the polymer composite (Fig. 3.1B) [15]. The piezo-resistive resistance is the resistance generated by the material intrinsically

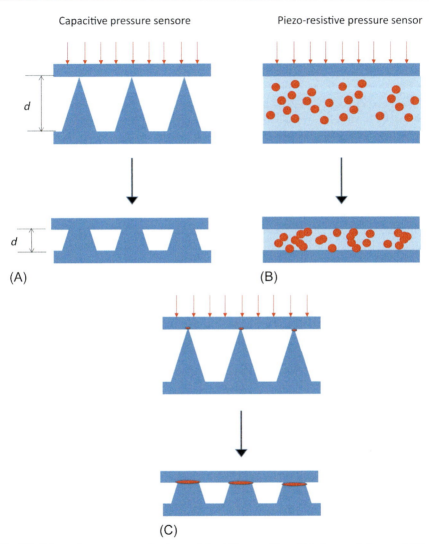

Fig. 3.1 Schematic sensory mechanisms of the (A) capacitive, (B) piezo-resistive, and (C) contact resistive sensors for measuring static pressure, respectively. Black arrows indicate applied pressure, red spheres indicate conductive fillers, and red areas indicate the contact area.

when the applied pressure is changed. However, the piezo-resistive resistance is easily changed by the influence of temperature, and it has the drawback of large electrical hysteresis and low-pressure sensitivity. It is not ideal to use a piezo-resistive pressure sensor as the main pressure-sensing approach [15]. In order to overcome these problems, many research groups have recently relied on contact resistance to improve the sensing effect of resistive pressure sensors. Contact resistance mainly exists on the interface between the conductor and electrode. When the contact area is increased, the resistance is decreased [14]. This method has low temperature sensitivity, thereby reducing the temperature effect. In addition, because the level of contact resistance mainly depends on the contact and not the thickness, the sensor can be fabricated with lightweight and ultrathin properties, which contributes greatly to its flexibility and stretchability.

Choong et al. used a pyramidal polydimethylsiloxane (PDMS) microstructure coated by a layer of PEDOT:PSS/PUD polymer mixture [12], and a conductive PEDOT:PSS film (~100 S/m) as the conductor and the electrode. When pressure is applied, the PDMS pyramidal structure will start to deform and increase the contact area of the upper and lower PEDOT:PSS electrodes. The sensing sensitivity of the sensor can reach to 4.88 kPa^{-1} when the pressure ranges from 0.37 to 5.9 kPa. Interestingly, this sensor has stretchability, and the researchers found that when the stretching ratio is increased, the sensitivity of the sensor is also increased, which may be related to the deformation of the pyramidal structure and the shortened distance between the electrodes. At 40% strain, the sensitivity of the sensor can reach to 10.3 kPa^{-1}. In their research report, Choong et al. determined the minimum detectable pressure to use in heartbeat detection is 23 Pa (93 mg). This study is the first report to use a resistive pressure sensor in the detection of human pulse signals (Fig. 3.2).

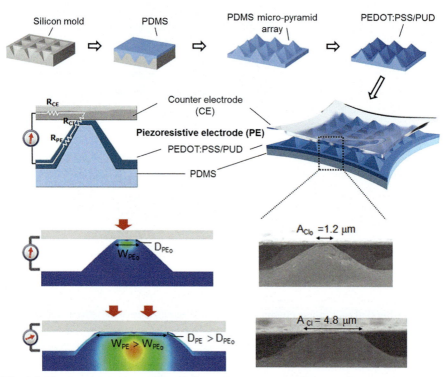

Fig. 3.2 The top picture shows the resistive pressure sensor, which uses the tapered structure PDMS and PEDOT:PSS/PUD to achieve the purpose of pressure detection. The bottom picture shows the deformation and contact area of the tapered structure under pressure [12]. Figure reproduced with permission from: John Wiley and Sons.

Tactile sensor based on capacitive structure

One-dimensional nanomaterials (such as metal nanowires [14], carbon fibers [7], and carbon nanotubes [13]) not only have good conducting properties, but also have good stretchability and adhesion ability, which is suitable for wearable electronics. On the resistive pressure sensor, Gong et al. adsorbed a large amount of gold nanowires on the paper and placed the gold nanometer substrate on the interdigital electrodes plated with PDMS to prepare a highly sensitivity pressure sensor [14]. When pressure is applied, the contact area between the gold nanowire and the interdigitated electrode increases to form a conductive channel. When the pressure is removed, the contact area decreases and the resistance increases. Based on this approach, the researchers obtained excellent sensing sensitivity ($1.14\,\text{kPa}^{-1}$) and fast response time (<17 ms). In addition, this approach can be used to analyze signals such as pressure, torsion, and sonic vibration (Fig. 3.3).

Pan et al. used a hollow ball structure to enable polypyrrole (PPy) to elastically deform and recover when external pressure is applied and released (Fig. 3.4A), thereby improving the contact stability of the pressure sensor and making the sensor

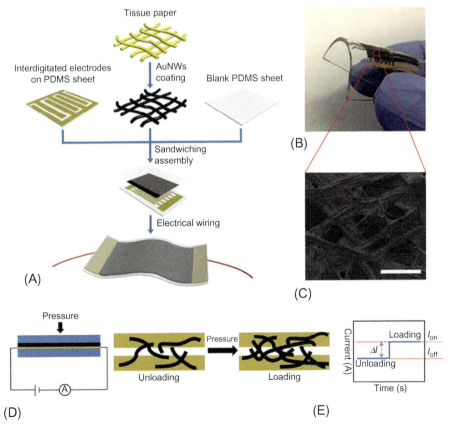

Fig. 3.3 One-dimensional gold nanometer pressure sensor [14]. Figure reproduced with permission from: Nature Publishing Grup.

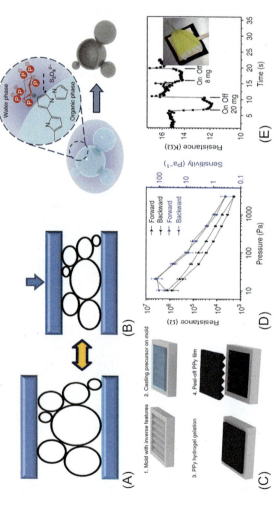

Fig. 3.4 (A) Schematic of the structural elasticity of the hollow sphere structure of PPy. (B) Schematic illustration of the interphase synthesis mechanism of PPy hydrogels with hollow-sphere microstructures. (C) Schematic process for the fabrication of micropatterned EMCP films. (D) Resistance response and pressure sensitivity of the EMCP pressure sensor, scale bar: 1 cm. (E) Transient response to the application and removal of a 20 mg weight and a flower petal (8 mg) on the micropatterned EMCP device [16]. Figure reproduced with permission from: Nature Publishing Group.

stable and its sensing capabilities reproducible. They made EMCP films by mixing and casting PPy gel precursors from two components in a Petri dish, and adopted a multi-phase reaction mechanism to achieve the PPy hollow sphere shape. By adjusting the microstructure of the sensor, they can pattern the surface of the EMCP film to further improve the sensitivity of the sensor. As shown in Fig. 3.4C, the sensor can be made more sensitive than human skin one by molding the microstructured surface topology (triangular cross-sectional contour of 0.5 mm height and 1 mm width) onto the PPy film before gelation. As shown in Fig. 3.4D, the microstructured PPy film can detect the placement and removal of a weight of 8 mg (e.g., petals) (approximately 0.8 Pa) [16].

Park et al. demonstrated the first scalable energy harvesting electronic skin (EHES) (Fig. 3.5), which can detect, distinguish, and collect various mechanical stimuli, owing to the scalability of its unique equipment architecture. In order to distinguish various tactile information such as normal pressure, strain, and bending, the researchers used the stretchability of the device to simultaneously measure changes in capacitance and sheet resistance due to lateral strain. In addition, this device can generate voltage and current in the range of tens of volts and tenths to several $\mu A\ cm^{-2}$ by harvesting the various mechanical stimuli mentioned previously. These features make EHES unique and practical in various electronic skin applications [17].

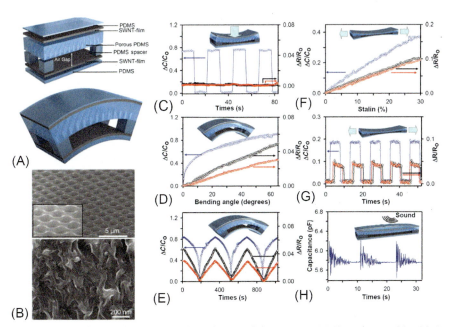

Fig. 3.5 (A) Device architecture. (B) SEM image of the porous PDMS surface and buckled SWNTs on PDMS surface. (C–H) Detection of various mechanical stimuli. Change in capacitance, change in top electrode resistance, and change in bottom electrode resistance are represented as blue circles, black diamonds, and red triangles, respectively [17].
Figure reproduced with permission from: John Wiley and Sons.

As the development of resistive pressure-sensing components is maturing, more and more scientists are trying to integrate pressure-sensing ability with various applications. For example, Chou et al. combined a stretchable resistive pressure sensor with a stretchable electrochromic device [13]. They used pyramidal microstructured PDMS as the substrate, but used carbon nanotubes as the conductive electrodes on the PDMS. They sprayed a layer of high-conductivity single-wall carbon nanotubes (SWNTs), and then used tape to remove the SWNTs on the top of the pyramidal microstructure. As shown in Fig. 3.6, when no pressure is applied, the resistance is about 2×10^6 kΩ, and when applying pressure of 22 kPa, the resistance quickly decreases to 50 kΩ. By combining the pressure sensor with the electrochromic material poly(3-hexyl thiophene) (P3HT), the color of the system could be changed by applying various pressure. Based on this design, the researchers successfully mimicked the function of chameleon skin for electronic skin applications.

In addition to electrochromic integration, a pressure sensor can also be combined with memory and transistor elements to form a multifunctional device [18]. The T. Someya research group reported memory function integrated with a pressure-sensing element [19]. The gate electrode of the field effect transistor was connected to the pressure sensor to enable the electronic device to record pressure change at the same time. The researchers uses a self-assembly monolayer as the dielectric layer, so the device could operate in a state of less than 3 V thereby reducing energy consumption. Unfortunately, they only used commercially available pressure-sensitive rubbers for the pressure sensors, so sensitivity information is not available.

In addition to the preceding applications, the Wang group of Georgia Tech and many research teams in the United States combined a triboelectric nanogenerator (TENG) with pressure sensing [5, 20]. They first made a tapered array on the surface of PDMS with a layer of gold coated on the PDMS, and a layer of silver nanowire and silver particles placed on the Al substrate (as shown in Fig. 3.7). When the upper and lower layers meet by applying external forces, the negative charge is transferred from the aluminum layer to the PDMS, while the positive charge is left on the aluminum layer. When the upper and lower layers are separated, the separated positive and negative charges form an open-loop potential. This open-loop potential is related to the distance between the upper and lower layers and the frictional charge density. The greater the applied pressure, the more changes in the potential. According to this property, the pressure change can be analyzed by measuring and analyzing the open-loop potential and short-circuit current. In this method, the sensitivity can reach to 0.31 kPa with a response time of less than 5 ms and a detection limit as low as 2.1 Pa. Notably, by this method, the short-circuit current is generated as a pulse signal, indicating that the static pressure and dynamic pressure changes of the device can be calculated. This is a report that can measure the dynamic pressure changes simultaneously. In addition, this triboelectric element can generate electricity by the object constantly moving backward and forward. This approach provides a good direction for self-powered electronic skin in the future [21].

The Z. Bao group developed an energy-efficient, skin-inspired mechanoreceptor by integrating a flexible organic transistor circuit and resistive pressure sensor (Fig. 3.8), which can directly convert the pressure into a digital frequency signal.

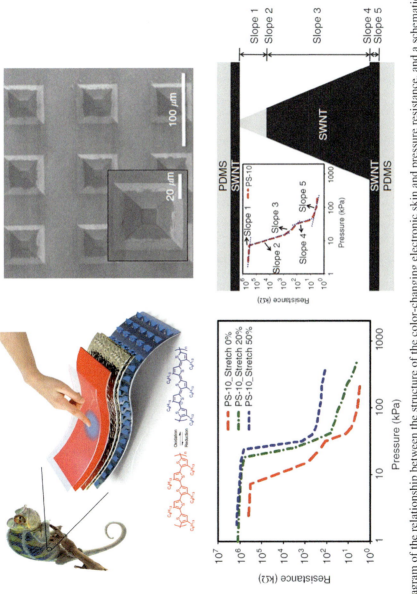

Fig. 3.6 Diagram of the relationship between the structure of the color-changing electronic skin and pressure resistance, and a schematic diagram of the mechanism explanation [13].
Figure reproduced with permission from: Nature Publishing Group.

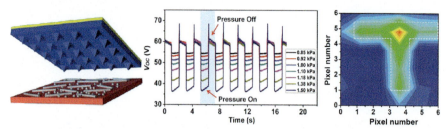

Fig. 3.7 Electronic skin with self-generated friction electrification and pressure sensing [20]. Reprinted with permission from L. Lin, Y. Xie, S. Wang, W. Wu, S. Niu, X. Wen, Z.L. Wang, ACS Nano, 7 (2013) 8266, Copyright (2013) American Chemical Society.

In order to achieve this concept, they developed a resistive pressure sensor with a wide impedance range. Carbon nanotubes and polyurethane elastomer were blended into a pyramidal microstructure, with the proportion of carbon nanotubes controlled to adjust the changes of the resistive sensor in order to achieve the required sensing characteristics for the mechanoreceptor [22].

3.3 Capacitive pressure sensors

To date, piezo-resistive devices are still the most popular device structure for tactile sensors due to their simple circuit design and good sensitivity. However, there are still some challenges for piezo-resistive sensors. The first challenge is the reliability of device performance. Many factors potentially influence electrical resistance changes, including moisture, humidity, impurities, temperature, contact resistance, and electrode brittleness. These factors make piezo-resistive sensors less reliable with large electrical hysteresis [23, 24]. This also causes piezo-resistive devices to be susceptible to the environment. Another challenge for piezo-resistive devices is power consumption. For a piezo-resistive sensor, although the low-resistance state can provide great sensitivity, it also leads to high current output. However, high current output is not preferred for wearable electronics because it results in greater power consumption (i.e., $P = I \times V$). This reduces the operation time of wearable devices. A capacitive sensor is an alternative high-performance pressure sensor. Compared to piezo-resistive pressure sensors, capacitive sensors have low power consumption because the current passing through a capacitor is near zero ideally. In the following section, we introduce the principles of capacitors and sensitivity calculations.

3.3.1 Principles of capacitive pressure sensors

A capacitor is an electronic device with a dielectric layer separated by two electrodes, known as a metal-insulator-metal (MIM) structure. The electrodes could be two parallel metallic plates or wires. Charges tend to accumulate at the interface between the insulator and electrodes. The capability of stored electrical charges is defined as

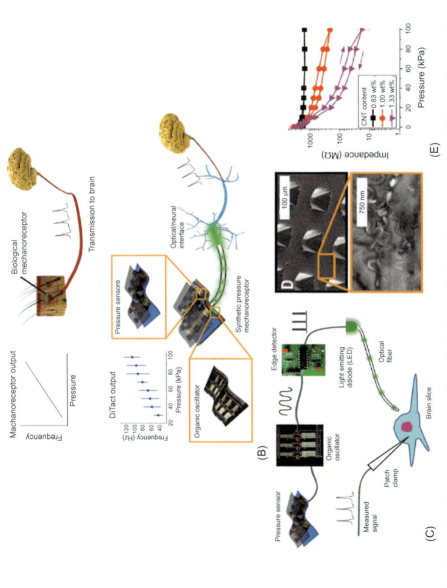

Fig. 3.8 (A) A biological mechanoreceptor system; (B) A skin-inspired artificial mechanoreceptor system; (C) Setup of the optoelectronic stimulation system for pressure-dependent neuron stimulation; (D) Scanning electron micrographs of the pyramids and the elastomer-CNT composite used to fabricate the piezoresistive sensors; and (E) Pressure response of sensors as a function of CNT concentration [22]. Figure reproduced with permission from AAAS.

capacitance, which is the ratio of the stored charges to the corresponding electrical voltages. Capacitance can be estimated from the following equation:

$$C = \frac{\varepsilon_0 kA}{d} \tag{3.2}$$

where ε_0 is the vacuum permittivity, k is the relative dielectric constant, A is the overlap area between two metallic conductors, and d is the thickness of the dielectric material [25]. According to this equation, capacitance is mainly controlled by three factors, including dielectric constant, thickness, and device area. Among these three factors, the change of the thickness is the main approach to sensing pressure (Fig. 3.1). When pressure is applied to the sensor, the capacitance increases owing to reduced dielectric thicknesses, leading to capacitance change. According to the capacitance changes, we find the relationship between capacitance and pressure for pressure sensors.

3.3.2 Microstructures of dielectric films

The key design of capacitive pressure sensors is the microstructure of a dielectric elastomer. PDMS is a well-known dielectric elastomer selected for capacitive pressure sensors owing to its stretchability, robustness, biocompatibility, and environmentally friendly features. These features make PDMS the most popular material candidate for the pressure-sensing capacitive dielectric layer. However, a capacitive sensor based on a flat PDMS film typically generates a very low-pressure sensitivity (<0.1 kPa). Moreover, due to its visco-elastic characteristics, the relaxation of the compressed PDMS films takes a few seconds. The slow relaxation time is not desirable for practical applications. To improve sensitivity and response time of capacitive sensors, a PDMS-based microscale elastic pyramid array is widely employed [3, 26]. Fig. 3.9 shows the detailed fabrication process of a PDMS pyramid array from a Si mold. The pyramid structure contains a large space and allows the deformation of PDMS. The pyramid-array dielectric film results in increased pressure sensitivity in the low-pressure regime (<1 kPa). The pressure sensitivity of the structured film is five times greater (0.55 kPa^{-1}) than that of an unstructured PDMS film (0.02 kPa^{-1}) and a line-structured film (0.1 kPa^{-1}) [5, 8]. Furthermore, the pyramid structure significantly reduces relaxation time ($\ll 1$ s). The fast response and relaxation times are essential for real-time health monitoring and signal resolution.

To improve the device performance of PDMS pyramid arrays, several structure concepts have been reported, as shown in Fig. 3.10. Park et al. demonstrated a porous pyramid dielectric layer (PPDL) to achieve ultrahigh sensitivity (greater than 40 kPa^{-1}) in a very low-pressure regime (100 Pa) (Fig. 3.10A) [27]. The porous structure can reduce elastic modulus of the pyramids. Furthermore, the pressures sensors showed great stretchability and thermal stability. They are insensitive to strain (up to 60%) and temperature changes (up to 100 °C). This great stability is important for a variety of applications. Pang et al. demonstrated another approach to improve tactile sensors. They placed a micro-hair structure inspired by a gecko underneath the PDMS

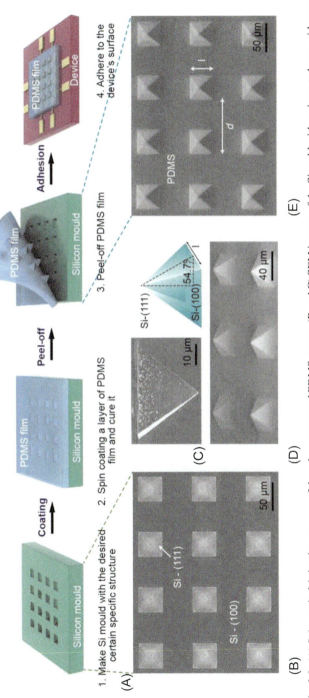

Fig. 3.9 (A) Schematic fabrication process of the microstructured PDMS arrays; (B and C) SEM images of the Si mold with an inverted pyramid array. (D) SEM images of the microstructured PDMS pyramid array. (E) Top-view SEM image of the micro-structure PDMS pyramid array [26]. Reproduced with permission from X. Wang, M. Que, M. Chen, X. Han, X. Li, C. Pan, Z. L. Wang, Adv. Mater. 29 (2017) 1605817, John Wiley and Sons.

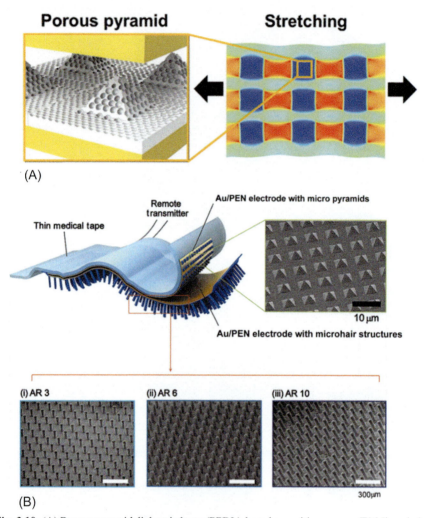

Fig. 3.10 (A) Porous pyramid dielectric layer (PPDL)-based capacitive sensor; (B) Micro-hairy structures underneath the PDMS pyramid arrays [9, 27].
Reproduced with permission from J.C. Yang, J.-O. Kim, J. Oh, S.Y. Kwon, J.Y. Sim, D.W. Kim, H.B. Choi, S. Park, ACS Appl. Mater. Interfaces, 11 (2019) 19472, Copyright (2017) American Chemical Society; Reproduced with permission from C. Pang, J.H. Koo, A. Nguyen, J.M. Caves, M.-G. Kim, A. Chortos, K. Kim, P.J. Wang, J.B.H. Tok, Z. Bao, Adv. Mater. 27 (2015) 634, John Wiley and Sons.

pyramid arrays (Fig. 3.10B) [9]. The structure significantly improved the adhesion between the rough surface of human skins and tactile sensors. The researchers observed that the sensors showed an increase of one order of magnitude in the signal-to-noise ratio. This indicates that tactile sensors can effectively contact with the rough surface of human skin.

Tactile sensor based on capacitive structure

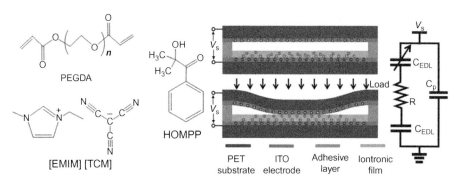

Fig. 3.11 Chemical structures of the ionic liquids and the polymer matrix. The air-gap structure of the ionic gel pressure sensor [29].
Reproduced with permission from B. Nie, R. Li, J. Cao, J.D. Brandt, T. Pan, Adv. Mater. 27 (2015) 6055, John Wiley and Sons.

In addition to the pyramid structures, the air-gap structure is another structure for pressure sensors. Viry et al. integrated an air gap with 150 μm in between the dielectric layers [28]. They found that the air-gap structure significantly improved sensitivity at low pressures (<2 kPa). Its low adhesion of the air gap facilitates mechanical detachment in between layers. This feature improves the signal-to-noise ratio and allows the device stable performance without the influence of the surroundings. Furthermore, the simple design of the air-gap structure does not require complicated processes. The researchers employed this air-gap structure to prepare three-axis tactile sensors. These sensors can sense a large range of pressures from very low pressure (100 Pa) to very high pressure (>200 kPa). However, the air-gap device's sensitivity is as low as 0.9 kPa^{-1}. This low sensitivity may be attributed to the low dielectric constant of the air (~1). To enhance sensitivity, Nie et al. utilized the air-gap structure and an ion gel polymer layer, which consists of an ionic liquid and polyethylene glycol diacrylate (PEGDA) (Fig. 3.11) [29]. Due to the high capacitance of the ion gel polymer (a few μF), the device sensitivity can be improved to 3 nF kPa^{-1}.

3.3.3 High capacitive dielectric materials for tactile sensors

As mentioned, PDMS is the most common used dielectric material for pressure sensors due to its commercial availability, non-toxicity, good stretchability, and biocompatibility. However, the disadvantage for PDMS is its relatively low dielectric constant (~2.4). The low dielectric constant causes low capacitive response and low sensitivity (0.01–0.3 kPa^{-1}) in pressure sensors. Therefore, the development of high k dielectrics is an important way toward high-sensitivity pressure sensors.

The addition of ionic liquids in dielectric elastomers is widely used for high-performance pressure sensors, due to their high capacitance (>1 μF), high ionic conductivity (10^{-3}–10^{-4} S/cm), wide electrochemical operation voltage ranges, and high boiling points. By adding the ionic liquid into a polymer matrix, the dramatic increase of capacitance enables a new direction for the development of soft electronics.

The ionic liquid can generate ultrahigh capacitance attributed to the formation of electrochemical double layers (EDLs). The capacitance of the EDLs can be estimated from the Helmholtz equation as follows: [30]

$$C = \frac{\varepsilon_0 k A}{\lambda} \tag{3.3}$$

Where ε_o is the vacuum permittivity, k is the effective dielectric constant, A is the cross-section area of the electrolyte between two electrodes, and λ is the thickness of the electrochemical double layer. Because the double-layer thickness is very thin, the double layer can provide a large capacitance in the order of a few μF. Recently, ionic liquids have been widely employed for capacitive pressure sensors [31, 32]. Cho et al. reported capacitive sensors based on pyramidal ionic gels. They used a thermoplastic fluorinated polymer, poly(vinylidenefluoride-*co*-hexafluoropropylene) (PVDF-HFP), as a host polymer, blended with the ionic liquid, 1-ethyl-3-methylimidazolium bis (trifluoromethylsulfonyl)amide ([EMI][TFSA]) (Fig. 3.12). The ion-gel micro-pyramids showed significant capacitance changes as a function of pressure. The pressure sensitivity can reach up to 41 kPa^{-1} at pressure less than 400 Pa, which is almost two orders of magnitude greater than that of its PDMS counterpart. In the range between

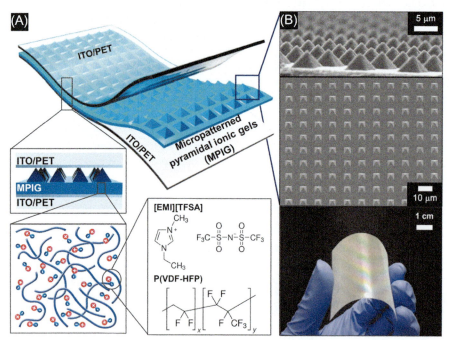

Fig. 3.12 Schematic device structures with micro-patterned ion-gel pyramids [33]. Reproduced with permission from S.H. Cho, S.W. Lee, S. Yu, H. Kim, S. Chang, D. Kang, I. Hwang, H.S. Kang, B. Jeong, E.H. Kim, S.M. Cho, K.L. Kim, H. Lee, W. Shim, C. Park, ACS Appl. Mater. Interfaces 9 (2017) 10128, Copyright (2017) American Chemical Society.

0.5 and 5 kPa, sensitivity still maintains a high value of 13 kPa^{-1}. The Kim group developed capacitive pressure sensors using polyurethane-based materials with the ionic liquid [EMIM$^+$][TFSI$^-$] to mimic the Piezo2 nanochannel in epidermal Merkel cells [34]. The ionic pairs can generate strong capacitive responses from a few nF to several tens µF when receiving mechanical deformation [31, 32, 34]. The devices act as capacitive artificial ionic mechanotransducers. After applying output stress, the ionic mechanotransducer tends to squeeze out the ionic pairs, leading to large capacitance changes, as shown in Fig. 3.13. The artificial ionic mechanotransducer exhibits high sensitivity of 25.8 nF kPa^{-1}. This work opens a new vision to mimic the behavior and tactile sensing ability of the mammalian physiology.

3.3.4 Electronic textiles with tactile sensing ability

Although the capacitive sensor showed great potential for thin-film pressure sensing applications, integrating it with textiles is rarely investigated. The main issue for the fabrication of pressure sensors on textiles is the highly rough surface of textiles. The voids on the textile result in an extremely rough surface. In principle, the capacitance of capacitive sensors is inversely proportional to the thickness of the dielectric film. However, the rough surface will result in a large deviation of the capacitive response. Furthermore, this significantly reduces the reliability and sensitivity of pressure sensors on the textiles. To avoid this issue, Lee et al. demonstrated a fiber-shaped pressure sensor consisting of a PDMS dielectric and a conductive Kevlar fiber, as shown in Fig. 3.14. Silver nanoparticles were grown on the surface of the fiber, leading to high conductivity. The PDMS dielectric layer was coated on the surface of the conductive Kevlar fiber for the fabrication of a pressure sensor. The fiber-shaped device based on PDMS-coated conductive fibers has a moderate sensitivity of 0.21 kPa^{-1} [4]. This sensitivity of PDMS-based fibers is relatively low, as compared to the PDMS film in the literature. This may be attributed to reduced contact area between the PDMS-coated fibers.

To improve the pressure sensitivity of capacitive sensors, tactile textiles using a photo-curable fluorinated elastomeric material (PVDF-HFP) have been demonstrated (Fig. 3.15) [35]. The fluorinated elastomer, PVDF-HFP, has been extensively investigated for polymer field-effect transistors [36], light-emitting diodes, and solar cells [37], due to its high dielectric constant. To enhance the capacitance response, PVDF-HFP is doped with quaternary ammonium salt, tetrabutylammonium perchlorate (TBAP). The addition of the ionic dopant significantly increases capacitive response of the dielectric layer. Unlike ionic liquids with a large EDL capacitance, TBAP does not have efficient diffusion in the polymer matrix. Instead, TBAP-doped polymer film shows an increased dielectric constant of the dielectric layer, thus leading to improved capacitance. Additionally, the fluorinated dielectric layer can be UV-cured using a transparent mold to avoid the thermal deformation of the polyester textiles underneath. This new fabrication processe can significantly increase the potential of pressure sensors on the textiles. This approach therefore offers a facile and fast approach to achieve reliable capacitive pressure sensors [35].

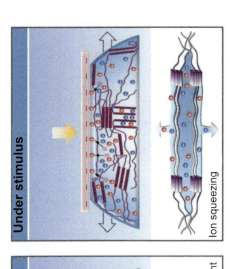

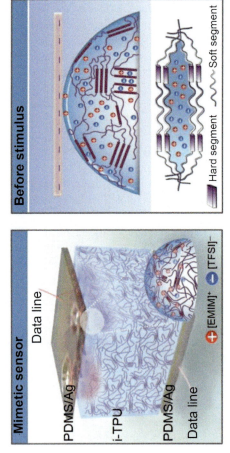
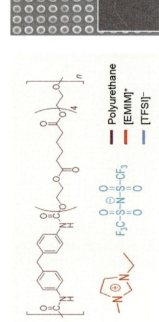

Fig. 3.13 Artificial mechanotransducer for tactile sensing [34].
Reproduced with permission from M.L. Jin, S. Park, Y. Lee, J.H. Lee, J. Chung, J.S. Kim, J.-S. Kim, S.Y. Kim, E. Jee, D.W. Kim, J.W. Chung, S.G. Lee, D. Choi, H.-T. Jung, D.H. Kim, Adv. Mater. 29 (2017) 1605973, John Wiley and Sons.

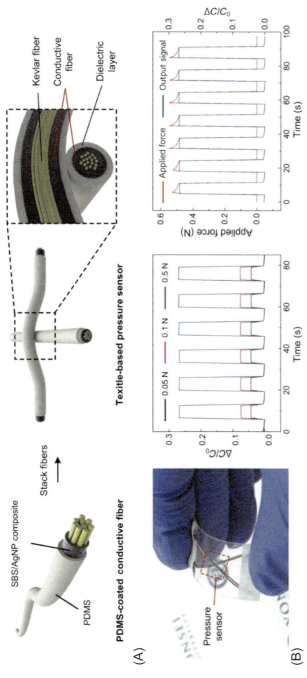

Fig. 3.14 (A) Schematic illustration of the fiber-shaped pressure-sensing devices; (B) Photograph of a cross-point fiber-shaped pressure sensor on a PET substrate; (C) Capacitive response of the pressure sensor with different loads of 0.05, 0.1, and 0.5 N. (D) Applied force and output signals of the fiber-shaped pressure sensors [4].

Reproduced with permission from L. Jaehong, K. Hyukho, S. Jungmok, S. Sera, K.J. Hoon, P. Changhyun, S. Seungbae, K.J. Hyung, J.Y. Hoon, K.D. Eun, L. Taeyoon, Adv. Mater. 27 (2015) 2433, John Wiley and Sons.

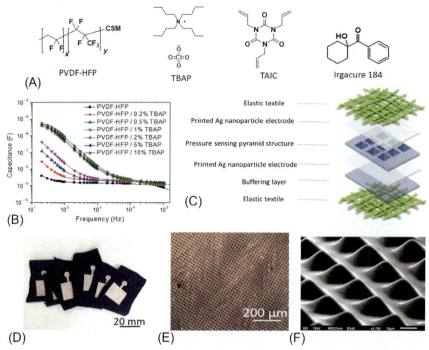

Fig. 3.15 (A) Chemical structures of photo-curable PVDF-HFP with the ionic dopant, TBAP; (B) Capacitance-frequency relationship of the TBAP-doped PVDF-HFP films with different dopant ratios. (C) The structure of the tactile-sensing textiles; (D) Photograph, (E) Optical microscopy image, and (F) SEM image of ion-doped pyramids [35].
Reproduced with permission from G.-T. Chen, C.-H. Su, S.-H. Wei, T.-L. Shen, P.-H. Chung, Q.-M. Guo, W.-J. Chen, Y.-F. Chen, Y.-C. Liao, W.-Y. Lee, Adv. Intell. Syst. (2020) 1900180, John Wiley and Sons

3.4 Perspective

The development of a soft tactile sensor is critical to the realization of wearable electronics and prosthetic electronic skin. Resistive and capacitive pressure sensors are promising technologies for the development of highly sensitive pressure sensors in low-pressure regimes. For both resistive and capacitive sensors, the microstructure design of the active layer significantly affects the electrical response and corresponding sensitivity. For resistive sensors, contact resistance at the interface between the active layer and electrodes plays an important role in resistance changes. Controlling the contact resistance is a promising approach toward highly sensitive resistive pressure sensors. In addition, resistance sensors are considered a relatively mature technology, so they have been integrated with various components to develop multifunctional systems. They can also be connected to circuit designs, such as the Wheatstone bridge, to increase sensitivity. For capacitive pressure sensors, in addition to the influence of microstructures, the addition of ionic electrolytes in the dielectric thin film dramatically enhances capacitive response. Although great progress has been made in the field of tactile sensors, there are still some challenges to overcome. For piezo-resistive devices, despite their high sensitivity and high output current, the devices exhibit large power consumption, which is not preferred for practical applications. For capacitive sensors, the dielectric film can be directly integrated with organic field-effect transistors (OFETs), leading to simplified fabrication processes and increased device density. There are many research works about the enhancement of capacitance by using ionic liquids. However, regarding slow ion diffusion, the ion-related dielectric materials require long relaxation time during the operation. Moreover, the capacitance in a high-frequency regime commonly shows a relatively low value, as compared to that in a low-frequency regime. Therefore, ion-doped capacitive sensors may be not suitable for high-frequency dynamic pressure measurement. For high-frequency applications, the development of high k elastomeric materials may be a possible way.

For both piezo-resistive and capacitive sensors, integration with OFETs is a promising circuit design for a large sensory array with reliable performance and low power consumption. In Chapter 4, we discuss the development of integrated arrays with OFETs and piezo-resistive or capacitive sensors. We believe that in the near future these types of pressure sensors can be combined with cross-disciplinary technologies such as printed electronic technology, stretchable materials, and other sensing elements to provide a strong interaction between robots (prosthetics) and humans.

References

[1] A. Chortos, J. Liu, Z.A. Bao, Nat. Mater. 15 (2016) 937.
[2] W. Xuewen, G. Yang, X. Zuoping, C. Zheng, Z. Ting, Adv. Mater. 26 (2014) 1336.
[3] S.C.B. Mannsfeld, B.C.K. Tee, R.M. Stoltenberg, C.V.H.H. Chen, S. Barman, B.V.O. Muir, A.N. Sokolov, C. Reese, Z. Bao, Nat. Mater. 9 (2010) 859.
[4] L. Jaehong, K. Hyukho, S. Jungmok, S. Sera, K.J. Hoon, P. Changhyun, S. Seungbae, K.J. Hyung, J.Y. Hoon, K.D. Eun, L. Taeyoon, Adv. Mater. 27 (2015) 2433.
[5] F.-R. Fan, L. Lin, G. Zhu, W. Wu, R. Zhang, Z.L. Wang, Nano Lett. 12 (2012) 3109.
[6] M. Stoppa, A. Chiolerio, Sensors 14 (2014) 11957.
[7] W. Qi, J. Muqiang, W. Chunya, Z. Yingying, Adv. Funct. Mater. 27 (2017) 1605657.
[8] G. Schwartz, B.C.K. Tee, J. Mei, A.L. Appleton, D.H. Kim, H. Wang, Z. Bao, Nat. Commun. 4 (2013) 1859.
[9] C. Pang, J.H. Koo, A. Nguyen, J.M. Caves, M.-G. Kim, A. Chortos, K. Kim, P.J. Wang, J. B.H. Tok, Z. Bao, Adv. Mater. 27 (2015) 634.
[10] I. Lee, H.J. Sung, Exp. Fluids 26 (1999) 27.
[11] A.V. Shirinov, W.K. Schomburg, Sensors Actuator A Phys. 142 (2008) 48.
[12] C. Chwee-Lin, S. Mun-Bo, L. Byoung-Sun, J. Sanghun, K. Dong-Su, K. Tae-Hyung, B. Jihyun, L.S. Hoon, B. Kyung-Eun, I. Jungkyun, J.Y. Jin, P.C. Eon, P. Jong-Jin, C. U-In, Adv. Mater. 26 (2014) 3451.
[13] H.-H. Chou, A. Nguyen, A. Chortos, J. W. F. To, C. Lu, J. Mei, T. Kurosawa, W.-G. Bae, J. B.H. Tok, Z. Bao, Nat. Commun. 6 (2015) 8011.

[14] S. Gong, W. Schwalb, Y. Wang, Y. Chen, Y. Tang, J. Si, B. Shirinzadeh, W. Cheng, Nat. Commun. 5 (2014) 3132.

[15] M.L. Hammock, A. Chortos, B.C.-K. Tee, J.B.-H. Tok, Z. Bao, Adv. Mater. 25 (2013) 5997.

[16] L. Pan, A. Chortos, G. Yu, Y. Wang, S. Isaacson, R. Allen, Y. Shi, R. Dauskardt, Z. Bao, Nat. Commun. 5 (2014) 3002.

[17] S. Park, H. Kim, M. Vosgueritchian, S. Cheon, H. Kim, J.H. Koo, T.R. Kim, S. Lee, G. Schwartz, H. Chang, Z. Bao, Adv. Mater. 26 (2014) 7324.

[18] C.-T. Chen, W.-Y. Lee, T.-L. Shen, H.-C. Wu, C.-C. Shih, B.-W. Ye, T.-Y. Lin, W.-C. Chen, Y.-F. Chen, Adv. Electron. Mater. 3 (2017) 1600548.

[19] T. Sekitani, T. Yokota, U. Zschieschang, H. Klauk, S. Bauer, K. Takeuchi, M. Takamiya, T. Sakurai, T. Someya, Science 326 (2009) 1516.

[20] L. Lin, Y. Xie, S. Wang, W. Wu, S. Niu, X. Wen, Z.L. Wang, ACS Nano 7 (2013) 8266.

[21] Y.-C. Lai, H.-M. Wu, H.-C. Lin, C.-L. Chang, H.-H. Chou, Y.-C. Hsiao, Y.-C. Wu, Adv. Funct. Mater. 29 (2019) 1904626.

[22] B.C.-K. Tee, A. Chortos, A. Berndt, A.K. Nguyen, A. Tom, A. McGuire, Z.C. Lin, K. Tien, W.-G. Bae, H. Wang, P. Mei, H.-H. Chou, B. Cui, K. Deisseroth, T.N. Ng, Z. Bao, Science 350 (2015) 313.

[23] T. Someya, Y. Kato, T. Sekitani, S. Iba, Y. Noguchi, Y. Murase, H. Kawaguchi, T. Sakurai, Proc. Natl. Acad. Sci. U. S. A. 102 (2005) 12321.

[24] M. Hussain, Y.-H. Choa, K. Niihara, J. Mater. Sci. Lett. 20 (2001) 525.

[25] M.E. Roberts, N. Queraltó, S.C.B. Mannsfeld, B.N. Reinecke, W. Knoll, Z. Bao, Chem. Mater. 21 (2009) 2292.

[26] X. Wang, M. Que, M. Chen, X. Han, X. Li, C. Pan, Z.L. Wang, Adv. Mater. 29 (2017) 1605817.

[27] J.C. Yang, J.-O. Kim, J. Oh, S.Y. Kwon, J.Y. Sim, D.W. Kim, H.B. Choi, S. Park, ACS Appl. Mater. Interfaces 11 (2019) 19472.

[28] L. Viry, A. Levi, M. Totaro, A. Mondini, V. Mattoli, B. Mazzolai, L. Beccai, Adv. Mater. 26 (2014) 2659.

[29] B. Nie, R. Li, J. Cao, J.D. Brandt, T. Pan, Adv. Mater. 27 (2015) 6055.

[30] S.H. Kim, K. Hong, W. Xie, K.H. Lee, S. Zhang, T.P. Lodge, C.D. Frisbie, Adv. Mater. 25 (2013) 1822.

[31] M.L. Jin, S. Park, J.-S. Kim, S.H. Kwon, S. Zhang, M.S. Yoo, S. Jang, H.-J. Koh, S.-Y. Cho, S.Y. Kim, C.W. Ahn, K. Cho, S.G. Lee, D.H. Kim, H.-T. Jung, Adv. Mater. 30 (2018) 1706851.

[32] V. Amoli, J.S. Kim, E. Jee, Y.S. Chung, S.Y. Kim, J. Koo, H. Choi, Y. Kim, D.H. Kim, Nat. Commun. 10 (2019) 4019.

[33] S.H. Cho, S.W. Lee, S. Yu, H. Kim, S. Chang, D. Kang, I. Hwang, H.S. Kang, B. Jeong, E. H. Kim, S.M. Cho, K.L. Kim, H. Lee, W. Shim, C. Park, ACS Appl. Mater. Interfaces 9 (2017) 10128.

[34] M.L. Jin, S. Park, Y. Lee, J.H. Lee, J. Chung, J.S. Kim, J.-S. Kim, S.Y. Kim, E. Jee, D.W. Kim, J.W. Chung, S.G. Lee, D. Choi, H.-T. Jung, D.H. Kim, Adv. Mater. 29 (2017) 1605973.

[35] G.-T. Chen, C.-H. Su, S.-H. Wei, T.-L. Shen, P.-H. Chung, Q.-M. Guo, W.-J. Chen, Y.-F. Chen, Y.-C. Liao, W.-Y. Lee, Adv. Intell. Syst. 2 (2020) 1900180.

[36] C. Lu, W.-Y. Lee, C.-C. Shih, M.-Y. Wen, W.-C. Chen, ACS Appl. Mater. Interfaces 9 (2017) 25522.

[37] P. Wang, S.M. Zakeeruddin, J.E. Moser, M.K. Nazeeruddin, T. Sekiguchi, M. Grätzel, Nat. Mater. 2 (2003) 402.

Tactile sensors based on organic field-effect transistors

Wen-Ya Lee*
Department of Chemical Engineering & Biotechnology, National Taipei University of Technology, Taipei, Taiwan
*Corresponding author: e-mail address: wenyalee@mail.ntut.edu.tw

4.1 Introduction

Owing to their mechanical flexibility, solution processability, and low-temperature process, organic field-effect transistors (OFETs) have been successfully used as active components for many electronic applications, including integrated circuits, organic light-emitting diodes (OLED), electronic skins, radio frequency identification tags, and sensors [1–8]. As compared to rigid silicon-based devices, OFETs can be mechanically flexible or even stretchable [3, 9]. The low-temperature process (<200°C) of OFETs allows plastic or elastic materials as substrates. Furthermore, their solution processability makes printing of large-area low-cost OFETs possible. These features make OFETs potentially applicable for flexible or stretchable electronic applications. Although the charge mobility of the OFETs is generally one or two orders magnitude lower than that of silicon-based devices, the operation speed of the OFETs is still sufficient for pressure sensors with the relatively slow speed (1 Hz to 1 kHz). Moreover, the tactile sensors integrated with OFETs can further reduce cross talk between adjacent devices and amplify the pressure-dependent electrical signals, leading to the enhancement of the pressure sensitivity in the sensor array. Therefore, the OFETs technology is a reasonable selection for the readout element in a tactile sensor array on flexible or stretchable substrates for potential applications in electronic skin. Before moving on to transistor-type pressure sensors, we will briefly introduce the operation principle of field-effect transistors.

4.2 Operation principle of field-effect transistors

The structure and electrical characteristics of the OFET are shown in Fig. 4.1. It is a three-terminal device consisting of electrodes, a semiconductor, and a dielectric layer. The electrodes are drain, source, and gate electrodes, respectively. The source electrode is always grounded (0 V). As gate voltage (V_G) is applied to the gate electrode, charges tend to accumulate at the interface between the semiconductor and the dielectric layer. Sequentially, the semiconductor will turn to a conducting channel near the interface for charge transport. On applying a drain-source voltage (V_{DS}), charges will be injected from the source electrode, which then move to the drain electrodes.

Functional Tactile Sensors. https://doi.org/10.1016/B978-0-12-820633-1.00008-5
© 2021 Elsevier Ltd. All rights reserved.

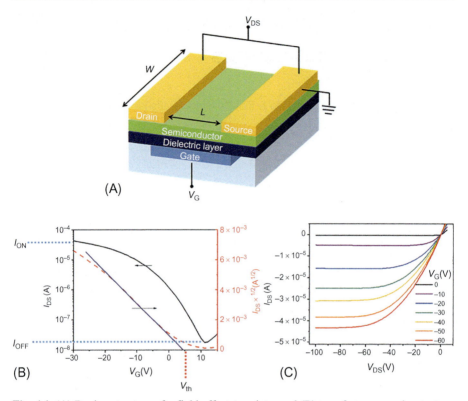

Fig. 4.1 (A) Device structure of a field-effect transistor and (B) transfer curve and output curves of the field-effect transistors.

The collected current is called the drain-source current (I_{DS}). Fig. 4.1B and C is the transfer and output curves of the FETs, respectively. Typically, there are two regimes in the output curve of FETs (Fig. 4.1C). When the drain-source voltage is small, i.e., $V_{DS} \ll V_G$, the current linearly increases with the applied V_{DS}. This current-voltage regime is called the linear regime. As $V_{DS} = V_G - V_{th}$, where V_{th} is the threshold voltage, the channel will reach a pinched-off point. In this pinched-off point, I_{DS} is saturated even with the increased V_{DS}. The output current is only modulated with the applied V_G. The current-voltage regime is called the saturation regime. In the saturation regime ($V_{DS} = V_G - V_{th}$), I_{DS} can be estimated according to the following equation [1]:

$$I_{DS} = \frac{W}{2L} C \mu_{sat} (V_G - V_{th})^2 \qquad (4)$$

On the other hand, the linear regime I_{DS} ($V_{DS} \ll V_G$) can be estimated from

$$I_{DS} = \frac{W}{2L} C \mu_{lin} (V_G - V_{th}) V_{DS} \qquad (5)$$

where μ_{sat} and μ_{lin} are the charge mobilities of the saturation and linear regimes, respectively, W and L are the channel width and length, respectively, V_{th} is the threshold voltage, and C is the capacitance per area of the dielectric layer.

Charge mobilities, threshold voltages, and ON/OFF current ratios are three critical parameters used to evaluate the device performance of the OFETs. The charge mobility is related to the operation speed of the OFETs. The device with a higher charge mobility shows faster operation speeds. The switching speed of the OFETs is proportional to the charge mobility and inversely proportional to the square of the channel lengths, i.e., $f \propto \frac{\mu}{L^2}$ [10]. For a high-performance OFET-based sensors, a high charge mobility is required. Another parameter is the threshold voltage. It is the minimum V_G required to create a conducting channel for charge transport. The threshold voltages can be extracted from $I_{DS}^{1/2} - V_G$ curve in the intercept of the x-axis (Fig. 4.1B). The threshold voltage can be affected by many factors, such as humidity, interface dipoles, defects, and dielectric capacitance. During the operation of an OFET, a shift of the threshold voltage can cause current hysteresis. The current hysteresis indicates that the forward-swept current is different from the backward-swept current. For a transistor, the current hysteresis is undesirable, because it represents the device instability. Therefore, a well-functioning transistor requires negligible current hysteresis. However, on the other hand, for an organic memory device, a larger current hysteresis is preferred, because this indicates more charges stored in the device. The third important parameter of OFETs is the ON/OFF current ratio, which is the ratio of the I_{DS} in the ON state and the OFF state (I_{ON}/I_{OFF}), which can be extracted from the transfer curve (Fig. 4.1B). The ON current is mainly attributed to the charge mobility of the semiconductor layer, while the OFF current is related to the gate current leakage and the bulk conductivity of the semiconductor layer. According to the equations, the output current is not only dependent on the capacitance, but also can be driven by the V_{DS} and V_G voltages. This means that the increase of both voltages and capacitances can further enhance current response in an OFET. Therefore, the OFET structure coupled with a piezo-resistive or capacitive sensor is an ideal architecture for the enhancement of electrical signals and sensitivity.

4.3 Tactile field-effect transistors

To realize the electronic skins, tactile sensory array with large area and high device density can provide sufficient resolution. The integration of each tactile sensor with an OFET is the efficient way toward the large sensory array, because the transistor enables to reduce cross talk and provide signal amplification. Moreover, the integration of the transistors and sensors makes an active-matrix circuit, which leads to a significant power saving. OFETs are capable of being integrated with capacitive or piezo-resistive tactile sensors, as shown in Fig. 4.2 [11]. A capacitive pressure sensors can be directly integrated with the dielectric layer in a transistor. On the other hand, a piezo-resistive sensor is required to be interconnected with an electrode in a transistor.

Piezo-resistive sensors have been commonly integrated with the source electrode of transistors in a tactile sensory array [12, 13]. Someya et al. have demonstrated a

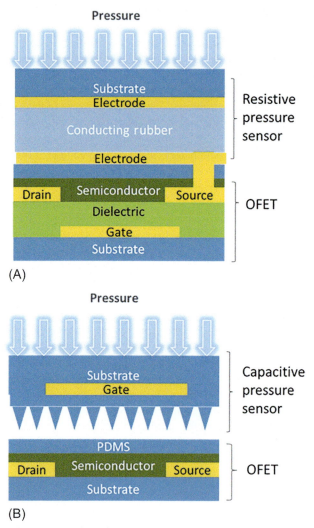

Fig. 4.2 Schematic representation of device structures of the OFET-based piezo-resistive and capacitive pressure sensors.

flexible pressure sensor with an OFET array. They used pentacene as an active layer of the OFET, and the pressure-sensitive conducting rubber was connected to the source electrodes for multiplexing. To connect a resistive sensor with a transistor, they first drilled a hole through the substrate and made an electronic connection between the two sides of the substrate. One side of the electrodes is used as the source electrode for the transistor and the other side is the electrode for the pressure-sensitive rubber layer. The pressure sensor array showed well-defined current response of a few μA at a pressure of 30 kPa, while the current is only tens of nA without pressure. To realize a large tactile sensory array for stretchable electronics, Martin et al. have demonstrated a stretchable active-matrix tactile sensor array on a prestretched ultrathin polymer foil, as shown in Fig. 4.3. The ultrathin electronic circuit looks like a feather and fully covers an uneven surface [14].

As mentioned in the last chapter, Mannsfeld and Tee et al. used elastic pyramids made by polydimethylsiloxane (PDMS) as a dielectric layer for capacitive pressure sensors. The pyramid structure led to a significant improvement in sensitivity ($0.55\ \mathrm{kPa}^{-1}$) in the low-pressure range ($<2\ \mathrm{kPa}$). Using a PDMS microstructured PDMS films, they fabricated an OFET with the pressure-sensing ability [15]. A rubrene single crystal as a semiconductor layer was placed on the top of the bottom-contact devices. The rubrene transistor was laminated with the structured PDMS/ITO/PET, as shown in Fig. 4.4. The tactile OFET showed similar fast response and relaxation when applying pressure. After a few years, Schwartz et al. further improved the device performance based on the pyramid structure, as shown in Fig. 4.5. They demonstrated a top-gate bottom-contact polymer-based OFETs with a piezo-capacitive PDMS pyramid layer [16]. Under applied pressure, the capacitance change of PDMS pyramid film resulted in the change of I_{DS}. I_{DS} can be affected by capacitance and gate voltages. Therefore, the output current is significantly amplified when giving a large V_{G}. The pressure sensitivity of $8.4\ \mathrm{kPa}^{-1}$ was achieved in the OFET-based pressure-sensitive active matrix, which is more than 10 times higher than the original capacitive sensor ($0.55\ \mathrm{kPa}^{-1}$) without coupling with a transistor. Due to its high sensitivity, the devices were demonstrated to monitor radial artery pulse waves. Compared with Mannesfeld's work in the tactile rubrene OFETs, Schwartz et al. exhibited much higher sensitivity. This may be attributed to the use of a PMDS interlayer on the top of the polymer semiconductor layer. The extra PDMS layer enables better contact with the PDMS pyramids and protection during the lamination.

4.4 Multifunctional transistor-based pressure sensors

The integration with transistors enables tactile sensors connect with other devices for the development of multifunctions devices [17]. In this section, we introduce several examples for the multifunction tactile OFETs. The first example is tactile transistor-type memory.

Sekitani et al. have reported a tactile OFET with a floating gate embedded in hybrid dielectrics [17]. The embedded floating gate causes nonvolatile memory effect in the OFETs. The hybrid dielectric of the OFET consisted of a self-assembly monolayer and a 4-nm-thick aluminum oxide. The ultrathin dielectric layer induces a high capacitance to drive the OFETs in the low voltage. The high capacitance allowed the devices to perform a threshold voltage shift in writing/erasing cycles under 6 V. Through connecting with a pressure-sensitive rubber sheet, the OFET-type memory can record the track of the applied pressure in a two-dimensional image. Chen et al. have developed another pressure memory device by integrating a paper-based piezo-resistive sensor with a chargeable donor-acceptor copolymer [18].

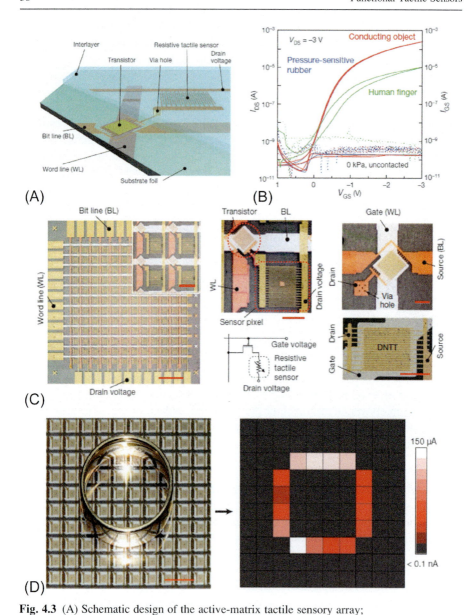

Fig. 4.3 (A) Schematic design of the active-matrix tactile sensory array; (B) pressure-dependent transfer curves of the active-matrix tactile sensory array; (C) circuit design of a resistive tactile sensor and a OFET; and (D) picture of the active-matrix tactile array and its corresponding current response with a metallic ring [14].
Reproduced with permission from M. Kaltenbrunner, T. Sekitani, J. Reeder, T. Yokota, K. Kuribara, T. Tokuhara, M. Drack, R. Schwödiauer, I. Graz, S. Bauer-Gogonea, S. Bauer, T. Someya, Nature 499 (2013) 458, Copyright © 2013, Springer Nature.

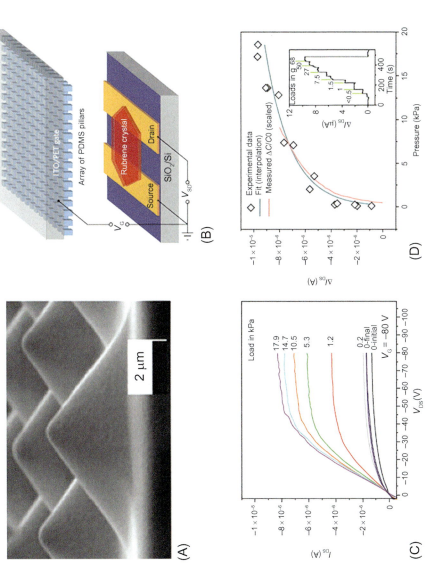

Fig. 4.4 (A) PDMS pyramids for capacitive pressure sensors; (B) schematic OFET structure for rubrene devices with structured PDMS dielectric layer; (C) pressure-dependent output curves of the rubrene OFETs; and (D) current response of the tactile transistor with the various pressures [15]. Reproduced with permission from S.C.B. Mannsfeld, B.C.K. Tee, R.M. Stoltenberg, C.V.H.H. Chen, S. Barman, B.V.O. Muir, A.N. Sokolov, C. Reese, Z. Bao, Nat. Mater. 9 (2010) 859, Copyright © 2010, Springer Nature.

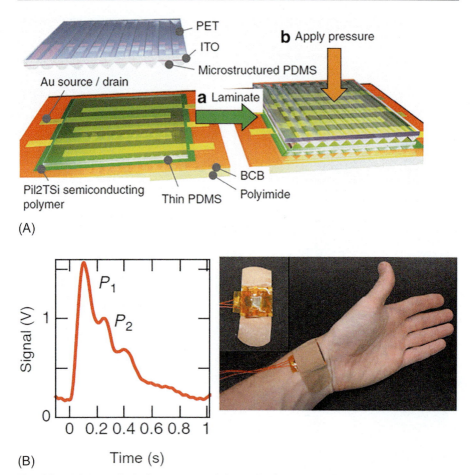

Fig. 4.5 (A) Schematic device structure of the tactile OFETs and (B) pulse wave of the radial artery of the hand performed from the PDMS pyramid OFET sensor [16].
Reproduced with permission from G. Schwartz, B.C.K. Tee, J. Mei, A.L. Appleton, D.H. Kim, H. Wang, Z. Bao, Nat. Commun. 4 (2013) 1859, Copyright © 2013, Springer Nature.

The piezo-resistive sensor has a porous structure, which utilizes a conductive polymer, PEDOT:PSS, and Ag nanowire composite. The ultrahigh sensitivity of 1089 kPa^{-1} can be achieved at the pressure below 1 kPa. The piezo-resistive pressure sensor was integrated with polymer-based nonvolatile transistor-type memory. The selenophene-diketopyrrolopyrrole donor-acceptor conjugated polymer (PSeDPP) acted as a chargeable semiconductor layer due to its ambipolarity. Therefore, the low-resistance state of the paper sensor can induce memory effects in the PSeDPP device. This tactile transistor memory is not only highly sensitive to pressure, but also can store the pressure information even after removing the pressure. The memory behaviors of this tactile transistor memory can record the trace of the applied touch (Fig. 4.6).

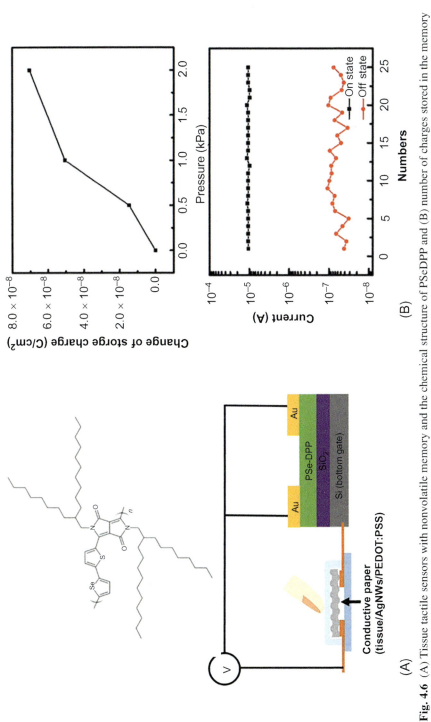

Fig. 4.6 (A) Tissue tactile sensors with nonvolatile memory and the chemical structure of PSeDPP and (B) number of charges stored in the memory under various pressures and the stability test of the ON/OFF state switching of write-read-erase-read cycles [18]. Reproduced with permission from C.-T. Chen, W.-Y. Lee, T.-L. Shen, H.-C. Wu, C.-C. Shih, B.-W. Ye, T.-Y. Lin, W.-C. Chen, Y.-F. Chen, Adv. Electron. Mater. 3 (2017) 1600548, John Wiley and Sons.

In addition to the tactile memory devices, researchers attempted to develop neuromorphologic electronics using the tactile sense. Bao and coworkers developed a bioinspired organic artificial afferent nerves consisting of tactile sensors, ring oscillators, and synaptic transistors [19, 20]. They used these devices to mimic the behaviors of a sensory nerve. The integrated circuit first collected tactile information (1–80 kPa) using carbon nanotube-based piezo-resistive pressure sensors as a mechanoreceptor. Then, the tactile information was converted by the ring oscillators into voltage pulses with different frequencies (0–100 Hz). The voltage pulses act as action potentials to simulate the synaptic transistors to induce learning behaviors. This demonstrates potential applications of the tactile sensors in neuromorphology engineering and artificial intelligent systems.

The Javey's group demonstrated an interactive electronic skin, which can visualize two-dimensional images to map the place with the applied pressure (Fig. 4.7) [21]. They combined flexible carbon nanotube OFETs, OLEDs, and pressure-resistant rubber sensors. In this integrated circuits, the active-matrix OLEDs exhibited instantaneous image response, where the surface was touched. The carbon nanotube OFETs showed a high mobility of 20 cm^2 V^{-1} s^{-1}, which sufficiently drove an OLED in a fast response time ($\sim$1 ms). However, the response time of the pressure-resistant rubber sheet is around 0.1 s. As a result, the switching speed of the interactive e-skin is limited by pressure-resistant rubber sheet. Despite the limited speed, this work still presented a pressure-sensitive active-matrix display involving the integration of pressure sensors, field-effect transistors, and OLED.

4.5 Perspective

To the realization of the applications of large-area low-power consumption electronic skins, pressure sensory arrays with an active matrix are necessary. OFETs are an essential element for active-matrix array. Recently, the charge mobilities of the polymer-based FETs have already achieved a high value over 10 cm^2 V^{-1} s^{-1} [22, 23]. This high-mobility feature enables development of high-speed sensor arrays. Although a simple passive matrix of pressure sensors was reported for large-area array [24], the passive-matrix design causes high-power consumption. Furthermore, additional electronic elements are required to reduce cross talks between devices for the passive-matrix array. In the active-matrix array, each sensor is integrated with one transistor. When applied pressure, the transistor can actively control the ON state of the sensor. This active-matrix design can save more power in the electronics with high-density sensors for high resolution.

Pressure-dependent resistive and capacitive sensors have made a great progress over the past decade. Many novel microstructures and pressure-sensitive materials have been reported in the literature [11–17, 21, 25–52]. However, a combination of the large-area sensory arrays with novel microstructures and materials is lagging behind a little. This might be attributed to the complicated manufacturing processes of multilayer OFET-based sensory arrays. We believe that

Tactile sensors based on organic field-effect transistors 63

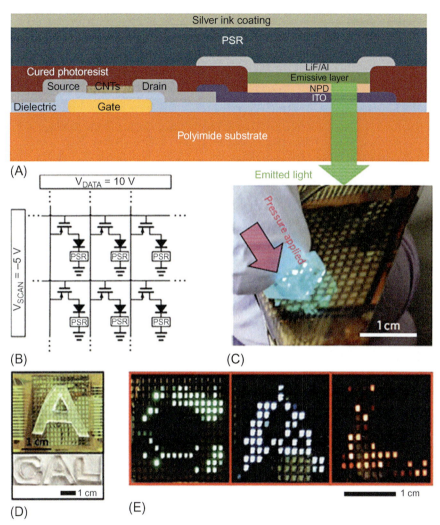

Fig. 4.7 User-interactive electronic skins [21]. (A) Schematic representation of device structures, (B) circuit designs of e-skin with the integration of an OFET, an OLED, and a pressure sensor; (C) photograph of the user-interactive e-skin when the devices is touched; (D) a PDMS stamp with CAL characters is pressed on the interactive e-skin; and (E) images of the interactive e-skin when pressing the PDMS stamp on the sensory array [21].
Reproduced with permission from C. Wang, D. Hwang, Z. Yu, K. Takei, J. Park, T. Chen, B. Ma, A. Javey, Nat. Mater. 12 (2013) 899, Copyright © 2013, Springer Nature.

the integration with OFETs is the only way toward the realization of the large-area electronic skins. For the electronic skin, softness is one of the key features. By taking advantage of the mechanical flexibility or stretchability of OFETs, artificial e-skin with a large area, high resolution, and stretchability could be developed in the near future.

References

[1] J. Zaumseil, H. Sirringhaus, Chem. Rev. 107 (2007) 1296.

[2] J. Xu, H.-C. Wu, C. Zhu, A. Ehrlich, L. Shaw, M. Nikolka, S. Wang, F. Molina-Lopez, X. Gu, S. Luo, D. Zhou, Y.-H. Kim, G.-J.N. Wang, K. Gu, V.R. Feig, S. Chen, Y. Kim, T. Katsumata, Y.-Q. Zheng, H. Yan, J.W. Chung, J. Lopez, B. Murmann, Z. Bao, Nat. Mater. 18 (2019) 594.

[3] J. Xu, S. Wang, G.-J.N. Wang, C. Zhu, S. Luo, L. Jin, X. Gu, S. Chen, V.R. Feig, J.W.F. To, S. Rondeau-Gagné, J. Park, B.C. Schroeder, C. Lu, J.Y. Oh, Y. Wang, Y.-H. Kim, H. Yan, R. Sinclair, D. Zhou, G. Xue, B. Murmann, C. Linder, W. Cai, J.B.-H. Tok, J.W. Chung, Z. Bao, Science 355 (2017) 59.

[4] J. Mei, D.H. Kim, A.L. Ayzner, M.F. Toney, Z. Bao, J. Am. Chem. Soc. 133 (2011), 20130.

[5] X. Gu, L. Shaw, K. Gu, M.F. Toney, Z. Bao, Nat. Commun. 9 (2018) 534.

[6] G. Giri, E. Verploegen, S.C.B. Mannsfeld, S. Atahan-Evrenk, D.H. Kim, S.Y. Lee, H.-A. Becerril, A. Aspuru-Guzik, M.F. Toney, Z. Bao, Nature 480 (2011) 504.

[7] Y. Diao, B.C.K. Tee, G. Giri, J. Xu, D.H. Kim, H.A. Becerril, R.M. Stoltenberg, T.H. Lee, G. Xue, S.C.B. Mannsfeld, Z. Bao, Nat. Mater. 12 (2013) 665.

[8] Z. Bao, J. Locklin (Eds.), Organic Field-Effect Transistors, CRC Press, Boca Raton, FL, 2007.

[9] J.Y. Oh, S. Rondeau-Gagné, Y.-C. Chiu, A. Chortos, F. Lissel, G.-J.N. Wang, B.C. Schroeder, T. Kurosawa, J. Lopez, T. Katsumata, J. Xu, C. Zhu, X. Gu, W.-G. Bae, Y. Kim, L. Jin, J.W. Chung, J.B.H. Tok, Z. Bao, Nature 539 (2016) 411.

[10] Y.-Y. Noh, N. Zhao, M. Caironi, H. Sirringhaus, Nat. Nanotechnol. 2 (2007) 784.

[11] W. Qi, J. Muqiang, W. Chunya, Z. Yingying, Adv. Funct. Mater. 27 (2017), 1605657.

[12] T. Someya, Y. Kato, T. Sekitani, S. Iba, Y. Noguchi, Y. Murase, H. Kawaguchi, T. Sakurai, Proc. Natl. Acad. Sci. U. S. A. 102 (2005), 12321.

[13] T. Someya, T. Sekitani, S. Iba, Y. Kato, H. Kawaguchi, T. Sakurai, Proc. Natl. Acad. Sci. U. S. A. 101 (2004) 9966.

[14] M. Kaltenbrunner, T. Sekitani, J. Reeder, T. Yokota, K. Kuribara, T. Tokuhara, M. Drack, R. Schwödiauer, I. Graz, S. Bauer-Gogonea, S. Bauer, T. Someya, Nature 499 (2013) 458.

[15] S.C.B. Mannsfeld, B.C.K. Tee, R.M. Stoltenberg, C.V.H.H. Chen, S. Barman, B.V.O. Muir, A.N. Sokolov, C. Reese, Z. Bao, Nat. Mater. 9 (2010) 859.

[16] G. Schwartz, B.C.K. Tee, J. Mei, A.L. Appleton, D.H. Kim, H. Wang, Z. Bao, Nat. Commun. 4 (2013) 1859.

[17] T. Sekitani, T. Yokota, U. Zschieschang, H. Klauk, S. Bauer, K. Takeuchi, M. Takamiya, T. Sakurai, T. Someya, Science 326 (2009) 1516.

[18] C.-T. Chen, W.-Y. Lee, T.-L. Shen, H.-C. Wu, C.-C. Shih, B.-W. Ye, T.-Y. Lin, W.-C. Chen, Y.-F. Chen, Adv. Electron. Mater. 3 (2017), 1600548.

[19] Y. Kim, A. Chortos, W. Xu, Y. Liu, J.Y. Oh, D. Son, J. Kang, A.M. Foudeh, C. Zhu, Y. Lee, S. Niu, J. Liu, R. Pfattner, Z. Bao, T.-W. Lee, Science 360 (2018) 998.

[20] B.C.-K. Tee, A. Chortos, A. Berndt, A.K. Nguyen, A. Tom, A. McGuire, Z.C. Lin, K. Tien, W.-G. Bae, H. Wang, P. Mei, H.-H. Chou, B. Cui, K. Deisseroth, T.N. Ng, Z. Bao, Science 350 (2015) 313.

[21] C. Wang, D. Hwang, Z. Yu, K. Takei, J. Park, T. Chen, B. Ma, A. Javey, Nat. Mater. 12 (2013) 899.

[22] Z. Yi, S. Wang, Y. Liu, Adv. Mater. 27 (2015) 3589.

[23] I. Kang, H.-J. Yun, D.S. Chung, S.-K. Kwon, Y.-H. Kim, J. Am. Chem. Soc. 135 (2013), 14896.

[24] T. Azuma, Sens. Mater. 13 (2001) 107.

[25] J.C. Yang, J.-O. Kim, J. Oh, S.Y. Kwon, J.Y. Sim, D.W. Kim, H.B. Choi, S. Park, ACS Appl. Mater. Interfaces 11 (2019), 19472.

[26] W. Xuewen, G. Yang, X. Zuoping, C. Zheng, Z. Ting, Adv. Mater. 26 (2014) 1336.

[27] X. Wang, M. Que, M. Chen, X. Han, X. Li, C. Pan, Z.L. Wang, Adv. Mater. 29 (2017), 1605817.

[28] P. Wang, S.M. Zakeeruddin, J.E. Moser, M.K. Nazeeruddin, T. Sekiguchi, M. Grätzel, Nat. Mater. 2 (2003) 402.

[29] C. Wang, W.-Y. Lee, D. Kong, R. Pfattner, G. Schweicher, R. Nakajima, C. Lu, J. Mei, T.H. Lee, H.-C. Wu, J. Lopez, Y. Diao, X. Gu, S. Himmelberger, W. Niu, J.R. Matthews, M. He, A. Salleo, Y. Nishi, Z. Bao, Sci. Rep. 5 (2015), 17849.

[30] L. Viry, A. Levi, M. Totaro, A. Mondini, V. Mattoli, B. Mazzolai, L. Beccai, Adv. Mater. 26 (2014) 2659.

[31] M. Stoppa, A. Chiolerio, Sensors 14 (2014), 11957.

[32] A.V. Shirinov, W.K. Schomburg, Sens. Actuators A Phys. 142 (2008) 48.

[33] M.E. Roberts, N. Queraltó, S.C.B. Mannsfeld, B.N. Reinecke, W. Knoll, Z. Bao, Chem. Mater. 21 (2009) 2292.

[34] C. Pang, J.H. Koo, A. Nguyen, J.M. Caves, M.-G. Kim, A. Chortos, K. Kim, P.J. Wang, J.B.H. Tok, Z. Bao, Adv. Mater. 27 (2015) 634.

[35] B. Nie, R. Li, J. Cao, J.D. Brandt, T. Pan, Adv. Mater. 27 (2015) 6055.

[36] C. Lu, W.-Y. Lee, C.-C. Shih, M.-Y. Wen, W.-C. Chen, ACS Appl. Mater. Interfaces 9 (2017), 25522.

[37] L. Lin, Y. Xie, S. Wang, W. Wu, S. Niu, X. Wen, Z.L. Wang, ACS Nano 7 (2013) 8266.

[38] I. Lee, H.J. Sung, Exp. Fluids 26 (1999) 27.

[39] Y.-C. Lai, H.-M. Wu, H.-C. Lin, C.-L. Chang, H.-H. Chou, Y.-C. Hsiao, Y.-C. Wu, Adv. Funct. Mater. 29 (2019), 1904626.

[40] D. Kong, R. Pfattner, A. Chortos, C. Lu, A.C. Hinckley, C. Wang, W.-Y. Lee, J.W. Chung, Z. Bao, Adv. Funct. Mater. 26 (2016) 4680.

[41] S.H. Kim, K. Hong, W. Xie, K.H. Lee, S. Zhang, T.P. Lodge, C.D. Frisbie, Adv. Mater. 25 (2013) 1822.

[42] M.L. Jin, S. Park, Y. Lee, J.H. Lee, J. Chung, J.S. Kim, J.-S. Kim, S.Y. Kim, E. Jee, D.W. Kim, J.W. Chung, S.G. Lee, D. Choi, H.-T. Jung, D.H. Kim, Adv. Mater. 29 (2017), 1605973.

[43] M.L. Jin, S. Park, J.-S. Kim, S.H. Kwon, S. Zhang, M.S. Yoo, S. Jang, H.-J. Koh, S.-Y. Cho, S.Y. Kim, C.W. Ahn, K. Cho, S.G. Lee, D.H. Kim, H.-T. Jung, Adv. Mater. 30 (2018), 1706851.

[44] L. Jaehong, K. Hyukho, S. Jungmok, S. Sera, K.J. Hoon, P. Changhyun, S. Seungbae, K.J. Hyung, J.Y. Hoon, K.D. Eun, L. Taeyoon, Adv. Mater. 27 (2015) 2433.

[45] M. Hussain, Y.-H. Choa, K. Niihara, J. Mater. Sci. Lett. 20 (2001) 525.

[46] S. Gong, W. Schwalb, Y. Wang, Y. Chen, Y. Tang, J. Si, B. Shirinzadeh, W. Cheng, Nat. Commun. 5 (2014) 3132.

[47] F.-R. Fan, L. Lin, G. Zhu, W. Wu, R. Zhang, Z.L. Wang, Nano Lett. 12 (2012) 3109.

[48] C. Chwee-Lin, S. Mun-Bo, L. Byoung-Sun, J. Sanghun, K. Dong-Su, K. Tae-Hyung, B. Jihyun, L.S. Hoon, B. Kyung-Eun, I. Jungkyun, J.Y. Jin, P.C. Eon, P. Jong-Jin, C. U-In, Adv. Mater. 26 (2014) 3451.

[49] H.-H. Chou, A. Nguyen, A. Chortos, J.W.F. To, C. Lu, J. Mei, T. Kurosawa, W.-G. Bae, J.B.H. Tok, Z. Bao, Nat. Commun. 6 (2015) 8011.

[50] A. Chortos, J. Liu, Z.A. Bao, Nat. Mater. 15 (2016) 937.

[51] S.H. Cho, S.W. Lee, S. Yu, H. Kim, S. Chang, D. Kang, I. Hwang, H.S. Kang, B. Jeong, E.H. Kim, S.M. Cho, K.L. Kim, H. Lee, W. Shim, C. Park, ACS Appl. Mater. Interfaces 9 (2017), 10128.

[52] V. Amoli, J.S. Kim, E. Jee, Y.S. Chung, S.Y. Kim, J. Koo, H. Choi, Y. Kim, D.H. Kim, Nat. Commun. 10 (2019) 4019.

Conductive composite-based tactile sensor

Haotian Chen[a] and Haixia Zhang[b,*]
[a]Laboratory for Soft Bioelectronic Interfaces, Centre for Neuroprosthetics, École Polytechnique Fédérale de Lausanne (EPFL), Geneva, Switzerland, [b]National Key Laboratory of Nano/Micro Fabrication Technology, Institute of Microelectronics, Peking University, Beijing, China
*Corresponding author: e-mail address: hxzhang@pku.edu.cn

5.1 Introduction

The field of stretchable electronics has developed rapidly in recent years due to their potential importance in diverse fields like health care, prosthesis, and robotics. The future development of electronics is expected to incorporate multifunctional electronic systems in human body or soft robots and imitate the function of the human tactile system, which requires the devices should be able to work stably under large strain or deformation. However, the combination of mechanical compliance and electrical conductivity is almost inaccessible to conventional, rigid silicon-based electronics. In order to solve this problem, scientists provided many efforts in the past 10 years, which can be summarized as two major approaches:

(1) making nonstretchable conductive materials become stretchable by new design and
(2) creating new stretchable and conductive materials.

There are several research groups that have made wonderful efforts in these two directions.

With the first strategy, Rogers et al. in Northwestern University have developed a number of techniques, which make silicon [1], gold [2], and some other conventionally used material [3] stretchable with delicate structures like serpentine [4], buckling [5], origami [6], kirigami [7], and so on. Despite its good working performance, the complex fabrication and high cost make it hard to prepare on a large scale.

As for the second approach, the rapid development of materials science in the recent days ranging from organic material to inorganic material with various forms like nanowire, nanotube, and two-dimensional (2D) materials provides more methods for stretchable devices, such as light-emitting diodes [8, 9], field-effect transistor [10], and tactile sensors [11] for electronic skin.

Compared to the first strategy, the second route is becoming more popular, especially as regards stretchable conductors and sensors due to its easy processing method and tunable properties.

Functional Tactile Sensors. https://doi.org/10.1016/B978-0-12-820633-1.00010-3
© 2021 Elsevier Ltd. All rights reserved.

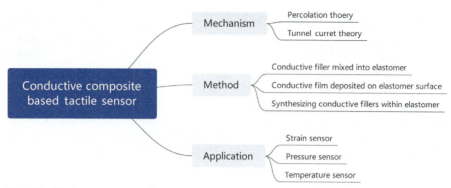

Fig. 5.1 Chapter structure outline.

As a result, in this chapter, we mainly discuss the stretchable conductive composite (SCC)-based tactile sensor. The outline of this chapter is shown in Fig. 5.1. First, a brief overview of this field is given. Second, we introduce the working mechanism of conductive composites like the conductor–insulator transition and percolation theory. The methods of preparing the conductive composites are presented in the third part. In the final part, some demonstrations of how to use SCC in various application fields including strain sensor, pressure sensor, temperature sensor, etc. are given.

5.2 Working mechanism

Firstly, we need to understand how SCC works as a conductive material and what kind of parameters will influence its properties. There are two main theories to explain it: percolation theory and tunnel current theory. In this section, we first review the classic percolation theory, which is widely used in all kinds of composites. Then, based on this theory, tunnel current theory, which plays a bigger role in the nanoscale, is introduced for conductive composites composed from nanoparticles, nanowires, nanotubes, and nanosheets.

5.2.1 Percolation theory

Through the incorporation of the conductive fillers into the elastomer matrix, the characteristics of the polymer, such as flexibility and stretchability, can be combined with the mechanical and conductive properties of the fillers. The electrical properties of SCC are dependent on the conductive networks formed by the conductive fillers. The increase of conductive fillers will no doubt enhance the conductivity of the material, however, leading to poor mechanical properties.

At the beginning process of mixing conductive fillers into elastomers, the SCC is not conductive at all due to the polymer blocking the conductive parts forming a connected path. As the content of conductive filler increases, a jump of conductivity will be found when a critical filler content is reached, as shown in Fig. 5.2. This specific

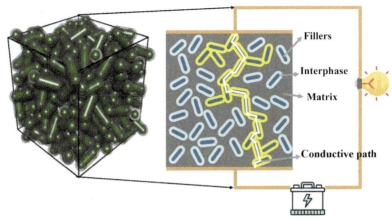

Fig. 5.2 Schematic view of the conductive composite with percolation path.
Reproduced with permission from Ref. W. Xu, P. Lan, Y. Jiang, D. Lei, H. Yang, Insights into excluded volume and percolation of soft interphase and conductivity of carbon fibrous composites with core-shell networks, Carbon 161 (2020) 392–402, copyright 2020 Elsevier.

content is termed as the "electrical percolation threshold" (p_c). Many parameters including the material of the conductive filler, the material of the elastomer, the geometry feature of the conductive filler, the dimension of the desired SCC, and so on can influence p_c [12].

The conductivity of SCC with an increasing amount of conductive filler can be determined by an equation according to the classical percolation theory [13]:

$$\sigma = \sigma_0 (p - p_c)^t \tag{5.1}$$

where σ is the conductivity of the SCC, p_c is the percolation threshold as mentioned, p is the filler content in SCC, σ_0 is a scaling factor proportional to the intrinsic conductivity of the filler, and t is the critical exponent related to the dimensionality of conductive network inside the SCC. Usually, t is taken as 1.3 and 2 for 2D and three-dimensional (3D) networks, respectively [14–16]. Actually, t values from 1.0 to 10 have been reported according to both the experimental and simulation results [17, 18]. The difference between the theoretical prediction and experimental measurement is still an important but unresolved issue.

5.2.2 Tunnel current theory

Initially, the insulator–conductor transition in SCC was considered to be only associated with the conductive networks throughout the whole sample. Currently, as more and more studies focus on the nanomaterial-based SCC, it is indicated that the electrons can flow through the insulating barrier via quantum mechanical tunneling between adjacent conductive regions. According to the numerical studies, the smaller the aspect ratio of the filler is, the more important role the tunneling current plays [19–21].

Kim et al. from the Hong Kong University of Science and Technology proposed a series of simple but effective models to calculate the p_c for spherical particles, 2D plates, and one-dimensional (1D) nanotubes or nanowires mixed within elastomers. Firstly, they assume each filler occupies a cubic unit cell (V_{c_sphere}, V_{c_plate}, or $V_{c\_nanotube}$, dependent on geometry feature of filler). For 2D and 1D fillers, the orientation and aspect ratio should be taken into consideration to create the cube. In this way, the average interparticle distance (D_{IP}) at which the electron tunneling happens can be regarded as the percolation threshold. In this way, they gave the equation for randomly distributed conductive fillers as follows [22, 23]:

$$V_{c\_sphere} = \frac{\pi D^3}{6(D + D_{IP})^3} \tag{5.2}$$

$$V_{c\_plate} = \frac{27\pi D^2 t}{4(D + D_{IP})} \tag{5.3}$$

$$V_{c_{nanotube}} = \frac{\xi\varepsilon\left(\frac{\pi D^3}{6}\right)}{(D + D_{IP})} + \frac{(1-\xi)27\pi D^2 l}{4(l + D_{IP})^3} \tag{5.4}$$

where D is the diameter of spherical particles, lateral diameter of plates, and cross-sectional diameter of nanotubes, t is the thickness of nanoplates, l is the length of nanotubes, ξ is the volume fraction of agglomerated nanotubes, and ε is the volume of nanotubes in an agglomerate.

Based on Eqs. (5.2)–(5.4), Fig. 5.3 shows how the p_c would change by changing according to size, thickness, and aspect ratio of the conductive fillers. They indicated for 3D fillers, that when the size of the particle increases, the p_c would also increase, while for 2D and 1D fillers, p_c will decrease as the size of filler increases.

5.3 Preparation methods

In order to fabricate the conductive composite, diverse materials with several different methods were studied. Here, we focus on the formation process and the relationship between conductive fillers and elastomers of these materials instead of the specific approaches or materials. In general, we divided all the conductive composites into three main categories: conductive filler mixed into elastomer, conductive film deposited on the elastomer surface, and synthesizing conductive fillers within an elastomer.

5.3.1 Conductive filler mixed into elastomer (3D)

Mixing conductive filler into elastomer is the most common and intuitive route to form a conductive composite. Metal and carbon-based nanomaterials are two kinds of appealing candidate to develop SCC.

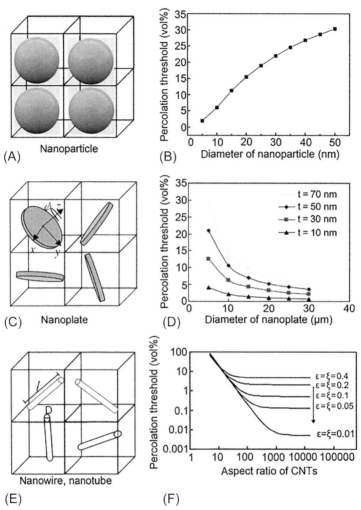

Fig. 5.3 (A), (C), and (E) show the schematic pictures of how the conductive filler can be located in a cell with different orientations. (B), (D), and (F) show the relationship between the predicted threshold percolation and sizes of different nanomaterials.
Reproduced with permission from Ref. M. Park, J. Park, U. Jeong, Design of conductive composite elastomers for stretchable electronics, Nano Today 9 (2014) 244–260, copyright 2014 Elsevier.

Ching-Ping Wong and coworkers from Georgia Institute of Technology proposed a facile and low-cost method to fabricate core–shell polystyrene/silver (PS@Ag) hybrid conductive particles dispersed in PDMS [24]. The core–shell structure, on the one hand, possesses high conductivity. On the other hand, it has lower density compared to pure silver. The microsphere shape helps to avoid aggregation in the polymeric matrix. PS@Ag hybrid particles were prepared using the as-prepared cross-linked

PS microspheres as polymer cores by an electroless plating method to coat sliver. Then, the particles were mixed with PDMS in a fast mixer at 2000 rpm for 5 min to get a viscous mixture, which could be cast into a mold to obtain the desired pattern (Fig. 5.4).

CNT is another common conductive filler material for fabricating conductive composites because of its comparatively high conductivity and low cost. A CNT-PDMS-based active sensor for sliding detection was proposed by our group from Peking University [25]. With the help of toluene, CNTs can be dispersed into PDMS uniformly with magnetic stirring for 4 h. After blending, toluene can evaporate away by itself. A 3D printing mold was created to shape the composite into different patterns, providing a customized way to develop various devices with this material (Fig. 5.5).

In addition to directly blending CNTs with PDMS, we developed another rapid but effective method [26]. First, a PDMS scaffold was prepared with cube sugar as the pore-creating agent. After it was prepared with sodium dodecylbenzenesulfonate (SDBS) as the surfactant, CNT inks were subsequently dropped onto the PDMS sponge. The resistance of the composite can be modulated by the dropping cycles. After rinsed by deionized water, the surfactant can be removed and CNT was coated onto the scaffold surface of PDMS uniformly (Fig. 5.6).

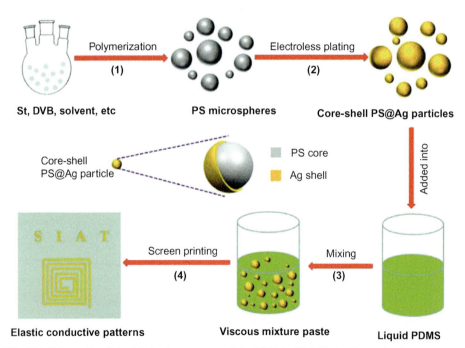

Fig. 5.4 Schematic of the fabrication process of the PS@Ag/PDMS elastic conductive patterns. Reproduced with permission from Ref. Y. Hu, T. Zhao, P. Zhu, Y. Zhu, X. Shuai, X. Liang, R. Sun, D. Daniel Lu, C.-P. Wong, Low cost and highly conductive elastic composites for flexible and printable electronics, J. Mater. Chem. C 4 (2016) 5839–5848, copyright 2016 RSC Publishing.

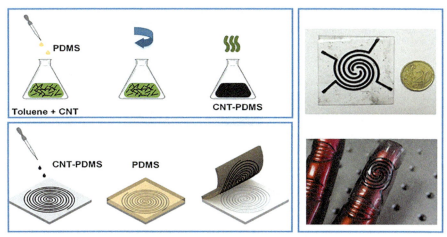

Fig. 5.5 Fabrication flow of prepare CNT-PDMS and application.
Reproduced with permission from Ref. H. Chen, L. Miao, Z. Su, Y. Song, M. Han, X. Chen, X. Cheng, D. Chen, H. Zhang, Fingertip-inspired electronic skin based on triboelectric sliding sensing and porous piezoresistive pressure detection, Nano Energy 40 (2017) 65–72, copyright 2017 Elsevier.

5.3.2 Conductive film deposited on elastomer surface (2D)

Although the directly mixing conductive filler with elastomer is a convenient method to achieve SCC, it is hard to make a thin device due to the viscosity of the composite. In order to solve this problem, the idea of depositing conductive film onto elastomer was proposed. Compared to embedding, deposition of conductive materials onto the surface of soft substrate not only provides an efficient way of fabricating thin film device, but also avoid the potential problem of nonuniform distribution of conductive fillers inside elastomer, which improve the stability and reliability.

Lacour and Suo from Princeton University first reported a method to deposit gold onto PDMS to form the stretchable interconnects by electron beam evaporation [27]. The thin film of gold is deposited with compressive stress, which causes the film to buckle. Under tensile strain, the buckling disappears when the external strain matches the as-deposited compressive strain in the film. In this case, the stretchable interconnect can sustain its function when stretched and relaxed by 12% strain (Fig. 5.7A). In addition, they presented another engineering method of prestretch and release PDMS substrate before and after sputtering gold to obtain stretchable gold strips, which can sustain good conductivity (38–70 Ω) under 15% strain (Fig. 5.7B) [28].

Compared to directly depositing metal onto the target surface, the solution-based fabrication method provides a simple and facile way to form a uniform film. Among various kinds of metal nanowires, silver nanowire (AgNW) is the most widely used material due to its incomparably high conductivity. Byun et al. from Sungkyukwan University proposed a stretchable AgNW/PDMS composite strain sensor using dispensing nozzle printing [29]. The electrical resistance could be modified by controlling the

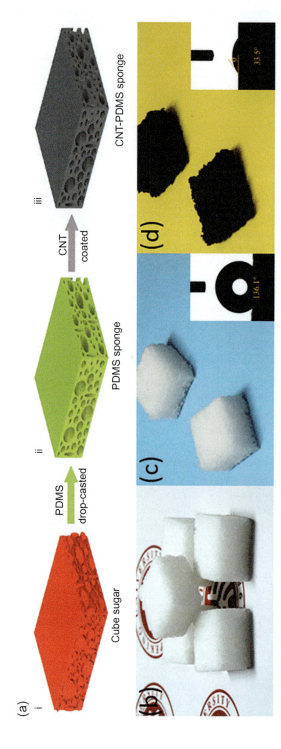

Fig. 5.6 CNT-PDMS sponge-based conductive composite. (A) Fabrication flow from commercial cube sugar to CNT-PDMS sponge. Photographs of cube sugar (B), PDMS sponge casted from cube sugar (C), and CNT-PDMS sponge (D), respectively. Reproduced with permission from Ref. Y. Song, H. Chen, Z. Su, X. Chen, L. Miao, J. Zhang, X. Cheng, H. Zhang, Highly compressible integrated supercapacitor–piezoresistance-sensor system with CNT-PDMS sponge for health monitoring, Small 13 (2017) 1702091, copyright 2017 WILEY-VCH Verlag GmbH & Co. KGaA, Weinheim.

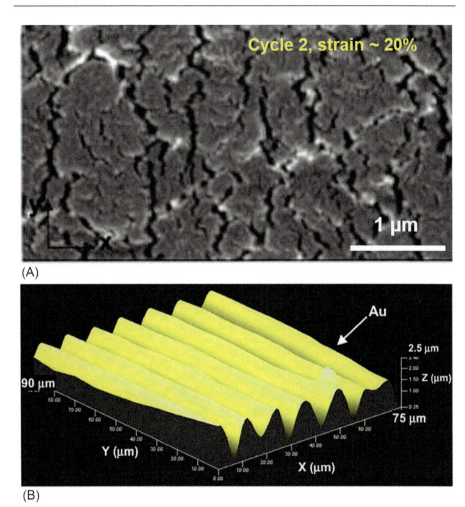

Fig. 5.7 Different formats of thin metal films on elastomeric substrates forming the stretchable conductive materials. (A) SEM image of gold microcracks at 20% elongation during the second stretching cycle. (B) 3D profile of a gold surface wave after release from 15% prestretch. Panel (A): Reproduced with permission from Ref. S.P. Lacour, D. Chan, S. Wagner, T. Li, Z. Suo, Mechanisms of reversible stretchability of thin metal films on elastomeric substrates, Appl. Phys. Lett. 88 (2006) 204103, copyright 2006 AIP Publishing, Panel (B): Reproduced with permission from Ref. S. Wagner, S.P. Lacour, J. Jones, P.I. Hsu, J.C. Sturm, T. Li, Z. Suo, Electronic skin: architecture and components, Physica E Low Dimens. Syst. Nanostruct. 25 (2004) 326–334, copyright 2004 Elsevier.

printing speed because the number of conductive fillers was proportional to the liquid ejection time (Fig. 5.8).

Despite its convenient way to design an arbitrary pattern, printing is not a good choice for large-scale fabrication. In this way, diverse coating methods were developed and expected to be integrated with traditional microfabrication processes.

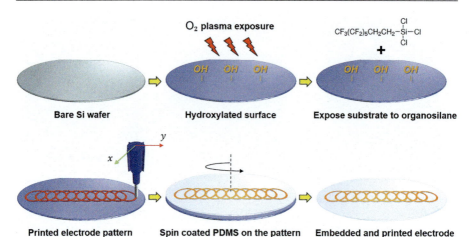

Fig. 5.8 Fabrication processes of stretchable printed AgNW/PDMS composite. Reproduced with permission from Ref. H. Lee, B. Seong, H. Moon, D. Byun, Directly printed stretchable strain sensor based on ring and diamond shaped silver nanowire electrodes, RSC Adv. 5 (2015) 28379–28384, copyright 2015 RSC Publishing.

Our group adopted the classic spin-coating method to directly deposit AgNW onto PDMS [30]. After obtaining the normal PDMS from wafer, the PDMS was kept in a prestrain state and treated it with oxygen plasma to form a hydrophilic surface. Then, AgNW solution (in ethanol) was spin coated 10 times to make a conductive network. After the AgNW/PDMS film was released, the sample was placed on the hot plate at 100°C to anneal. Finally, another layer of PDMS was also spin coated to encapsulate the conductive path (Fig. 5.9A).

In addition to direct spin coating, Jong-Woong Kim's group from Korea Electronics Technology Institute developed a method to transfer patterned AgNW film onto a specific polymer [31]. First, they spin coated the AgNW solution (in isopropanol) onto the Kapton film and heated it on a hot plate at 60°C for 10 min to remove the remaining organic solvent. The tandem compound pattern was prepared using lithography, after which the AgNWs were etched with an acid solution (chromium etchant). After stripping the photoresist, they could obtain a photolithography-leveled high-resolution AgNW pattern. In order to transfer AgNW from Kapton to a new substrate (PDMS/polyurethane urea), a roll laminator was utilized to help AgNW adhere to the elastomer surface. After the transferring, three cycles of intense-pulsed-light irradiation with 2.0 kV voltage and 500 μs pulse duration were applied to the composites to further enhance the adhesion between the nanowires and polyurethane urea (Fig. 5.9B).

Besides 2D metal nanomaterial, carbon nanotube (CNT) as a popular 2D carbon-based nanomaterial was also adopted to develop SCC by lots of groups. Park et al. from KAIST demonstrated a sandwich-structured strain sensor based on highly purified, solution-processed, 99% metallic CNT-PDMS thin film with 92% transparency [32]. After it was treated by oxygen plasma, the PDMS substrate was functionalized with an amine-terminated group to hold CNTs using a 0.1 g/mL poly-L-lysine

Conductive composite-based tactile sensor

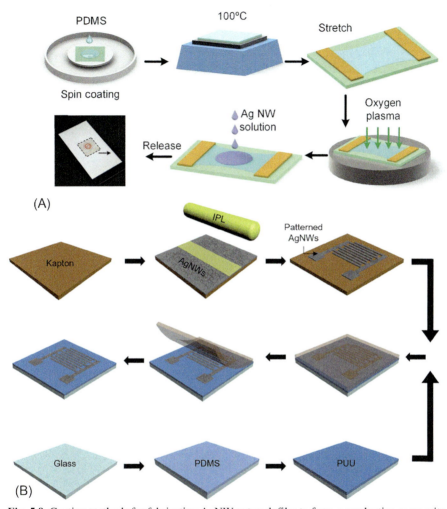

Fig. 5.9 Coating methods for fabricating AgNW network film to form a conductive composite. (A) Fabrication flow of spin-coating AgNW on prestretched PDMS. (B) Fabrication procedure for AgNW/PUU/PDMS-based pressure sensors by the transferring method.
Panel (A): Reproduced with permission from Ref. M. Shi, H. Wu, J. Zhang, M. Han, B. Meng, H. Zhang, Self-powered wireless smart patch for healthcare monitoring, Nano Energy 32 (2017) 479–487, copyright 2017 Elsevier. Panel (B): Reproduced with permission from Ref. B. You, C. Jong Han, Y. Kim, B.-K. Ju, J.-W. Kim, A wearable piezocapacitive pressure sensor with a single layer of silver nanowire-based elastomeric composite electrodes, J. Mater. Chem. A 4 (2016) 10435–10443, copyright 2016 RSC Publishing.

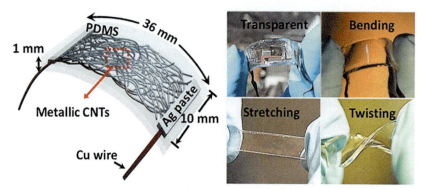

Fig. 5.10 Schematic illustration of CNT-PDMS composite strain sensor in the form of a sandwich structure and photographs showing its state of bending, stretching, and twisting. Reproduced with permission from Ref. J. Lee, M. Lim, J. Yoon, M.S. Kim, B. Choi, D.M. Kim, D.H. Kim, I. Park, S.-J. Choi, Transparent, flexible strain sensor based on a solution-processed carbon nanotube network, ACS Appl. Mater. Interfaces 9 (2017) 26279–26285, copyright 2017 American Chemical Society.

solution. The CNT network was created by spray coating of 99% metallic CONT solution with concentration of 0.01 mg/mL directly onto PDMS substrate. The density of the CNT network can be easily controlled by adjusting the times of CNT spray. Finally, the composite film was annealed at 90°C for an hour (Fig. 5.10).

As different materials have their own electromechanical properties, there is always a challenge in combination of sensitivity and dynamic range when they were made into sensors. Lipomi et al. from UCSD reported a highly sensitive and robust SCC-based strain sensor containing three-layered materials: graphene, palladium, and PEDOT:PSS [33]. In order to get this multilayer composite, first a granular film of palladium was evaporated on top of single-layer graphene supported on copper foil, followed by a sputter deposition of 50 nm aluminum through a stencil. The Al film served as an etch resist, which could define the serpentine structure. Following the etching process, poly(methyl methacrylate) (PMMA) was deposited on the copper/Gr/Pd substrate. After copper was removed, the graphene/Pd/PMMA layer was transferred by a water bath onto a PDMS layer. Then, the PEDOT:PSS was deposited by spray coating once the PMMA dissolved in acetone to release the soft part. The detailed process flow is shown in Fig. 5.11. The advantage of stacking these materials in a structured blend instead of mixing them into one layer is that its high sensitivity (0.001% strain) can be retained with a wide dynamic range (86%).

5.3.3 Synthesizing conductive fillers within elastomer

The main issue for above two methods is that conductive fillers tend to aggregate at low-volume fraction before the percolation threshold can be reached. In-situ polymerization is an effective way to solve the problem as the polymer chain and fillers can be

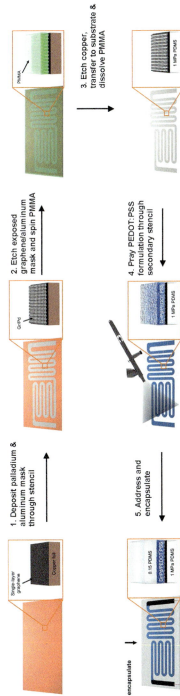

Fig. 5.11 Schematic illustration of process used to fabricate the structured blend of Gr/Pd/PEDOT:PSS for use as a wearable strain sensor. Reproduced with permission from Ref. J. Ramírez, D. Rodriquez, A.D. Urbina, A.M. Cardenas, D.J. Lipomi, Combining high sensitivity and dynamic range: wearable thin-film composite strain sensors of graphene, ultrathin palladium, and PEDOT:PSS, ACS Appl. Nano Mater. 2 (2019) 2222–2229, copyright 2019 American Chemical Society.

dispersed and grafted on the molecular scale, which ensures an excellent filler dispersion and good interfacial strength between filler and elastomer matrix.

This approach usually starts by mixing a metal salt in a cross-linkable prepolymer, or in a polymer gel that can be swollen by the metal salt. The metal precursors absorbed in the cross-linked gels are reduced to form metal nanoparticles by dipping in a reducing agent solution or exposure to a reductant vapor. This method not only enhanced the volume fraction due to the good miscibility of metal precursors with elastomer, but also reduced the aggregation because the movement of in situ synthesized nanoparticles was restricted by the elastomer matrix.

Jeong et al. from Yonsei University demonstrated a conductive composite mat of silver nanoparticles and rubber fibers that allows the formation of highly stretchable circuits through a fabrication process that is compatible with any substrate and scalable for large-area application [34]. First, a nonwoven mat of electrospun styrene-block-butadiene-block-styrene (SBS) fibers was collected and then dipped in a silver precursor solution ($AgCF_3COO$ in ethanol). The precursor and the solvent are absorbed by the fibers, thus making the fiber mat swollen. After drying, the precursor is reduced by a hydrazine hydrate ($N_2H_4 \cdot 4H_2O$) solution, which produce silver nanoparticles inside the fiber and silver shells at the surfaces of the fibers. The elasticity was maintained at much the same level with the as-spun fiber mat, which was attributed to the curved contour pathway of the fibers. The geometrical confinement of the nanoparticles in the microscale rubber dimension helped the composite retain a high conductivity (2500 S cm^{-1}) at 100% strain (Fig. 5.12).

Despite all the advantages of the in situ synthesizing method to obtain the SCC, the main problem is the material limitation. Currently, most groups only focus on the metal material and the kind of polymer to be mixed has its own limitation to cooperate with the metal precursor, which restricts its application fields.

5.4 Applications

Due to its integration of stretchability and conductivity, SCC is widely utilized in soft sensors. By modulating the material and content of conductive fillers, the SCC can obtain different properties as desired. Furthermore, with the help of patterning techniques, SCC can be shaped into diverse structures and shapes to further enhance the performance as a sensor. In this case, many research studies focus on how to better use SCC's unique properties to achieve a high-performance sensors. Among all tactile sensors, strain sensor, pressure sensor, and temperature sensor are the most common and important three sensors for constructing a tactile system, which is our main focus in this chapter.

5.4.1 Strain sensor

SCC is the ideal material for fabricating resistive sensors because its resistance will change obviously under deformation because the microscaled conductive network can be easily changed under all kinds of external mechanical stimuli, thus resulting into

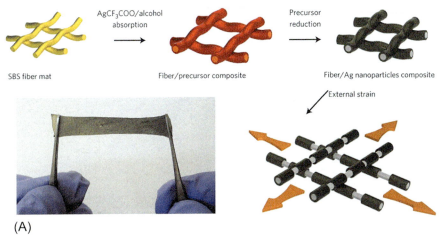

Fig. 5.12 Schematic of the processes to synthesize conductive fillers inside an elastomer. Reproduced with permission from Ref. M. Park, J. Im, M. Shin, Y. Min, J. Park, H. Cho, S. Park, M.-B. Shim, S. Jeon, D.-Y. Chung, J. Bae, J. Park, U. Jeong, K. Kim, Highly stretchable electric circuits from a composite material of silver nanoparticles and elastomeric fibres, Nat. Nanotechnol. 7 (2012) 803–809, copyright 2012 Springer Nature.

the resistance signal. Usually, the filler content should be above the percolation threshold to ensure a stable sensing behavior. The sensing mechanism can be explained as follows. When the strain is small, the connections between neighboring fillers are still there while the tunneling current contributes more to the resistance change. When the strain increases, the distance is beyond the tunneling distance so that the tunnel current almost disappears between original neighboring fillers and the conductive network rebuild in this period, generating much more change of resistance.

CNTs are a widely utilized conductive filler for resistive composite strain sensors due to their excellent mechanical and electrical properties. A representative work integrates randomly oriented and well-aligned CNTs, acting as sensitive and stretchable conductive elements, respectively, into a continuous changing structure [35]. Three types of CNT films (CNT networks, CNT arrays, and gradient CNTs) were studied as strain sensors. The anisotropic CNT strain sensor can be stretched up to 270% with a sensitivity of gauge factor (GF) of 0.85. On the contrary, the sensitivity of the isotropic CNTs strain sensors is ultrahigh (GF = 175), but it cannot work above the applied strain of 40%. The gradient CNTs strain sensors, in which the anisotropic and isotropic domains are combined, exhibit both high sensitivity (GF = 13.5, strain from 200% to 500% with a linearity of 0.98; GF = 7.7, strain from 0% to 200% with a linearity of 0.96), and a large working range (0%–570%) (Fig. 5.13).

Percolation networks formed by metal can offer comparable performance. Park et al. from KAIST presented a highly flexible, stretchable, and sensitive strain sensors based on the nanocomposite of silver nanowire (AgNW) network embedded between two PDMS layers [36]. The AgNW-based strain sensors show strong piezoresistivity

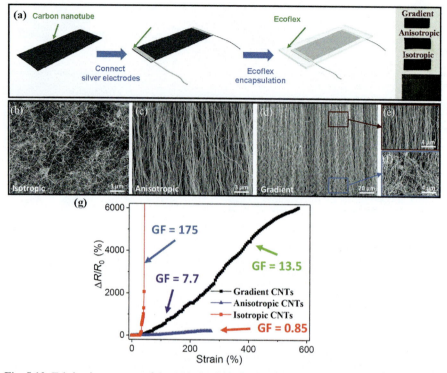

Fig. 5.13 Fabrication process of three kinds of CNTs strain sensors, their SEM characterization and their working performances. (A) Fabrication of the strain sensors with different CNTs. SEM images of isotropic CNTs (B), anisotropic CNTs (C), gradient CNTs (D), anisotropic part of gradient CNTs (E), and isotropic part of gradient CNTs (F). (G) Relative resistance response of isotropic, anisotropic, and gradient CNTs strain sensors under different applied strain. Reproduced with permission from Ref. B. Liang, Z. Lin, W. Chen, Z. He, J. Zhong, H. Zhu, Z. Tang, X. Gui, Ultra-stretchable and highly sensitive strain sensor based on gradient structure carbon nanotubes, Nanoscale 10 (2018) 13599–13606, copyright 2018 RSC Publishing.

with tunable gauge factors in the ranges of 2–14 and a high stretchability up to 70% (Fig. 5.14).

5.4.2 Pressure sensor

Pressure sensor is another commonly used soft mechanical sensor. Most resistive sensors transduce pressure through changes in contact resistance between two electrodes. A recent work from the TL Ren group at Tsinghua University adopted abrasive paper as a template and reduced graphene oxide as conductive material to fabricate a randomly distributed microstructure [37]. The PDMS was coated on the abrasive paper to obtain a micropatterned flexible substrate. After dip coating the graphene oxide (GO), a high temperature was applied to reduce the GO. Finally, a face-to-face package was used to

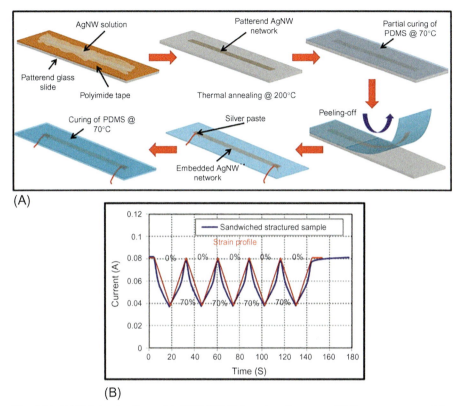

Fig. 5.14 (A) Fabrication flow of AgNW-PDMS-based strain sensor. (B) Response curve of the strain sensor under stretch/release cycles with 70% strain.
Reproduced with permission from Ref. M. Amjadi, A. Pichitpajongkit, S. Lee, S. Ryu, I. Park, Highly stretchable and sensitive strain sensor based on silver nanowire–elastomer nanocomposite, ACS Nano 8 (2014) 5154–5163, copyright 2014 American Chemical Society.

prepare pressure sensors with different roughness surfaces. The sensitivity of the pressure sensor is as high as 25.1 kPa^{-1} with a linearity range up to 2.6 kPa (Fig. 5.15).

Another interesting work reported by our group integrated the microstructure and porous structure together to enhance the pressure sensitivity [38]. A CNT-PDMS composite was first mixed with sugar microparticles before curing and then the mixture was cast into the desired mold (with micropyramid or microcone) to have the outer shape. After some simulation and calculation, the sensor with sharper angle of microstructure and higher porosity could obtain the highest sensitivity (Fig. 5.16).

5.4.3 Temperature sensor

Due to the thermal expansion of the polymer matrix with increase of temperature, the change in the conductive network will cause resistance change. In this way, the SCC can also serve as a temperature sensor.

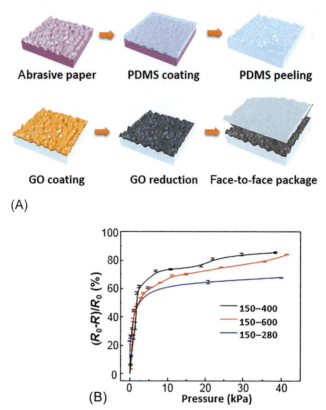

Fig. 5.15 (A) Fabrication process of the graphene pressure sensor. (B) Relative resistance variation vs the pressure for the pressure sensors using different roughnesses of PDMS. Reproduced with permission from Ref. Y. Pang, K. Zhang, Z. Yang, S. Jiang, Z. Ju, Y. Li, X. Wang, D. Wang, M. Jian, Y. Zhang, R. Liang, H. Tian, Y. Yang, T.-L. Ren, Epidermis microstructure inspired graphene pressure sensor with random distributed spinosum for high sensitivity and large linearity, ACS Nano 12 (2018) 2346–2354, copyright 2018 American Chemical Society.

The positive temperature coefficient (PTC) effect is a typical temperature–resistivity behavior of solid-state conductive composite caused by the thermal expansion polymer matrix, which may break the conductive path. Heaney from Raychem Corporation presented experimental data on resistance vs temperature behavior of disordered carbon black in polyethylene (PE) with melting temperature of 130°C [39]. In order to verify the assumption that thermal expansion causes the resistance change, he observed the experimental data of resistance–temperature, thickness–temperature, and resistance–thickness, respectively. The results well explained the relationship between the resistance, volume change, and temperature (Fig. 5.17).

A CNT-PU composite with negative temperature coefficient (NTC) was reported by Li and coworkers from Sichuan University [40]. By designing close cell foam, they

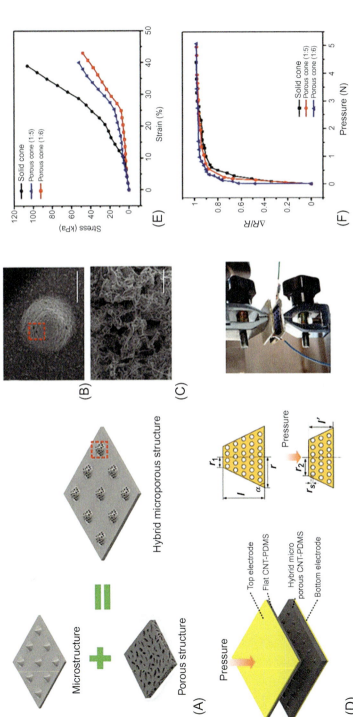

Fig. 5.16 (A) Schematic picture of hybrid microporous structure. (B) and (C) show the SEM images of the composite. (D) Schematic picture and photo of how the hybrid structure deformed under pressure. (E) The strain–stress curve of the sensor. (F) The response curve of resistance change at different pressures.

Reproduced with permission from Ref. H. Chen, Y. Song, H. Guo, L. Miao, X. Chen, Z. Su, H. Zhang, Hybrid porous micro structured finger skin inspired self-powered electronic skin system for pressure sensing and sliding detection, Nano Energy 51 (2018) 496–503, copyright 2018 Elsevier.

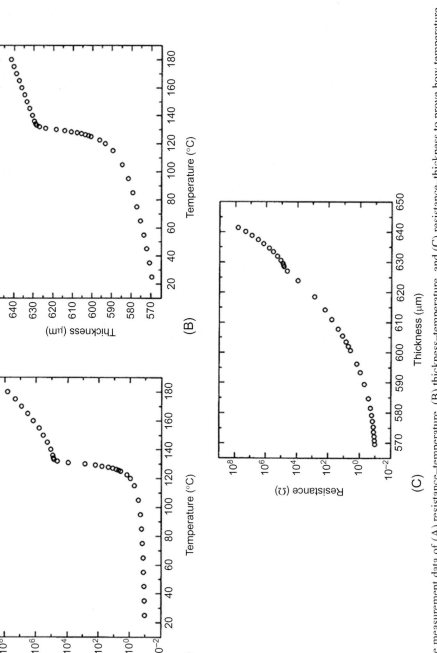

Fig. 5.17 The measurement data of (A) resistance–temperature, (B) thickness–temperature, and (C) resistance–thickness to prove how temperature changes the resistance.
Reproduced with permission from Ref. M.B. Heaney, Resistance-expansion-temperature behavior of a disordered conductor–insulator composite, Appl. Phys. Lett. 69 (1996) 2602–2604, copyright 1996 AIP Publishing.

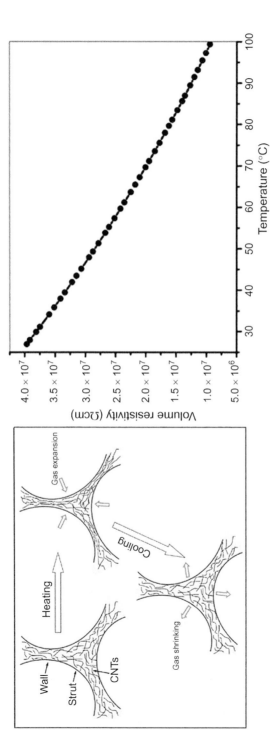

Fig. 5.18 Schematic diagram of the microstructural changes in CNT/sPU foam composites during heating/cooling cycles. Reproduced with permission from Ref. Z.-D. Xiang, T. Chen, Z.-M. Li, X.-C. Bian, Negative temperature coefficient of resistivity in lightweight conductive carbon nanotube/polymer composites, Macromol. Mater. Eng. 294 (2009) 91–95, copyright 2009 WILEY-VCH Verlag GmbH & Co. KGaA, Weinheim.

use the CNT-PU composite to wrap a large volume of CO_2. When the composite is heated, the closed gas will expand, which will squeeze the cell walls. As the cell walls are getting thinner, the curved CNTs in the wall become more straight and the conductive network turns from a disordered pattern into a more ordered pattern. Thus, more conductive paths are created, which decreases the resistance of the composite (Fig. 5.18).

5.5 Summary

In conclusion, because of the great potential in areas of health care and smart robotics, soft tactile sensors are a fast-developing field. Stretchable conductive composite integrates the properties from conductive fillers and elastomer matrix, which present one of the most promising approaches and provide diverse choices for constructing specific stretchable electronics. Here, we start by introducing the working mechanism of SCC based on percolation theory. Especially, the most updated tunnel current theory is also included. Then, the method of preparing SCC is summarized into three main categories: conductive filler mixed into an elastomer, conductive film deposited on the elastomer surface, and synthesizing conductive fillers within an elastomer. Finally, several typical tactile sensors including strain sensor, pressure sensor, and temperature sensor are demonstrated. In summary, it is believed that SCC is expected to play a more and more important role in stretchable sensors and more functional devices like optoelectronics and energy devices.

References

[1] D.-Y. Khang, H. Jiang, Y. Huang, J.A. Rogers, A stretchable form of single-crystal silicon for high-performance electronics on rubber substrates, Science 311 (2006) 208–212.

[2] K.-I. Jang, S.Y. Han, S. Xu, K.E. Mathewson, Y. Zhang, J.-W. Jeong, G.-T. Kim, R.C. Webb, J.W. Lee, T.J. Dawidczyk, R.H. Kim, Y.M. Song, W.-H. Yeo, S. Kim, H. Cheng, S.I. Rhee, J. Chung, B. Kim, H.U. Chung, D. Lee, Y. Yang, M. Cho, J.G. Gaspar, R. Carbonari, M. Fabiani, G. Gratton, Y. Huang, J.A. Rogers, Rugged and breathable forms of stretchable electronics with adherent composite substrates for transcutaneous monitoring, Nat. Commun. 5 (2014) 1–10.

[3] C. Dagdeviren, Y. Su, P. Joe, R. Yona, Y. Liu, Y.-S. Kim, Y. Huang, A.R. Damadoran, J. Xia, L.W. Martin, Y. Huang, J.A. Rogers, Conformable amplified lead zirconate titanate sensors with enhanced piezoelectric response for cutaneous pressure monitoring, Nat. Commun. 5 (2014) 1–10.

[4] J.A. Fan, W.-H. Yeo, Y. Su, Y. Hattori, W. Lee, S.-Y. Jung, Y. Zhang, Z. Liu, H. Cheng, L. Falgout, M. Bajema, T. Coleman, D. Gregoire, R.J. Larsen, Y. Huang, J.A. Rogers, Fractal design concepts for stretchable electronics, Nat. Commun. 5 (2014) 1–8.

[5] J. Song, H. Jiang, Y. Huang, J.A. Rogers, Mechanics of stretchable inorganic electronic materials, J. Vac. Sci. Technol. A 27 (2009) 1107–1125.

[6] Z. Yan, F. Zhang, J. Wang, F. Liu, X. Guo, K. Nan, Q.W. Lin, M. Gao, D. Xiao, Y. Shi, Y. Qiu, H. Luan, J.H. Kim, Y. Wang, H. Luo, M. Han, Y. Huang, Y. Zhang, J.A. Rogers, Controlled mechanical buckling for origami-inspired construction of 3D microstructures in advanced materials, Adv. Funct. Mater. 26 (2016) 2629–2639.

[7] Y. Zhang, Z. Yan, K. Nan, D. Xiao, Y. Liu, H. Luan, H. Fu, X. Wang, Q. Yang, J. Wang, W. Ren, H. Si, F. Liu, L. Yang, H. Li, J. Wang, X. Guo, H. Luo, L. Wang, Y. Huang, J.A. Rogers, A mechanically driven form of Kirigami as a route to 3D mesostructures in micro/nanomembranes, Proc. Natl. Acad. Sci. U. S. A. 112 (2015) 11757–11764.

[8] Y. Wang, C. Zhu, R. Pfattner, H. Yan, L. Jin, S. Chen, F. Molina-Lopez, F. Lissel, J. Liu, N.I. Rabiah, Z. Chen, J.W. Chung, C. Linder, M.F. Toney, B. Murmann, Z. Bao, A highly stretchable, transparent, and conductive polymer, Sci. Adv. 3 (2017), e1602076.

[9] C. Larson, B. Peele, S. Li, S. Robinson, M. Totaro, L. Beccai, B. Mazzolai, R. Shepherd, Highly stretchable electroluminescent skin for optical signaling and tactile sensing, Science 351 (2016) 1071–1074.

[10] J. Xu, S. Wang, G.-J.N. Wang, C. Zhu, S. Luo, L. Jin, X. Gu, S. Chen, V.R. Feig, J.W.F. To, S. Rondeau-Gagné, J. Park, B.C. Schroeder, C. Lu, J.Y. Oh, Y. Wang, Y.-H. Kim, H. Yan, R. Sinclair, D. Zhou, G. Xue, B. Murmann, C. Linder, W. Cai, J.B.-H. Tok, J.W. Chung, Z. Bao, Highly stretchable polymer semiconductor films through the nanoconfinement effect, Science 355 (2017) 59–64.

[11] S. Wang, J. Xu, W. Wang, G.-J.N. Wang, R. Rastak, F. Molina-Lopez, J.W. Chung, S. Niu, V.R. Feig, J. Lopez, T. Lei, S.-K. Kwon, Y. Kim, A.M. Foudeh, A. Ehrlich, A. Gasperini, Y. Yun, B. Murmann, J.B.-H. Tok, Z. Bao, Skin electronics from scalable fabrication of an intrinsically stretchable transistor array, Nature 555 (2018) 83–88.

[12] H. Deng, L. Lin, M. Ji, S. Zhang, M. Yang, Q. Fu, Progress on the morphological control of conductive network in conductive polymer composites and the use as electroactive multifunctional materials, Prog. Polym. Sci. 39 (2014) 627–655.

[13] M. Park, J. Park, U. Jeong, Design of conductive composite elastomers for stretchable electronics, Nano Today 9 (2014) 244–260.

[14] Z.-M. Dang, Y.-H. Lin, C.-W. Nan, Novel ferroelectric polymer composites with high dielectric constants, Adv. Mater. 15 (2003) 1625–1629.

[15] M. Panda, V. Srinivas, A.K. Thakur, On the question of percolation threshold in polyvinylidene fluoride/nanocrystalline nickel composites, Appl. Phys. Lett. 92 (2008), 132905.

[16] F. He, S. Lau, H.L. Chan, J. Fan, High dielectric permittivity and low percolation threshold in nanocomposites based on poly(vinylidene fluoride) and exfoliated graphite nanoplates, Adv. Mater. 21 (2009) 710–715.

[17] W. Bauhofer, J.Z. Kovacs, A review and analysis of electrical percolation in carbon nanotube polymer composites, Compos. Sci. Technol. 69 (2009) 1486–1498.

[18] S. Vionnet-Menot, C. Grimaldi, T. Maeder, S. Strässler, P. Ryser, Tunneling-percolation origin of nonuniversality: theory and experiments, Phys. Rev. B 71 (2005), 064201.

[19] N. Hu, Y. Karube, C. Yan, Z. Masuda, H. Fukunaga, Tunneling effect in a polymer/carbon nanotube nanocomposite strain sensor, Acta Mater. 56 (2008) 2929–2936.

[20] Y. Konishi, M. Cakmak, Nanoparticle induced network self-assembly in polymer–carbon black composites, Polymer 47 (2006) 5371–5391.

[21] X. Jing, W. Zhao, L. Lan, Effect of particle size on electric conducting percolation threshold in polymer/conducting particle composites, J. Mater. Sci. Lett. 19 (2000) 377–379.

[22] J. Li, J.-K. Kim, Percolation threshold of conducting polymer composites containing 3D randomly distributed graphite nanoplatelets, Compos. Sci. Technol. 67 (2007) 2114–2120.

[23] J. Li, P.C. Ma, W.S. Chow, C.K. To, B.Z. Tang, J.-K. Kim, Correlations between percolation threshold, dispersion state, and aspect ratio of carbon nanotubes, Adv. Funct. Mater. 17 (2007) 3207–3215.

[24] Y. Hu, T. Zhao, P. Zhu, Y. Zhu, X. Shuai, X. Liang, R. Sun, D. Daniel Lu, C.-P. Wong, Low cost and highly conductive elastic composites for flexible and printable electronics, J. Mater. Chem. C 4 (2016) 5839–5848.

[25] H. Chen, L. Miao, Z. Su, Y. Song, M. Han, X. Chen, X. Cheng, D. Chen, H. Zhang, Fingertip-inspired electronic skin based on triboelectric sliding sensing and porous piezoresistive pressure detection, Nano Energy 40 (2017) 65–72.

[26] Y. Song, H. Chen, Z. Su, X. Chen, L. Miao, J. Zhang, X. Cheng, H. Zhang, Highly compressible integrated supercapacitor–piezoresistance-sensor system with CNT–PDMS sponge for health monitoring, Small 13 (2017), 1702091.

[27] S.P. Lacour, S. Wagner, Z. Huang, Z. Suo, Stretchable gold conductors on elastomeric substrates, Appl. Phys. Lett. 82 (2003) 2404–2406.

[28] S.P. Lacour, J. Jones, Z. Suo, S. Wagner, Design and performance of thin metal film interconnects for skin-like electronic circuits, IEEE Electron Device Lett. 25 (2004) 179–181.

[29] H. Lee, B. Seong, H. Moon, D. Byun, Directly printed stretchable strain sensor based on ring and diamond shaped silver nanowire electrodes, RSC Adv. 5 (2015) 28379–28384.

[30] M. Shi, H. Wu, J. Zhang, M. Han, B. Meng, H. Zhang, Self-powered wireless smart patch for healthcare monitoring, Nano Energy 32 (2017) 479–487.

[31] B. You, C.J. Han, Y. Kim, B.-K. Ju, J.-W. Kim, A wearable piezocapacitive pressure sensor with a single layer of silver nanowire-based elastomeric composite electrodes, J. Mater. Chem. A 4 (2016) 10435–10443.

[32] J. Lee, M. Lim, J. Yoon, M.S. Kim, B. Choi, D.M. Kim, D.H. Kim, I. Park, S.-J. Choi, Transparent, flexible strain sensor based on a solution-processed carbon nanotube network, ACS Appl. Mater. Interfaces 9 (2017) 26279–26285.

[33] J. Ramírez, D. Rodriquez, A.D. Urbina, A.M. Cardenas, D.J. Lipomi, Combining high sensitivity and dynamic range: wearable thin-film composite strain sensors of graphene, ultrathin palladium, and PEDOT:PSS, ACS Appl. Nano Mater. 2 (2019) 2222–2229.

[34] M. Park, J. Im, M. Shin, Y. Min, J. Park, H. Cho, S. Park, M.-B. Shim, S. Jeon, D.-Y. Chung, J. Bae, J. Park, U. Jeong, K. Kim, Highly stretchable electric circuits from a composite material of silver nanoparticles and elastomeric fibres, Nat. Nanotechnol. 7 (2012) 803–809.

[35] B. Liang, Z. Lin, W. Chen, Z. He, J. Zhong, H. Zhu, Z. Tang, X. Gui, Ultra-stretchable and highly sensitive strain sensor based on gradient structure carbon nanotubes, Nanoscale 10 (2018) 13599–13606.

[36] M. Amjadi, A. Pichitpajongkit, S. Lee, S. Ryu, I. Park, Highly stretchable and sensitive strain sensor based on silver nanowire–elastomer nanocomposite, ACS Nano 8 (2014) 5154–5163.

[37] Y. Pang, K. Zhang, Z. Yang, S. Jiang, Z. Ju, Y. Li, X. Wang, D. Wang, M. Jian, Y. Zhang, R. Liang, H. Tian, Y. Yang, T.-L. Ren, Epidermis microstructure inspired graphene pressure sensor with random distributed spinosum for high sensitivity and large linearity, ACS Nano 12 (2018) 2346–2354.

[38] H. Chen, Y. Song, H. Guo, L. Miao, X. Chen, Z. Su, H. Zhang, Hybrid porous micro structured finger skin inspired self-powered electronic skin system for pressure sensing and sliding detection, Nano Energy 51 (2018) 496–503.

[39] M.B. Heaney, Resistance-expansion-temperature behavior of a disordered conductor–insulator composite, Appl. Phys. Lett. 69 (1996) 2602–2604.

[40] Z.-D. Xiang, T. Chen, Z.-M. Li, X.-C. Bian, Negative temperature coefficient of resistivity in lightweight conductive carbon nanotube/polymer composites, Macromol. Mater. Eng. 294 (2009) 91–95.

Mechanoluminescent materials for tactile sensors

Dengfeng Peng[a,*] and Sicen Qu[b]
[a]School of Physics and Optoelectronic Engineering, Shenzhen University, Shenzhen, China,
[b]Department of Physical Education, Shenzhen University, Shenzhen, China
*Corresponding author. e-mail address: pengdengfeng@szu.edu.cn

6.1 Background

A myriad of materials in nature, either living or nonliving, exhibit luminescence. For example, jellyfish glow with the bioluminescent protein aequorin under mechanical stimulation. In recent years, researchers have worked to imitate stimuli luminescence [1] or color change [2] for making tactile sensors. There are mainly two kinds of tactile sensors made of luminous materials: electroluminescent tactile sensors (ETS) and mechanoluminescent tactile sensors.

6.1.1 Direct-current-driven electroluminescent tactile sensors

The first type of ETS is an LED tactile sensor based on semiconductors that are maintained by direct current. In 2013, Wang et al. reported a light-emitting screen tactile sensor based on capacitive "flexible electronic skin" (Fig. 6.1) [3]. The brightness of the LED responds to the amplitude of the pressure, that is, the greater the pressure applied, the greater the brightness exhibited. With an array of 16 × 16 pixels, the "electronic skin" can be used as an interactive wallpaper for display on wristwatches or integrated into a car dashboard to achieve touch control. The electronic skin enables the more accurate touch of the robot.

In the same year, Pan and Wang's group developed an ETS pressure sensor based on a p-GaN/n-ZnO nanowire LED array grown on a sapphire substrate [4]. The sensors achieved an extraordinarily high resolution of 2.7 μm, which exceeds the tactile resolution of human skin. In 2015, the same team reported a tactile pressure sensor based on a PEDOT:PSS/ZnO LED array grown on organic flexible substrates [5, 6] (Fig. 6.2). The organic/inorganic hybrid sensor offers a flexible platform for more application scenarios to sense strain distribution in real time. The LED-based tactile sensors have opened up a new way to direct sensing and visualization of tactile pressure.

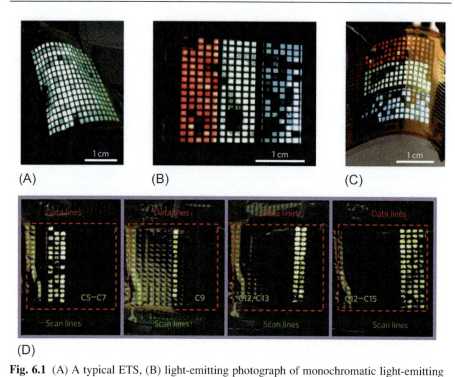

Fig. 6.1 (A) A typical ETS, (B) light-emitting photograph of monochromatic light-emitting ETS for electronic skin under bending conditions, (C) light-emitting photos of polychromatic skin under normal stretching and bending conditions, respectively, and (D) unit-selective light-emitting physical picture of monochromatic light-emitting skin collected [3].
Reprinted with permission from C. Wang, D. Hwang, Z. B. Yu, K. Takei, J. Park, T. Chen, B. Ma, A. Javey, User-interactive electronic skin for instantaneous pressure visualization, Nat. Mater. 12 (2013) 899. Copyright © 2013, Nature Publishing Group.

6.1.2 Alternating electric field-driven electroluminescent tactile sensors

In 2016, Robert et al. reported a super-elastic "skin" (Fig. 6.3) with a stretchability of around 600% [1]. One of the important parts of this artificial skin is the electroluminescent elastic layer embedded with zinc sulfide electroluminescent powders. This artificial skin can sense tactile pressure. The capacitance of the device changes according to the deformation when the elastomer is deformed, resulting in the vibration of luminance. Based on a similar principle, the researchers prepared a multicolor luminous unit for the flexible artificial skin by conjunction of photolithography and a transfer printing process [7]. The prototype of the tactile sensor requires transparently conductive hydrogel as the electrode to supply the core layer. The commercial ZnS electroluminescent phosphor used in the reported skin should be driven by a high voltage/frequency alternating current (2500 V, 700 Hz), which can be solved in later research by adding optimized dielectric materials [8].

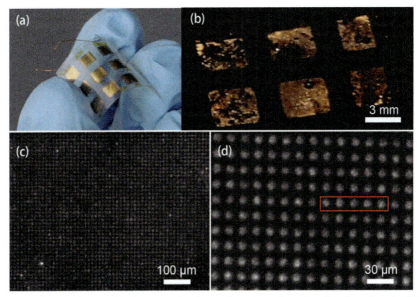

Fig. 6.2 High-resolution visual pressure/strain sensing system based on the piezoelectric effect. Images (A) and (B) are digital photographs of a fabricated ETS when unlit and electrically lit, respectively. Images (C) and (D) are optical images of a ZnO NW/p-polymer LED array ETS device. The brightness of the piezoelectric ZnO nanowire array varies with pressure, so the pressure can be visualized by luminous intensity [6].
Reprinted with permission from R. R. Bao, Z. L. Wang, et al., Flexible and controllable piezo-phototronic pressure mapping sensor matrix by ZnO NW/p-polymer LED array, Adv. Funct. Mater. 25 (2015) 2884. Copyright © 2015, Wiley VCH.

The aforementioned two types of ETS require electrodes and a power supply system connected by wires to electrodes. The arrangement and the interface contact between wires and electrodes limits the stability and safe operability of the device, and the tactile users should warrant that there is no short circuit in case of power supply when put to use in real applications. Therefore, it is important to find light-emitting materials and composites that do not need additional power supply and that can emit light directly under human touch. Once refined, it is possible to develop a simple and wearable light-emitting tactile sensor.

6.2 Mechanoluminescence materials for tactile sensors

Mechanoluminescence (ML) is a phenomenon characterized by photon emission in response to mechanical stimuli (Fig. 6.4) such as pinching, pressing, rubbing, squeezing, stretching, etc. [9–14].

Different from photoluminescence and electroluminescence, ML is a kind of luminescence phenomenon excited directly by external mechanical energy. As a type of smart luminescent material, ML crystals have been used in applications of battery-free

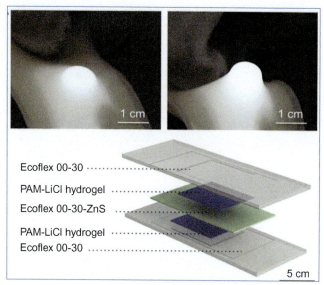

Fig. 6.3 Stretchable flexible ETS and skin structure composition; stretchable elastic light-emitting film based on ZnS electroluminescent materials [1].
Reprinted with permission from C. Larson, B. Peele, S. Li, S. Robinson, M. Totaro, L. Beccai, B. Mazzolai, R. Shepherd, Highly stretchable electroluminescent skin for optical signaling and tactile sensing, Science 351 (2016) 1071. Copyright © 2016, Nature Publishing Group.

Fig. 6.4 Scheme of ML material in response to finger touch stimulus.

light source, stress sensing, and artificial skin. Some ML materials can maintain high brightness (120 cd/m^2) emission under repeated dynamic stress ($\sim$10 kPa over 100,000 cycles) [15], showing unique advantages in the field of visual tactile sensors.

Francis Bacon, first recorded the phenomenon of stress luminescence of artificial objects in 1605. He observed that artificial sugar produces visible luminescence when cut or scraped. Natural substances such as ores and manmade substances such as aromatic compounds and alkaloids exhibit ML behaviors. However, most of them have

ML characteristics only under certain conditions, for instance impact by high pressure and the luminous intensity is usually very weak because this kind of crystal does not have enough of an effective luminous center. They emit light only when they are fractured or plastic deformed, therefore they are destructive luminescence materials. In recent years, with the rapid development of advanced functional luminescent materials and the continuous innovation of photoelectric detection technology and intelligent sensor devices, ML materials have been developed to show repeatable performances.

Xu from the National Institute of Advanced Industrial Science and Technology (AIST) in Japan has been dedicated to the research of repeatable ML materials and their applications since the 1990s. She creatively applied these materials to new light sources, biological imaging, passive detection of dynamic stress of electronic skin and artificial bone, detection of building structures, real-time sensing and monitoring, and other major fields. Based on the investigation and summary of the literature, the recoverable ML materials (inorganics) are as shown in Fig. 6.5. Among them, the most frequently reported ML materials are $SrAl_2O_4$:Eu(SAO-E) and ZnS:Mn/Cu. Photographs of both ML materials are shown in Fig. 6.6 [16, 17]. Xu prepared a high brightness ML thin film based on SAO-E that can produce bright and visible stress luminescence under slight drawing force, as shown in Fig. 6.6 (*up panel*).

In order to understand the properties and mechanism of ML materials, it is important to know the classifications of stress luminescence. Chandra classified ML luminescence into 23 types, covering all kinds of ML materials forms under various conditions. The generation process of ML is complex; usually one ML material with multiple luminous properties. Generally, ML material has one or more stress-induced excitation modes. In 2014, Chandra combined his long-term research experiences of

Doped simple oxide:	ZrO_2:Ti^{4+}
Doped chalcogenides:	ZnS:Mn^{2+};ZnS:Cu^{2+};ZnS:Mn^{2+}/Cu^{2+}; ZnS:Al^{3+}/Cu^2; ZnS:Al^{3+}/Mn^{2+}/Cu^{2+}; CdS:Mn^{2+}, $ZnTe$: Mn^{2+}
Doped oxysulfide:	$CaZnOS$: Mn^{2+}/Sm^{3+}/Er^{3+}; $BaZnOS$: Mn^{2+}
Doped aluminates:	$SrAl_2O_4$: Eu^{2+}/Ce^{3+}/Ho^{3+}/Dy^{3+}/Er^{3+}; $ZnAl_2O_4$:Mn^{2+}; $CaYAl_3O_7$: Eu^{2+}/Ce^{3+}; $CaAl_2SiO_7$:Ce^{3+}
Doped silicates	$Ca_2MgSi_2O_7$: Eu^{2+}; Eu^{2+}/Dy^{3+}; $CaAl_2Si_2O_8$:Eu^{2+} $SrCaMgSi_2O_7$: Eu^{2+}; $SrBaMgSi_2O_7$:Eu^{2+}; $Sr_2Mg_2Si_2O_7$:Eu^{2+}
Doped titanates:	$CaTiO_3$-$BaTiO_3$: Pr^{3+}; $CaTiO_3$:Pr^{3+}
Doped niobates:	$CaNb_2O_6$: Pr^{3+}/ Eu^{3+}/Tb^{3+};$LiNbO_3$:Pr^{3+}
Doped phosphates:	$CaZr(PO_4)_2$: Eu^{2+}; $SrMg_2(PO_4)_2$:Eu^{2+}
Doped stannates:	$Sr_{n+1}Sn_nO_{3n+1}$:Sm^{3+}/Nd^{3+}
Doped tungstates:	$Ca_2Gd_2W_3O_{14}$:Eu^{3+}
Doped oxynitrides:	$BaSi_2O_2N_2$: Eu^{2+}; $SrSi_2O_2N_2$: Eu^{2+}; $CaSi_2O_2N_2$: Eu^{2+}

Fig. 6.5 Representatives of repeatable mechanoluminescent materials.

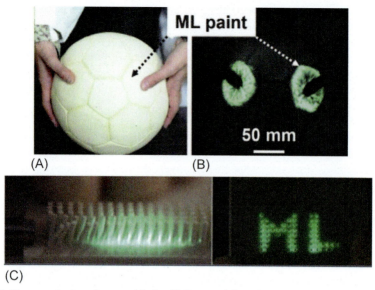

Fig. 6.6 Photographs based on two kinds of ML material coatings and composite devices. (A) ML photos of SAO coating on a volleyball's surface; (B) ML photos of finger touchees; and (C) ML photos of wind-driven ZnS:Cu/PDMS arrays; the right is an ML pattern driven under wind perpendicular to the array [16, 17].
Panels (A) and (B): Reprinted with permission from L. S. Liu, C. N. Xu, A. Yoshida, D. Tu, N. Ueno, S. Kainuma, Scalable elasticoluminescent strain sensor for precise dynamic stress imaging and onsite infrastructure diagnosis, Adv. Mater. Technol. 4 (1) (2019) 1800336. Copyright © 2019, Wiley VCH; panel (C): Reprinted with from S. M. Jeong, S. Song, K. I. Joo, J. Kim, S. H. Hwang, J. Jeong, H. Kim, Bright, wind-driven white mechanoluminescence from zinc sulphide microparticles embedded in a polydimethylsiloxane elastomer, Energy Environ. Sci. 7 (2014) 3338.

stress luminescence and further summarized all of them into two categories according to the excitation mode of ML objects caused by mechanical force, namely deformation luminescence (DML) and triboluminescence, as shown in Fig. 6.7. DML can be subdivided into three types: plastic-ML, fracto-ML, and elastico-ML. Among them, elastico-ML sometimes accompanies piezoelectric effect, so it is also referred to as piezoluminescence [18].

The ML intensity of most elastico-ML materials decreases with increased cycle times of the applied mechanical force. After one cycle release, it's necessary to use ultraviolet (UV) ray to irradiate because UV ray has high photon energy, which can compensate the ML emission and recover with equal luminescent intensity as same as the first cycle. The ML objects recovered by UV light are called UV-recoverable ML materials. Most of the ML materials systems listed in Fig. 6.5 are the typically UV-recoverable ML inorganics, such as $SrAl_2O_4:Eu^{2+}$ [14] (Fig. 6.8A), $CaTiO_3$-$BaTiO_3:Pr^{3+}$ [19, 20], $CaZnOS:Cu^+/Mn^{2+}$ [21, 22], and Li/$NaNbO_3:Pr^{3+}$ [23, 24].

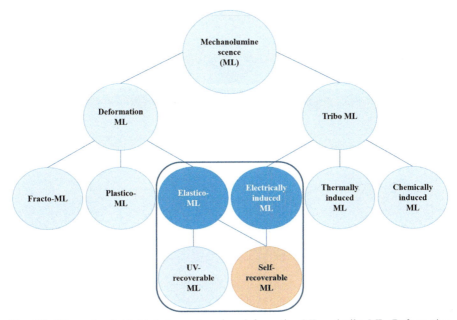

Fig. 6.7 ML can be divided into two categories: deformation ML and tribo ML. Deformation ML can be subdivided into three categories: fracto-ML, plastic-ML, and elastic-ML. Tribo ML can be divided into three categories: field (electric field) electroluminescence, thermoluminescence, and chemiluminescence. Elastico-ML can be divided into two forms: UV-recovery ML and self-recovery ML. Electrically induced ML, or electric-field-induced luminescence, is also a kind of self-recovery ML.

It is found that ZnS-based elastico-ML materials have recoverable ML capability under continuous mechanical force without external irradiation energy compensation, directly converting mechanical energy to light emission and showing repeatable luminescence without decreasing intensity. For example, compared to the aforementioned SAO, ZnS-doped luminescent ions can repeat more than 10^6 times of dynamic light emission under mechanical force, and still maintain ML performance [15].

It is found that SAO luminescence has attenuation that needs to be compensated by UV irradiation. ZnS does not need irradiation, but it can maintain luminous intensity well to a certain extent. This is due to the polymer material wrapped with SAO undergoing mechanical fatigue under 100,000 times of external force. The luminescence of these two systems is derived from luminescent ions, for example, the self-recovery ML of the ZnS:Cu matrix originates from the luminescent centers of Cu ions, while UV-recoverable ML of SAO-E comes from Eu ions. The difference between the luminescent properties of the two systems is determined by the properties of the matrix itself, such as structural properties, defects, and so on.

Due to the special stress luminescent repeatability of zinc sulfide, it has attracted much attention. At the same time, a series of related researches on zinc sulfide composites have been carried out. For example, based on the composite elastomer of

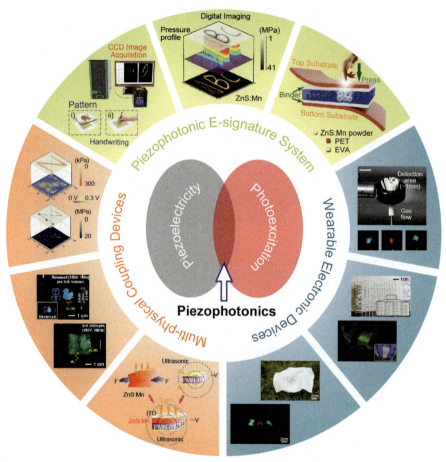

Fig. 6.8 Application scenarios of mechanoluminescence for advanced optoelectronic devices, such as artificial skin, stress sensing, battery-free display, and weareble and multiintegrated devices and so on [10].
Reprinted from X. D. Wang, D. F. Peng, B. L. Huang, C. F. Pan, Z. L. Wang, Piezophotonic effect based on mechanoluminescent materials for advanced flexible optoelectronic applications, Nano Energy 55 (2019) 389.

doped ZnS materials, the ZnS-based wind driving ML soft array was designed. They dance under the fluctuations of the wind and shine vividly (Fig. 6.6). By taking advantage of self-recovery ML properties, and the merit of directly pressure-to-optical signal onverted and output, ZnS-based ML materials have proved to be promising for use in many fields, including wearable electronics, e-signatures, and multiphysical coupling devices [10], as shown in Fig. 6.8.

If applied to the sensor, we need a reliable linear relationship and a fast response speed. Therefore, in addition to repeatability, the linear relationship between the luminous characteristics and the stress and response time of the stress–light transition are also crucial. The relationship between luminous intensity and applied pressure of ZnS:Mn is linear and might be suitable for pressure detection and visualization, as shown in Fig. 6.9 [18].

Before making devices, it is important to choose the material form (film, powder, large single crystal, or ceramic microcrystalline). Xu has conducted a lot of research and accumulated data on the properties of doped ZnS. Powders and bulks as well as single crystal films can emit ML light. Xu sintered a ceramic target of ZnS:Mn^{2+} (~1.5%) in a vacuum and then prepared a high brightness elastico-ML film via magnetron sputtering method. The relationship between luminous intensity and stress is linear, which is suitable for stress detection. For the first time, the concept of ZnS:Mn-based ML material used in artificial skin is proposed [13] (the main experimental data is shown in Fig. 6.10). The film material is based on a rigid glass substrate, which needs to overcome difficulty of bending and stretching of a soft skin.

In order to make future devices more flexible and easy to process, Jeong et al. demonstrated the self-recovery ML properties of a ZnS-PDMS (polydimethylsiloxane) flexible composite that can still maintain good luminescent intensity under hundreds of thousands of cycles of stretching (Fig. 6.11). They mixed phosphors with different

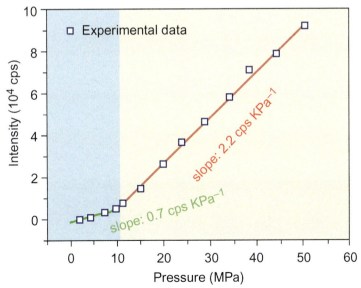

Fig. 6.9 Linear relationship of intensity and applied pressure (*lift figure*), the decay curve of ML under one mechanical pulse [12].
Reprinted from X.D. Wang, H.L. Zhang, R.M. Yu, L. Dong, D.F. Peng, A. H. Zhang, Y. Zhang, H. Liu, C.F. Pan, Z.L. Wang, Dynamic pressure mapping of personalized handwriting by a flexible sensor matrix based on the mechanoluminescence process, Adv. Mater. 2714 (2015) 2324.

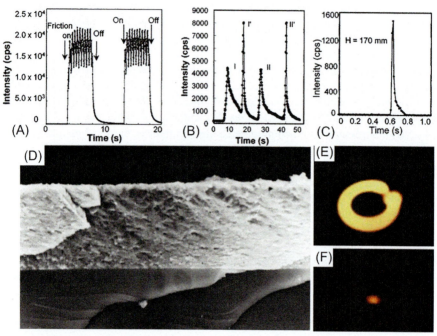

Fig. 6.10 Panels (A), (B), and (C) show ML-reproducible responses of ZnS:Mn film to friction under the same conditions. Panel (D) is a photograph of the ML of the ZnS:Mn film on ITO substrate under the impact of a free-falling ball; panel (E) is a ML photograph of the sliding force of a small steel bar; and panel (F) is a cross-section SEM picture of the sputtered film [13]. Reprinted with permission from C. N. Xu, T. Watanabe, M. Akiyama, X. G. Zheng, Artificial skin to sense mechanical stress by visible light emission. Appl. Phys. Lett. 74 (1999) 1236, Copyright © 1999, AIP Publishing.

Fig. 6.11 (A) SEM of ZnS:Mn ML film sample; (B) ML photographs under stretchable force; and (C) cross-section SEM of composited film, showing the ML powder is about 10 μm and concentrated on one side of the film [25].
Reprinted with permission from S. M. Jeong, S. Song, S. K. Lee, N.Y. Ha, Color manipulation of mechanoluminescence from stress-activated composite films, Adv. Mater. 25 (2013) 6194. Copyright © 2013, 2015, Wiley VCH.

luminescent color, and tuned the luminescent color by varying the mixing ratio. They found that the color of luminescence changes when tuning the action frequency. As the frequency increases, the wavelength becomes shorter, and vice versa. The ZnS:Cu/Mn ML powders used had an average particle size of 25 μm. The ML particle (ZnS:Cu/Mn) encapsulated PDMS composites, namely, the stretchable LED was made in advance through the pattern confining method, that is, the pattern/logo contains ML particles, while the frame of the matrix does not. Therefore, under the activation of stretching, twisting, or kneading, the device displays patterns showing real-time luminescence. As such, it is expected to have advantages of being flexible and stretchable in soft tactile sensors.

In addition to PDMS complex, there are many polymer materials that can be used for composites and packaging of ML powders. Pan et al. developed polyethylene terephthalate (PET) packaging ML particles (ZnS:Mn). They fixed the particles between two PET films, as shown in Fig. 6.12. Under the action of sliding, it was possible to distinguish the pressure of handwriting. This device also has flexible characteristics; it can bend but not stretch. In order to improve the luminous intensity, the researchers realized the optimization of the unit structure (Fig. 6.13), for example, using the mold to solidify the pyramid stress contact unit translation (the fabrication process is shown in Fig. 6.14). ML powder can also be made a hybrid with TPU and other polymers, and can be 3D printed into various patterns [24,26].

Doped zinc sulfide is a commercial electroluminescent phosphor that can be driven by electric fields. Therefore, the triboelectric field can also excite the light emission of the doped ZnS powders when the powder experiences friction. Zhu superposed the triboelectric effect to ML to enhance luminescence intensity [28]. They developed a kind of device that uses triboelectric-triggered luminescence. The light-emission properties belong to a type of self-recovery ML, so the stability of these devices is also very appreciable, as shown in Fig. 6.14. Dong et al. reported an intrinsically stretchable and transparent alternating-current electroluminescent device driven with a triboelectric nanogenerator [29]. Kim proposed a new origin of ML of an elastomer composite film through the study of the microstructure of Cu-doped ZnS particles encapsulated by a 500-nm Al layer [30].

Fang and Hu developed a multifunctional tactile skin based on PDMS and ML phosphors (Fig. 6.15) that could sense both gentle optical and electrical signals of tactile pressure with high sensitivities [27,30]. Interestingly, Song Yanlin has greatly improved the corresponding sensitivity by adding nano-sized silicon oxide in PDMS and ZnS powders [31], which are found to emit light driven by a girl's face skin as shown in Fig. 6.16. These interesting findings pave the way for the development of ML skin tactile sensors.

To optimize the structure, Peng and Jeong independently prepared hybrid fabrics using ZnS powders and PDMS [33–35] by taking the advantages of textile technologies. It was found that ML fabrics greatly improve the flexibility of the device, which has an important application in directional-strain detection and wearable things. Fig. 6.17 shows the scheme, and the real ML picture of the ML emissions for stretchable ML fabrics. The fiber also has electroluminescent properties, which increases its operability and multifunctionality.

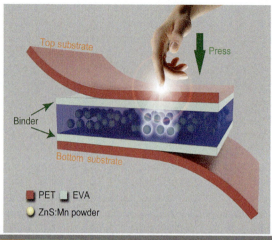

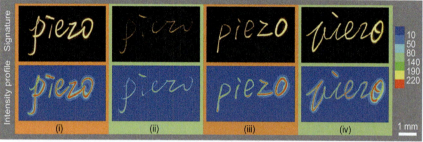

Fig. 6.12 Schematic structure of PET-dominated packaging ML devices (*top panel*). Demonstrations of recording the handwriting (scribing) habits of four volunteers by devices (*top panel*). Pressure evenly distributed to every point during the scribing process, with difference in force (i)–(iv) [12].
Reprinted with permission from X. D. Wang, H. L. Zhang, R. M. Yu, L. Dong, D. F. Peng, A. H. Zhang, Y. Zhang, H. Liu, C. F. Pan, Z. L. Wang, Dynamic pressure mapping of personalized handwriting by a flexible sensor matrix based on the mechanoluminescence process, Adv. Mater. 2714 (2015) 2324. Copyright © 2015, Wiley VCH.

Organic polymer materials also have flexible properties. Some organic polymers have the characteristics of ML emission. In recent years, organic ML materials have received much attention [36–41]. For example, Li et al. [42] proved that organic ML materials can be used for monitoring physiological parameters of the human body. They used pure organic ML luminogens (tPE-2-Th and tPE-3-Th) to monitor heartbeat (Fig. 6.18). Due to the high sensitivity and good performance of these materials, they can also be used in soft tactile sensors.

The aforementioned examples prove that ML materials have unique advantages in smart tactile sensors. Flexible optical fiber sensor units with one end of the optical fiber being an elastico-ML fiber, as shown in Fig. 6.19, have been proposed. In these, we encapsulate the ML material into an independent optical fiber. By touching the elastic-ML contactor at the end of the optical fiber, the optical fiber is guided to

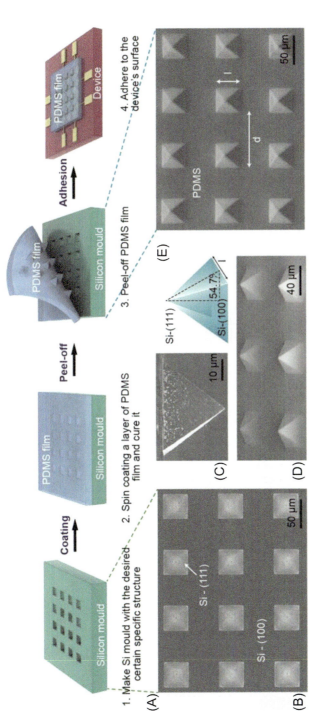

Fig. 6.13 (A) Schematic fabrication process of microstructured PDMS films for incorporation with ZnS phosphors. (B) and (C) SEM images of the mold with regular inverted pyramid by using UV photolithography and wet etching process. (D) and (E) SEM images of the PDMS film with pyramid structure lifted off from the Si mold [27].
Reprinted with permission from X. D. Wang, M. L. Que, M. X. Chen, X. Han, X. Y. Li, C. F. Pan, Z. L. Wang, Full dynamic-range pressure sensor matrix based on optical and electrical dual-mode sensing, Adv. Mater. 29 (15) (2017) 1605817. Copyright © 2019, Wiley VCH.

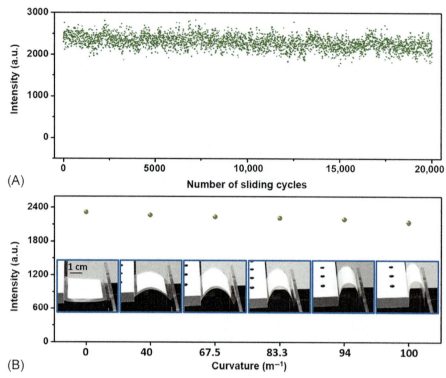

Fig. 6.14 (A) The light intensity of ZnS:Cu is almost constant under 2000 cycles of sliding force, the contact pressure is 60 kPa, and the light threshold pressure is 10 kPa [28] and (B) the light intensity basis of the composite film under different bending degrees is consistent, which shows that ZnS:Cu self-recovery stress light-emitting material has good light-emitting repeatability and stability.
Reprinted with permission from X. Y. Wei, X. D. Wang, S. Y. Kuang, L. Su, H. Y. Li, Y. Wang, C. F. Pan, Z. L. Wang, G. Zhu, Dynamic triboelectrification-induced electroluminescence and its use in visualized sensing. Adv. Mater. 28 (2016) 6656. Copyright © 2019, Wiley VCH.

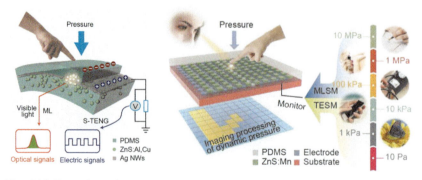

Fig. 6.15 Two schematic structures of an integrated device [25,32].
Reprinted with permission from S. M. Jeong, S. Song, S. K. Lee, N.Y. Ha, Color manipulation of mechanoluminescence from stress-activated composite films, Adv. Mater. 25 (2013) 6194.; H. Q. Fang, X. D. Wang, Q. Li, D. F. Peng, Q. F. Yan, C. F. Pan, A stretchable nanogenerator with electric/light dual-mode energy conversion, Adv. Eng. Mater. 6 (2016) 1600829. Copyright © 2016, 2017, Wiley VCH.

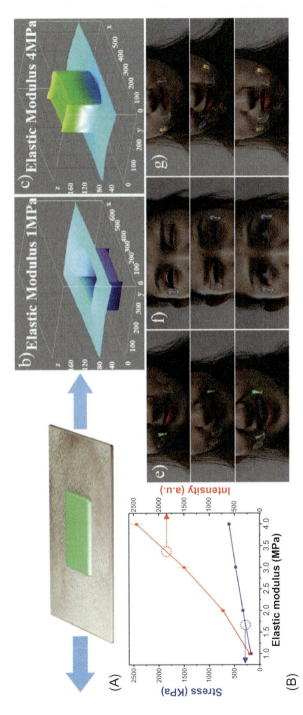

Fig. 6.16 The skin-driving ML exhibition. (A) An ML film stretching finite element analysis simulating (FEA) on skin. (B) and (C) 3D representation of FEA for pressure distribution ML film with 1 MPa (half of human facial skin) elastic modulus attached skin (B), and 4 MPa elastic modulus attached skin (C). (D) The relation of elastic modulus corresponding to stress by FEA and ML intensity by experiments. The skin-driven various color ML response to lips corner (E), canthus (F), and cheek (G) muscle movements [33]. Reprinted with permission from X. Qian, Z. R. Cai, M. Su, F. Y. Li, W. Fang, Y. D. Li, X. Zhou, Q. Y. Li, X. Q. Feng, W. B. Li, X. T. Hu, X. D. Wang, C. F. Pan, Y. L. Song, Printable skin-driven mechanoluminescence devices via nanodoped matrix modification, Adv. Mater. 30 (2018) 1800291. Copyright © 2018, Wiley VCH.

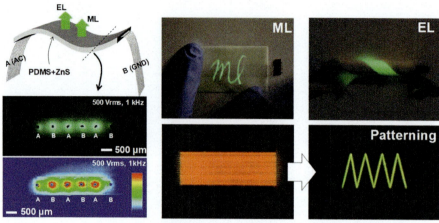

Fig. 6.17 Fiber ZnS@PDMS, which emits light under the action of deformation, can be fabricated into a directional array, so it can respond to strain in different directions [33]. Reprinted with permission from X. Qian, Z. R. Cai, M. Su, F. Y. Li, W. Fang, Y. D. Li, X. Zhou, Q. Y. Li, X. Q. Feng, W. B. Li, X. T. Hu, X. D. Wang, C. F. Pan, Y. L. Song, Printable skin-driven mechanoluminescence devices via nanodoped matrix modification, Adv. Mater. 30 (2018) 1800291. Copyright©2020, Elsevier Ltd.

Fig. 6.18 Organic ML materials for detecting heartbeat. [42].
Reprinted with permission from C. Wang, Y. Yu, Y. H. Yuan, C. Y. Ren, Q. Y. Liao, J. Q. Wang, Z. F. Chai, Q. Q. Li, Z. Li, Heartbeat-sensing mechanoluminescent device based on a quantitative relationship between pressure and emissive intensity, Matter 2 (1) (2020) 181. Copyright © 2020, Elsevier Ltd.

analyze the tactile information. At the same time, the optical fiber can also be combined into an array, so we use a number of optical fiber units to sense 2D and 3D information. For example, information on the location and strength of the stress as well as the roughness of the contact surface and so on. At the same time, we combine corresponding algorithms for data feedback and machine learning methods to research integrated devices.

Mechanoluminescent materials for tactile sensors   107

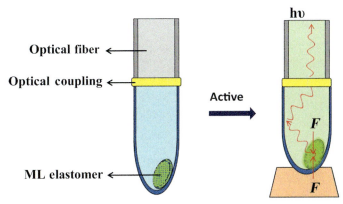

Fig. 6.19 The scheme of a flexible unit of an optical fiber tactile sensor based on ML materials; the ML light-emitting unit under touch. The position of touch and the magnitude of tactile force are fed back by the ML position and brightness.

6.3 Conclusions and prospective

This chapter introduced the phenomenon, classification, materials, and composite form of ML. Taking self-recovery ML material as an example, it presented representative work of ML composite materials. The possible selection scheme for tactile integrated device unit was provided. This kind of material emits light under touch, so it has unique advantages in tactile sensing. However, it does have some challenges.

A particle can act as an independent sensor without electrodes or wires. The sensing and visualization performances are mainly dependent on particle properties such as particle size distribution, crystal orientation, composition, ML intensity, and color. The study of these properties and luminescence of particles, combined with theoretical calculation, leads to development of new structures and materials with better performance. How to fix and package single particles is another problem.

The ultimate goal of ML is use as a new light source [43]. In the future, the material system may be an approach to integrating organic and organic-inorganic composites and macromolecules, while the luminous stability and repeatability of organic materials need to be guaranteed. In addition, the research of material genetic engineering, multidimensional printing technology, and machine learning will help promote the development of ML materials. Nowadays, more and more researchers are engaged in the field of ML materials and related sensor applications [43–71].

At present, the ML process can only be used for dynamic pressure sensing due to the intrinsically dynamic-stimulus properties of the ML materials. If we develop an ML material with memory function, it may be used for static pressure recording. Smet recently reported a kind of ML material that can be stored for several hours, and then released by other stimulus regulation [69].

Acknowledgments

This work was supported by the Natural Science Foundation of China (No. 61875136), the Guangdong Natural Science Foundation (No. 2020A1515010316), Shenzhen Fundamental Research Project (No. 201708183000260), Scientific Research Foundation as Phase II construction of high level University for the Youth Scholars of Shenzhen University (No. 000002110223).

References

[1] C. Larson, B. Peele, S. Li, S. Robinson, M. Totaro, L. Beccai, B. Mazzolai, R. Shepherd, Highly stretchable electroluminescent skin for optical signaling and tactile sensing, Science 351 (2016) 1071.

[2] H.H. Chou, A. Nguyen, A. Chortos, A chameleon-inspired stretchable electronic skin with interactive colour changing controlled by tactile sensing, Nat. Commun. 6 (2015) 8011.

[3] C. Wang, D. Hwang, Z.B. Yu, K. Takei, J. Park, T. Chen, B. Ma, A. Javey, User-interactive electronic skin for instantaneous pressure visualization, Nat. Mater. 12 (2013) 899.

[4] C.F. Pan, Z.L. Wang, et al., High-resolution electroluminescent imaging of pressure distribution using a piezoelectric nanowire LED array, Nat. Photonics 7 (2013) 752.

[5] C.F. Wang, C.F. Pan, Z.L. Wang, et al., Enhanced emission intensity of vertical aligned flexible ZnO nanowire/p-polymer hybridized LED array by piezo-phototronic effect, Nano Energy 11 (2014) 2211.

[6] R.R. Bao, Z.L. Wang, et al., Flexible and controllable piezo-phototronic pressure mapping sensor matrix by ZnO NW/p-polymer LED array, Adv. Funct. Mater. 25 (2015) 2884.

[7] S. Li, B.N. Peele, C.M. Larson, H.C. Zhao, R.F. Shepherd, A stretchable multicolor display and touch interface using photopatterning and transfer printing, Adv. Mater. 28 (2016) 9770.

[8] Y.J. Tan, H. Godaba, G. Chen, S.T.M. Tan, G.X. Wan, G. Li, P.M. Lee, Y.Y. Cai, S. Li, R. F. Shepherd, J.S. Ho, B.C.K. Tee, A transparent, self-healing and high-k dielectric for low-field-emission stretchable optoelectronics, Nat. Mater. 19 (2020) 182.

[9] D.F. Peng, B. Chen, F. Wang, Recent advances in doped mechanoluminescent phosphors, ChemPlusChem 80 (2015) 1209.

[10] X.D. Wang, D.F. Peng, B.L. Huang, C.F. Pan, Z.L. Wang, Piezophotonic effect based on mechanoluminescent materials for advanced flexible optoelectronic applications, Nano Energy 55 (2019) 389.

[11] J.C. Zhang, X.S. Wang, G. Marriott, C.-N. Xu, Trap-controlled mechanoluminescent materials, Prog. Mater. Sci. 103 (2019) 678.

[12] X.D. Wang, H.L. Zhang, R.M. Yu, L. Dong, D.F. Peng, A.H. Zhang, Y. Zhang, H. Liu, C. F. Pan, Z.L. Wang, Dynamic pressure mapping of personalized handwriting by a flexible sensor matrix based on the mechanoluminescence process, Adv. Mater. 2714 (2015) 2324.

[13] C.N. Xu, T. Watanabe, M. Akiyama, X.G. Zheng, Artificial skin to sense mechanical stress by visible light emission, Appl. Phys. Lett. 74 (1999) 1236.

[14] C.N. Xu, Y. Liu, M. Akiyama, K. Nonaka, X.G. Zheng, Visualization of stress distribution in solid by mechanoluminescence, Proc. SPIE 4448 (2001) 399.

[15] S.M. Jeong, S. Song, S.K. Lee, B. Choi, Mechanically driven light-generator with high durability, Appl. Phys. Lett. 102 (2013) 051110.

[16] L.S. Liu, C.N. Xu, A. Yoshida, D. Tu, N. Ueno, S. Kainuma, Scalable elasticoluminescent strain sensor for precise dynamic stress imaging and onsite infrastructure diagnosis, Adv. Mater. Technol. 4 (1) (2019) 1800336.

[17] S.M. Jeong, S. Song, K.I. Joo, J. Kim, S.H. Hwang, J. Jeong, H. Kim, Bright, wind-driven white mechanoluminescence from zinc sulphide microparticles embedded in a polydimethylsiloxane elastomer, Energy Environ. Sci. 7 (2014) 3338.

[18] G. Alzetta, N. Minnaja, S. Santucci, Piezoluminescence in zinc-sulphide phosphors, Il Nuovo Cimento (1955–1965) 23 (1962) 910.

[19] X.S. Wang, C.N. Xu, H. Yamada, K. Nishikubo, X.G. Zheng, Electro-mechano-optical conversions in Pr^{3+}-doped $BaTiO_3$-$CaTiO_3$ ceramics, Adv. Mater. 17 (2005) 1254.

[20] H.F. Zhao, X.S. Wang, X.M. Feng, Elastico-mechanoluminescence in non-piezoelectric CaTiO3:Pr^{3+}. in: Conference: 2016 Joint IEEE International Symposium on the Applications of Ferroelectrics, European Conference on Application of Polar Dielectrics, and Piezoelectric Force Microscopy Workshop (ISAF/ECAPD/PFM), 2016. https://doi.org/10.1109/ISAF.2016.7578068.

[21] D. Tu, C.N. Xu, Y. Akihito, Mechanism of mechanical quenching and mechanoluminescence in phosphorescent CaZnOS:Cu, Light Sci. Appl. 4 (2015) 356.

[22] J.C. Zhang, C.N. Xu, S. Kamimura, Y. Terasawa, H. Yamada, X.S. Wang, An intense elastico-mechanoluminescence material CaZnOS:Mn^{2+} for sensing and imaging multiple mechanical stresses, Opt. Express 21 (2013) 12976.

[23] D. Tu, C.N. Xu, LiNbO$_3$:Pr^{3+}: a multi-piezo material with simultaneous piezoelectricity and sensitive piezoluminescence. Adv. Mater. 29 (2017) 1606914. https://doi.org/10.1002/adma.201606914.

[24] J.C. Zhang, C.P. Yi, F.Z. Li, Z.Z. Hong, W. He, X.F. Liu, J.R. Qiu, Achieving thermo-mechano-opto-responsive bitemporal colorful luminescence via multiplexing of dual lanthanides in piezoelectric particles and its multidimensional anticounterfeiting, Adv. Mater. 30 (2018) 1804644.

[25] S.M. Jeong, S. Song, S.K. Lee, N.Y. Ha, Color manipulation of mechanoluminescence from stress-activated composite films, Adv. Mater. 25 (2013) 6194.

[26] X.D.K. Patel, B.E. Cohen, L. Etgar, S. Magdassi, Fully 2D and 3D printed anisotropic mechanoluminescent objects and their application for energy harvesting in the dark, Mater. Horiz. 5 (2018) 708–714.

[27] X.D. Wang, M.L. Que, M.X. Chen, X. Han, X.Y. Li, C.F. Pan, Z.L. Wang, Full dynamic-range pressure sensor matrix based on optical and electrical dual-mode sensing, Adv. Mater. 29 (15) (2017) 1605817.

[28] X.Y. Wei, X.D. Wang, S.Y. Kuang, L. Su, H.Y. Li, Y. Wang, C.F. Pan, Z.L. Wang, G. Zhu, Dynamic triboelectrification-induced electroluminescence and its use in visualized sensing, Adv. Mater. 28 (2016) 6656.

[29] X.C. Wang, J.L. Sun, L. Dong, C.F. Lv, K.K. Zhang, Y.Y. Shang, T. Yang, J.Z. Wang, C. X. Shan, Stretchable and transparent electroluminescent device driven by triboelectric nanogenerator, Nano Energy 58 (2019) 410–418.

[30] S. Wook Shin, J.P. Oh, C.W. Hong, E.M. Kim, J.J. Woo, G.S. Heo, J.H. Kim, Origin of mechanoluminescence from Cu-doped ZnS particles embedded in an elastomer film and its application in flexible electro-mechanoluminescent lighting devices, ACS Appl. Mater. Interfaces 8 (2) (2016) 1098–1103.

[31] Y.L. Zhang, Y.S. Fang, J. Li, Q.H. Zhou, Y.J. Xiao, K. Zhang, B.B. Luo, J. Zhou, B. Hu, Dual-mode electronic skin with integrated tactile sensing and visualized injury warning, ACS Appl. Mater. Interfaces 9 (2017) 37493–37500.

[32] H.Q. Fang, X.D. Wang, Q. Li, D.F. Peng, Q.F. Yan, C.F. Pan, A stretchable nanogenerator with electric/light dual-mode energy conversion, Adv. Eng. Mater. 6 (2016) 1600829.

[33] X. Qian, Z.R. Cai, M. Su, F.Y. Li, W. Fang, Y.D. Li, X. Zhou, Q.Y. Li, X.Q. Feng, W. B. Li, X.T. Hu, X.D. Wang, C.F. Pan, Y.L. Song, Printable skin-driven mechanoluminescence devices via nanodoped matrix modification, Adv. Mater. 30 (2018) 1800291.

[34] S.K. Song, B. Song, C.H. Cho, S.K. Lim, S.M. Jeong, Textile-fiber-embedded multi-luminescent devices: a new approach to soft display systems, Mater. Today 32 (2020) 46.

[35] S.M. Jeong, S. Song, H.J. Seo, W.M. Choi, S.H. Hwang, S.G. Lee, S.K. Lim, Battery-free, human-motion-powered light-emitting fabric: mechanoluminescent textile, Adv. Sustain. Syst. 1 (2017) 1700126.

[36] L.S. Zhan, Z.X. Chen, S.L. Gong, Y.P. Xiang, F. Ni, X. Zeng, G.H. Xie, C.L. Yang, A simple organic molecule realizing simultaneous TADF, RTP, AIE, and mechanoluminescence: understanding the mechanism behind the multifunctional emitter, Angew. Chem. Int. Ed. 131 (2019) 17815.

[37] B.J. Xu, J.J. He, Y.X. Mu, Q.Z. Zhu, S.K. Wu, Y.F. Wang, Y. Zhang, C.J. Jin, C.C. Lo, Z. G. Chi, A. Lien, S.W. Liu, J.R. Xu, Very bright mechanoluminescence and remarkable mechanochromism using a tetraphenylethene derivative with aggregation-induced emission, Chem. Sci. 6 (2015) 3236.

[38] D.S. Mukherjee, D.P. Thilagar, Renaissance of organic triboluminescent materials, Angew. Chem. Int. Ed. 58 (2019) 7922.

[39] F. Yang, Y. Yuan, R.P. Sijbesma, Y.L. Chen, Sensitized mechanoluminescence design toward mechanically induced intense red emission from transparent polymer films, Macromolecules 53 (3) (2020) 905.

[40] L. Kan, H.G. Cheng, B. Li, X.Y. Zhang, Q. Wang, H. Wei, N. Ma, Anthracene dimer crosslinked polyurethanes as mechanoluminescent polymeric materials, New J. Chem. 43 (2019) 2658–2664.

[41] Y. Hirai, T. Nakanishi, Y. Kitagawa, K. Fushimi, T. Seki, H. Ito, Y. Hasegawa, Triboluminescence of lanthanide coordination polymers with face-to-face arranged substituents, Angew. Chem. Int. Ed. 129 (2017) 7277.

[42] C. Wang, Y. Yu, Y.H. Yuan, C.Y. Ren, Q.Y. Liao, J.Q. Wang, Z.F. Chai, Q.Q. Li, Z. Li, Heartbeat-sensing mechanoluminescent device based on a quantitative relationship between pressure and emissive intensity, Matter 2 (1) (2020) 181.

[43] V.K. Chandra, B.P. Chandra, P. Jha, Self-recovery of mechanoluminescence in ZnS:Cu and ZnS:Mn phosphors by trapping of drifting charge carriers, Appl. Phys. Lett. 103 (2013) 161113.

[44] M.C. Wong, L. Chen, M.K. Tsang, Y. Zhang, J.H. Hao, Magnetic-induced luminescence from flexible composite laminates by coupling magnetic field to piezophotonic effect, Adv. Mater. 27 (2015) 4488.

[45] Y. Zhang, G. Gao, H.L.W. Chan, J.Y. Dai, Y. Wang, J.H. Hao, Piezo-phototronic effect-induced dual-mode light and ultrasound emissions from ZnS:Mn/PMN-PT thin-film structures, Adv. Mater. 24 (2012) 1729.

[46] B.L. Huang, Energy harvesting and conversion mechanisms for intrinsic upconverted mechano-persistent luminescence in CaZnOS, Phys. Chem. Chem. Phys. 18 (2016) 25946.

[47] B.L. Huang, D.F. Peng, C.F. Pan, Energy relay center for doped mechanoluminescence materials: a case study on Cu-doped and Mn-doped CaZnOS, Phys. Chem. Chem. Phys. 19 (2017) 1190.

[48] K.F. Wang, L.R. Ma, X.F. Xu, S.Z. Wen, J.B. Luo, Triboluminescence dominated by crystallographic orientation, Sci. Rep. 6 (2016) 26324.

[49] H.L. Zhang, D.F. Peng, W. Wang, L. Dong, C.F. Pan, Mechanically induced light emission and infrared-laser-induced upconversion in the Er-doped CaZnOS multifunctional piezoelectric semiconductor for optical pressure and temperature sensing, J. Phys. Chem. C 119 (2015) 28136.

[50] L.J. Li, K.L. Wong, P.F. Li, M.Y. Peng, Mechanoluminescence properties of Mn^{2+}-doped BaZnOS phosphor, J. Mater. Chem. C 4 (2016) 8166.

[51] W. Wang, D.F. Peng, H.L. Zhang, X.H. Yang, C.F. Pan, Mechanically induced strong red emission in samarium ions doped piezoelectric semiconductor CaZnOS for dynamic pressure sensing and imaging, Opt. Commun. 395 (2017) 24.

[52] L.J. Li, L. Wondraczek, L.H. Li, Y. Zhang, Y. Zhu, M.Y. Peng, C.B. Mao, CaZnOS:Nd3+ emits tissue-penetrating near-infrared light upon force loading, ACS Appl. Mater. Interfaces 10 (2018) 14509.

[53] X. Wu, X.J. Zhu, P. Chong, J.L. Liu, L.N. Andre, K.S. Ong, K. Brinson Jr., A.I. Mahdi, J. C. Li, L.E. Fennod, H.L. Wang, G.S. Hong, Sono-optogenetics facilitated by a circulation-delivered rechargeable light source for minimally invasive optogenetics, Proc. Natl. Acad. Sci. U. S. A. 116 (2019) 26332.

[54] S. Wu, S. Zeng, Z.F. Wang, F. Wang, H. Zhou, J.C. Zhang, Z.P. Ci, L.Y. Sun, Efficient mechanoluminescent elastomers for dual-responsive anticounterfeiting device and stretching/strain sensor with multimode sensibility, Adv. Funct. Mater. 28 (2018) 1803168.

[55] J.C.G. Bünzlia, K.L. Wong, Lanthanide mechanoluminescence, J. Rare Earths 36 (2018) 1.

[56] Y.Y. Du, Y. Jiang, T.Y. Sun, J.X. Zhao, B.L. Huang, D.F. Peng, F. Wang, Mechanically excited multicolor luminescence in lanthanide ions, Adv. Mater. 31 (2019) 1807062.

[57] D.F. Peng, Y. Jiang, B.L. Huang, Y.Y. Du, J.X. Zhao, X. Zhang, R.H. Ma, S. Golovynskyi, B. Chen, F. Wang, A ZnS/CaZnOS heterojunction for efficient mechanical to optical energy conversion by conduction band offset, Adv. Mater. 32 (2020) 201907747.

[58] J. Li, C.N. Xu, D. Tu, X.N. Chai, X.S. Wang, L.S. Liu, E. Kawasaki, Tailoring bandgap and trap distribution via Si or Ge substitution for Sn to improve mechanoluminescence in $Sr_3Sn_2O_7:Sm^{3+}$ layered perovskite oxide, Acta Mater. 145 (2018) 462.

[59] X.Y. Fu, S.H. Zheng, J.P. Shi, Y.N. Liu, X.Y. Xu, H.W. Zhang, Investigation of the cyan phosphor $Ba_2Zr_2Si_3O_{12}:Eu^{2+}$, Dy^{3+}: mechanoluminescence properties and mechanism, J. Alloys Compd. 766 (2018) 221.

[60] H.M. Chen, L.W. Wu, F. Bo, J.K. Jian, L. Wu, H.W. Zhang, L.R. Zheng, Y.F. Kong, Y. Zhang, J.J. Xu, Coexistence of self-reduction from Mn^{4+} to Mn^{2+} and elastico-mechanoluminescence in diphase $KZn(PO_3)_3:Mn^{2+}$, J. Mater. Chem. C 7 (2019) 7096.

[61] S.S. Zhang, S. Wang, T. Hu, S.H. Xuan, H. Jiang, X.L. Gong, Study the safeguarding performance of shear thickening gel by the mechanoluminescence method, Compos. Part B Eng. 180 (2020) 107564.

[62] C.J. Chen, Y.X. Zhuang, D. Tu, X.D. Wang, C.F. Pan, R.J. Xie, Creating visible-to-near-infrared mechanoluminescence in mixed-anion compounds $SrZn_2S_2O$ and SrZnSO, Nano Energy 68 (2020) 104329.

[63] F.L. Wang, F.L. Wang, X.D. Wang, S.C. Wang, J.F. Jiang, Q.L. Liu, X.T. Hao, L. Han, J. J. Wang, C.F. Pan, H. Liu, Y.H. Sang, Mechanoluminescence enhancement of ZnS:Cu, Mn with piezotronic effect induced trap-depth reduction originated from PVDF ferroelectric film, Nano Energy 63 (2019) 103861.

[64] K.P. Min, J. Kim, K.D. Song, G.W. Kim, A G-fresnel optical device and image processing based miniature spectrometer for mechanoluminescence sensor applications, Sensors 19 (16) (2019) 3528.

[65] T.A. Arica, G. Topcu, A. Pala, M.M. Demir, Experimental apparatus for simultaneous measurement of triboelectricity and triboluminescence, Measurement 152 (2020) 107316.

[66] D.O. Olawale, T. Dickens, W.G. Sullivan, O.I. Okoli, J.O. Sobanjo, B. Wang, Progress in triboluminescence-based smart optical sensor system, J. Lumin. 131 (2011) 1407.

[67] R.S. Fontenot, K.N. Bhat, W.A. Hollerman, M.D. Aggarwal, Triboluminescent materials for smart sensors, Mater. Today 14 (2011) 292.

[68] N. Terasaki, Innovative first step toward mechanoluminescent ubiquitous light source for trillion sensors, Sens. Mater. 28 (2016) 827.

[69] R.R. Petit, S.E. Michels, A. Feng, P.F. Smet, Adding memory to pressure-sensitive phosphors, Light Sci. Appl. 8 (1) (2019) 1.

[70] J. Zhang, L.K. Bao, H.Q. Lou, J. Deng, A. Chen, Y.J. Hu, Z.T. Zhang, X.M. Sun, H. S. Peng, Flexible and stretchable mechanoluminescent fiber and fabric, J. Mater. Chem. C 5 (2017) 8027.

[71] Q.Q. Guo, B.X. Huang, C.H. Lu, T. Zhou, G.H. Su, L.Y. Jia, X.X. Zhang, A cephalopod-inspired mechanoluminescence material with skin-like self-healing and sensing properties, Mater. Horiz. 6 (2019) 996.

Mechanophores in polymer mechanochemistry: Insights from single-molecule experiments and computer simulations

7

Wenjin Li[*]
Institute for Advanced Study, Shenzhen University, Shenzhen, China
[*]Corresponding author. e-mail address: liwenjin@szu.edu.cn

7.1 Introduction

Mechanochemistry, wholly different from traditional thermochemistry, photochemistry, and electrochemistry, focuses on how mechanical force affects the reactivity of molecules. The term mechanochemistry first appeared in Ostwald's books. However, Ostwald's version of mechanochemistry is that mechanical force just generate heat, which then accelerates chemical reactions [1]. Today, we know that mechanical force can directly induce chemical reactions. For instance, mechanical force can break down a polymer chain in the middle to form two radicals [2].

With the advent of various approaches, especially the single-molecule force spectroscopy (SMFS) [3, 4], our insights into the way that the mechanical force accelerates chemical reactions was greatly advanced. One of the intensively studied case is the thiol-disulfide exchange. Fernandez and coworkers engineered a disulfide bond into immunoglobulin domain of titin, and observed an exponentially increasing rate of the disulfide reduction by dithiothreitol when subjected to stretching forces from atomic force microscopy (AFM) in a range of less than 600 pN [5]. Recently, a computational study extended the studied force to 2000 pN, and discovered that the transition state was shifted toward the reactant at high forces [6]. When catalyzed by a variety of reducing agents, the activation length of disulfide reduction varied from 0.23 to 0.46 Å [7], indicating significant differences among the structures of transition states under the presence of various agents. Furthermore, when catalyzed by an enzyme (thioredoxin), the disulfide reduction was observed to be slowed down in a low-force range while the reaction was accelerated in the regime of higher forces [8, 9].

Mechanical force is a vector and thus directional. For forces of the same magnitude, their effects on a process may vary if their directions are different. Such phenomenon has been observed in protein unfolding studies. For example, Carrion-Vazquez et al. have studied the mechanical stability of polyubiquitin chains by constant velocity pulling with AFM [10]. They applied force either between its Lys48 and C terminus

---

Functional Tactile Sensors. https://doi.org/10.1016/B978-0-12-820633-1.00005-X
© 2021 Elsevier Ltd. All rights reserved.

or between the N and C terminus of the same domain, and found that under the same pulling velocity the average unfolding force in the former case was much lower than that in the latter one. In addition, the unfolding force in the former case showed a stronger dependence to the pulling rate. Later, Li and Mokarov [11] have studied the dependence of mechanical stability of ubiquitin along four pulling coordinates with molecular dynamic (MD) simulation, and they have found an even lower unfolding force when the Lys11 and C terminus are extended. Recently, a near-zero unfolding force has been observed by Best et al. when a force is applied to residues 24 and 52 [12]. The mechanical stability was observed as well to be dependent on the direction of the applied force in another study, in which a β-sheet protein, E2lip3, was stretched along two distinct coordinates [13].

Apparently, the force dependence of chemical reactions is considerably more complex than originally expected, and our understanding in the mechanochemical behavior of chemical reactions remains very limited. However, there exist many efforts in applying our knowledge on mechanochemistry to the development of new materials. One main concept in material design is the incorporation of mechanophores (force-sensitive chemical groups that undergo chemical reactions in response to mechanical forces) into polymer chains. For instances, the colorless spiropyran can be transformed by mechanical force into a colored product via a ring-opening reaction [14]. This mechanophore was then used by Sottos and coworkers to synthesize a mechanoresponsive material, in which the stress-induced damages were visualized via a color change upon tensile loading [15]. Bielawski and coworkers [16] recently observed that force-induced cycloreversion converted triazoles with high fidelity into their azide and alkyne constituents, which is another example that can be employed to develop mechanoresponsive materials.

The chemical reactivity of many mechanophores was reported to be tuned by mechanical forces of various magnitudes and directions, similar to the above-mentioned phenomena on biomolecules. More importantly, mechanical force was observed to change the reaction pathway of mechanophores. Pathways, which are thermochemically forbidden, can be accessible under mechanical force. For instance, under thermal activation, the Woodward-Hoffmann (WH) rules [17, 18] predict that both *trans*- and *cis*-isomer of 1,2-disubstituted benzocyclobutene proceed along a conrotatory pathway of ring opening. However, the *cis*-isomer was observed to reaction via a disrotatory ring opening when subjected to external forces [19], which was also supported by computational results from the Martinez's group [20].

In the following, we will first describe the theory of mechanochemistry briefly in Section 7.2. The common experimental approaches, especially single-molecule techniques, and the computational methods are then introduced in Sections 7.3 and 7.4, respectively. In Sections 7.5 and 7.6, recent mechanochemical studies on typical examples of covalent and organometallic mechanophores, respectively, are summarized with particular attention paid to the results from single-molecule experiments and theoretical calculations. In addition, the effect of polymer chain on the mechanochemical behavior of mechanophore is the subject of Section 7.7. Finally, the perspectives of mechanophores in polymer mechanochemistry are discussed in Section 7.8.

7.2 Brief theory of mechanochemistry

Zhurkov has proposed a theoretical model to quantify the kinetics of the breakdown of solids under stress [21, 22]. The model is based on the experimental study of polymer midchain scission in viscous flow. The rate of chain-rupture K is given by adding a term $\alpha\sigma$ into the Arrhenius equation:

$$K = K_0 \exp\left[-(E_a - \alpha\sigma)/RT\right] \tag{7.1}$$

where σ is the tensile stress from viscous flow and in the unit of N m^{-2} and α is a coefficient in the unit of m^3 mol^{-1}. Thus, their product $\alpha\sigma$ gives the unit of energy and is interpreted as the mechanical work. K_0 is the frequency factor and E_a is the activation energy. With Eq. (7.1) as the prototype, Bell has introduced the mechanical force F explicitly and developed a new model given by Eq. (7.2) [23]:

$$r(F) = r_0 \exp\left[\beta F \Delta x\right] \tag{7.2}$$

where $\beta = k_B T$, r_0 is the force-free rate constant. Δx is the activation length defined as the distance between reactant and transition state along the coordinate of external force. For the first time, Bell's model links the macroscopic force with a microscopy quantity Δx, which is obviously of physical and chemical significance. It assumes that the mechanical force does not alter the reaction pathway and the energy barrier is lowered by mechanical force linearly. Bell's model predicts a force-independent transition state, in contrast to the observation in the theoretical study of the disulfide reduction by dithiothreitol [6]. In addition, it failed to predict the experimental observations in studies such as protein unfolding by AFM and nanopore unzipping of DNA hairpins [24–26]. By applying the Kramers theory [27] in the high-barrier limit to single-well models of free-energy surfaces, Hummer and coworkers have derived a unified formula for the force-dependent rate constant [24, 28, 29]:

$$r(F) = r_0 \left(1 - vF\Delta x/\Delta G^{\ddagger}\right)^{1/v-1} \exp\left\{\beta\Delta G^{\ddagger}\left[1 - \left(1 - vF\Delta x/\Delta G^{\ddagger}\right)^{1/v}\right]\right\} \tag{7.3}$$

in which $\Delta G^{\ddagger}$ is the apparent free energy of activation and v specifies the shape of the free-energy surface. This Dudko-Hummer model, also known as the cusp model, takes into account the effect of the applied force on the structure of transition state.

Other theoretical models include the Boulatov's Taylor expansion [30], Makarov's extended Bell theory [31], and a theoretical model for the prediction of the rupture probabilities of molecular bonds in series as proposed by Gaub and coworkers [32].

7.3 Single-molecule approaches

7.3.1 Introduction

Mechanical force can be applied to molecules interested and the kinetic or dynamic properties of the reaction are then recorded by a variety of approaches. These methods can be divided into ensemble approaches and single-molecule measurements according to the amount of samples [an ensemble of reactants (e.g., solid powder) or a single molecule] used in the experiments.

In the early stage of the study in mechanochemistry, the reactants are always a massive of molecules, which are, for example, in the form of dry powder or solute in a solvent. That is to say, an ensemble of molecules is under the influence of mechanical forces generated by milling [33, 34], mastication [35], and viscous flow [36, 37] among many others. The applied forces are in arbitrary directions and vary in magnitude as well. In ensemble measurement, the mechanical force is not directly monitored, and other quantities correlated with the mechanical force (e.g., the strain rate in viscous flow) are measured instead. With the increasing advances in method development, new ensemble methods such as ultrasonic irradiation [38] and molecular tensile machines with bottlebrush macromolecules [39, 40] have been proposed. Ultrasound is now commonly used to apply force selectively on chemical bonds within polymers. For example, ultrasound has been applied to break weak peroxide [41] and azo [42] bonds in covalent polymers, and weak coordinative palladium-phosphorus bond in Palladium(II) coordination polymers [43]. Importantly, single-molecule methods such as SMFS [44] and molecular force probe [45] have been recently developed as well.

7.3.2 Single-molecule force spectroscopy

In order to quantify the mechanochemical effects of force on a mechanophore at the single-molecule level, techniques such as AFM, optical tweezers, and magnetic tweezers were developed to enable the pulling of a single molecule at a time with a measurable force [3, 46–48]. Among these three techniques, AFM is the major one applied intensively in the studies of polymer mechanochemistry. Thus, only the principle and applications of AFM are briefly introduced here.

In AFM, one end of a mechanophore-functionalized polymer is connected to the tip of a cantilever, and the other linked to the substrate on a piezoelectric positioner (Fig. 7.1A). Upon moving the positioner downwards, the polymer chain is pulled and the generated force results in a displacement of a well-calibrated cantilever, which is then derived from the position of a laser beam reflected by the cantilever. The applied force is quantified by the product of the displacement of the cantilever and its spring constant, while the extension of the polymer chain is given by the distance from the tip of the cantilever to the surface of the positioner.

There are two popular modes to pull the polymer chain with AFM: force-probe and force-clamp experiments. In force-probe experiments, the piezoelectric positioner moves at constant velocity and the change of the generated force with the extension of the polymer chain is recorded to give the force-extension curve. Initially, force-probe AFM was utilized to enforce protein unfolding, where the force-extension curve possesses a characteristic sawtooth-like pattern [4, 49, 50]. In force-clamp experiments, the movement of the piezoelectric positioner is controlled by a feedback of the measured force to ensure a constant force, and the extension of the polymer chain is monitored as a function of time. The typical force spectroscopy for the forced unfolding of a polyprotein chain is a staircase-like curve [51, 52]. For mechanophores such as metal-ligand coordination complexes between Pd pincer complexes and pyridine ligands [53, 54], their mechanochemical activation results in the rupture of

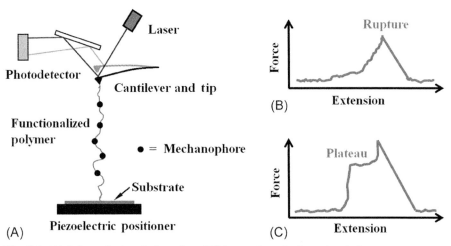

Fig. 7.1 (A) Schematic description of an AFM setup in a single-molecule force spectroscopy experiment, here considering the example of the mechanophore-functionalized polymer. (B) Typical force-extension curve for a functionalized polymer with a mechanophore whose mechanochemical activation results in the rupture of the polymer. (C) Typical force-extension curve for a functionalized polymer with a mechanophore, whose mechanochemical activation results in an extension but not a breaking of the mechanophore. There is a plateau region that indicates the mechanochemical activation events of the mechanophore.

the polymer. A typical force-extension curve for such mechanophores is shown in Fig. 7.1B, in which the rupture of the polymer is indicated by a sudden drop of force and the stretching of the polymer backbone gives the nonlinear increase in force before rupture. The mechanochemical activation of mechanophores like cyclopropane derivatives [55, 56] results in an extension instead of a breaking of the mechanophore. However, the extension is typically several angstroms and the resulted force drop is not large enough to be visible in the force-extension curve. For these mechanophores, the force-extension curve presents a plateau before the final rupture of the polymer (Fig. 7.1C), in contrast to the sawtooth-like signature in the unfolding of a polyprotein chain. The plateau region is attributed to the mechanochemical activation events of the mechanophore, and the force at which the plateau is observed (named the plateau force) provides an additional feature to quantify the force dependence of the mechanophore.

7.3.3 Molecular force probe

Molecular force probe developed by Boulatov and coworkers is a novel way to apply forces along two fixed atoms of small molecules [45, 57]. In molecular force probe, the mechanophore is embedded into a macrocycle, which consists of a stiff stilbene, two linkers, and the mechanophore (Fig. 7.2). Stiff stilbene can undergo isomerization from its Z-isomer to the E-isomer upon absorption of light, which results in a great increase of the C6-to-C6′ distance from ∼5.1 to ∼8.3 Å [59]. Upon the photoisomerization of stiff stilbene, the macrocycle exerts internal forces along its C6,

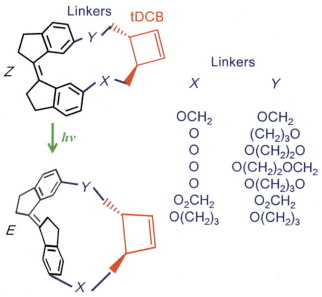

Fig. 7.2 Examples of molecular force probes employed in Refs. [45, 58]. Macrocycles consist of the mechanophore tDCB (*red*), two linkers (named *X* and *Y*, *blue*), and stiff stilbene (*black*). Upon absorption of light, the macrocycles swith from their compact Z-isomers (*top*) to the strained E-isomers (*bottom*). A variety of linkers with different length and geometry are listed at the right.

$C6'$ axis to the linkers and the mechanophore [45, 57]. The restoring force applied on the mechanophore can be derived from quantum chemical calculations. By varying the length and geometry of the linker molecules, a variety of restoring forces can then be generated. Thus, the mechanochemical reactivity of a mechanophore can be quantitatively studied using the molecular force probe.

This molecular approach was first applied to investigate the ring-opening reaction of *trans*-3,4-dimethylcyclobutene (tDCB) into the corresponding 2,4-hexadiene derivative (Fig. 7.2). The rate of ring opening was shown to increase when tDCB is subjected to the restoring force present in the macrocycles, with smaller macrocycles inducing higher rates [45, 57]. Computations at the level of density functional theory (DFT) showed that the restoring forces generated upon the photoisomerization of stiff stilbene were in the sub-nN regime. A recent study with force distribution analysis (FDA) indicated a significantly different geometry of force application in these macrocycles as compared to the one in SMFS experiments [58].

7.4 Computational approaches

7.4.1 *Force-probe and force-clamp molecular dynamics*

AFM experiments with constant velocity pulling can be simulated with force-probe molecular dynamics (FPMD) [60]. In order to reproduce the dynamic details at the atomic level of the rupture process of the chemical bond under external forces, the atom of the molecule that is connected to the cantilever through a polymer linker in AFM experiments is taken as the pulling point (PP) and is subjected to a harmonic potential given by.

$$V_{\text{spring}} = K_0[Z_{\text{PP}}(t) - Z_{\text{cant}}(t)]^2/2 \tag{7.4}$$

in which K_0 is the force constant, $Z_{\text{PP}}(t)$ is the position of the PP, and $Z_{\text{cant}}(t)$ is the cantilever position. This harmonic spring potential approximates the elastic cantilever and is moved with constant velocity v along the desired direction. The spring potential is thus centered at $Z_{\text{cant}}(t) = Z_{\text{cant}}(0) + vt$, where $Z_{\text{cant}}(0)$ is the starting position of the spring at $t = 0$, where it is relaxed. The force exerted on the PP is thus given by

$$F_{\text{PP}} = K_0(Z_{\text{PP}}(t) - Z_{\text{cant}}(t)) \tag{7.5}$$

and is recorded to give the force-extension curve, in which the largest force observed gives the rupture force. The resulted force profiles provide direct comparison to experimental ones. The FPMD simulations with classical force fields have been intensively applied to large biological systems such as the avidin-biotin complex [60, 61], titin immunoglobulin domains [62], and titin kinase [63].

AFM experiments carried out under constant pulling force [51, 64] can be also simulated with MD by keeping the force exerted on the PP to be constant. Such force-clamp molecular dynamics (FCMD) simulations have been applied in the study of titin immunoglobulin domain [65] and the mechanochemical behavior of a thiol/disulfide exchange in a protein [6], just to name a few. FPMD and FCMD are oftentimes collectively called as steered MD.

7.4.2 Constrained geometries simulate external force (CoGEF)

To simulate the force-extension curve resulted from SMFS measurement under constant velocity, Beyer proposed the "constrained geometries simulate external force (CoGEF)" method [66]. In CoGEF, the potential energy surface (PES) is obtained by a relaxed potential energy scan along the pulling direction, which is usually approximated by the distance r between the two atoms on which the external force is exerted. More specifically, starting from the equilibrium geometry with no force exerted, a geometry optimization is carried out by allowing all the coordinates to be relaxed other than the distance r, which is increased by a small amount and kept fixed. Such constrained geometry optimizations are repeated by taking the geometry from the previous step as the starting structure and increasing r by a small step. The small step is usually a fixed amount, but can also be varied in each step. The CoGEF method will generate a potential curve as a function of the distance r, called CoGEF potential. An example of the CoGEF potential calculated for the *cis*-dicyano-substituted cyclobutane under stretching is given in Fig. 7.3 [66]. Two characteristic parameters can be obtained from the CoGEF potential: the rupture force (F_{max}) that is given by the maximum slope along the CoGEF potential curve toward dissociation and the breaking point distance (R_{max}) that is defined as the distance corresponding to the rupture force [67]. The CoGEF method has been employed to evaluate the mechanical

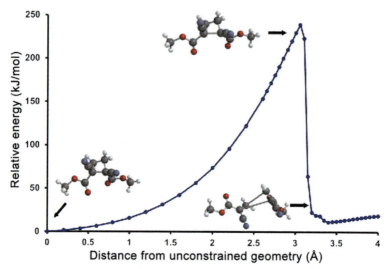

Fig. 7.3 CoGEF potential of the *cis*-dicyano-substituted cyclobutane mechanophore (see *inset*) when two terminal methyl ester groups are pulled apart. The portrayed structures from left to right correspond to the geometries of the mechanophore at equilibrium, under the rupture force, and when being stretched far beyond the breaking point.
Adapted with permission from M. J. Kryger, A. M. Munaretto, J. S. Moore, Structure–mechanochemical activity relationships for cyclobutane mechanophores, J. Am. Chem. Soc. 133 (2011) 18992–18998. Copyright 2011 American Chemical Society.

strength of various covalent bonds, such as H—H, H—F, H—Cl in diatomic molecules and O—H, O—F, O—Cl in triatomic molecules [66]. It has also been systematically tested in predicting the force-induced rupture of bonds like Si—C, C—C, and C—N in simple polyatomic molecules [67].

7.4.3 External force explicitly included (EFEI)

Marx's group proposed a method named "external force explicitly included (EFEI)" to study molecules stretched under constant forces in computer simulations [68]. For a given PES $V(\mathbf{q})$ as a function of the positional coordinates $\mathbf{q}$ of a system, the PES will be altered when a constant force $\mathbf{F}$ is applied on a mechanical coordinate $\mathbf{r}(\mathbf{q})$. In case of that the force acts on two atoms of the system, $\mathbf{r}(\mathbf{q})$ is simply the distance between the two atoms. A force-transformed PES can be defined as [68]

$$V_{EFEI}(\mathbf{q}, \mathbf{F}) = V(\mathbf{q}) - \mathbf{F} \times \mathbf{r}(\mathbf{q}) \tag{7.6}$$

At the stationary point of the force-transformed PES, that is, the first derivative of $V_{EFEI}(\mathbf{q}, \mathbf{F})$ with respect to $\mathbf{q}$ vanishes for a given constant $\mathbf{F}$, $V_{EFEI}(\mathbf{q}, \mathbf{F})$ is proved to be the Legendre transform of the corresponding CoGEF potential. Thus, EFEI provides the same structure as the one from CoGEF when the mechanical coordinate is fixed in a given value. Importantly, searching the stationary points in the force-tilted PES $V_{EFEI}(\mathbf{q}, \mathbf{F})$, structures such as reactant, transition, and product state structures can be determined using the same searching approaches (for instance, Fukui's intrinsic reaction coordinate concept [69]) as those traditionally applied to the original PES. Thus, the reactant, transition, and product state structures as well as the activation energies as a function of the external force can be well investigated.

7.4.4 Force distribution analysis approaches

The knowledge of how the external force is distributed in each degree of freedom in a molecule can facilitate our understanding of the mechanochemical properties of mechanophores. For example, such knowledge help us to identify which part of a molecule is more susceptible to the external force than others, and to reveal the way that the external force is selectively transducted to a bond in a particular geometry of a molecule. To obtain such knowledge, several FDA approaches have been developed [58, 65, 70].

7.4.4.1 Classical force distribution analysis

Classical FDA was developed to analyze the change of internal forces for a system simulated with classical force fields when the system was under external perturbations such as the external force pulling [65, 71]. As the averaged atomic force on each atom converges to zero in equilibrium, the FDA method use forces for each atom pair to estimate the change of internal forces. The change of such pairwise force for each atom pair i,j (ΔF_{ij}) is defined as the difference between the pairwise force in the perturbed state after force application (F_{ij}^{pert}) and the one before the application of force (F_{ij}^{ref}). Thus,

$$\Delta F_{ij} = F_{ij}^{Pert} - F_{ij}^{ref} \tag{7.7}$$

In classical FDA, the pairwise forces can be easily derived from the nonpolarizable force field, in which bond interaction, van der Waals interaction and Coulombic interaction are generally pairwise interactions. Angular and dihedral interactions can be transformed into pairwise interactions by intuitive approximations [71]. Recent efforts from the same group introduced time-resolved FDA to enable the monitoring of pairwise forces during conformational changes [72].

7.4.4.2 Force-matching force distribution analysis (FM-FDA)

The classical FDA is not applicable to quantum mechanically described systems, as it is not trivial to extract the classical pairwise forces from the quantum Hamiltonian due to the delocalization of and the correlation between electrons. Thus, a new FDA approach was proposed based on force matching (FM) [58].

FM proposed by Ercolessi and Adams aimed at the construction of an effective force field from ab initio MD simulations [73]. Upon extension and improvement from the group of Voth [74–76], the method evolved to be a new method, called multi-scale coarse-graining (MS-CG). In MS-CG, an effective force field was defined by a set of L parameters η_1, η_2, ..., η_L, which can be determined when the differences between the forces obtained by the effective force field and the ones supplied by ab initio MD simulations. More specifically, for a trajectory with M configurations and N atoms in each configuration, the effective force field was optimized by minimizing the following objective function:

$$\chi^2 = \frac{1}{3MN} \sum_{i=1}^{N} \sum_{j=1}^{M} \left| F_{ij}^{\text{ref}} - F_{ij}^{\text{mat}}(\eta_1, \eta_2, ..., \eta_L) \right|^2 \tag{7.8}$$

where F_{ij}^{mat} is the matched force on atom i in configuration j calculated from the effective force field and F_{ij}^{ref} is the corresponding reference forces obtained from ab initio MD simulations. Generally, Eq. (7.8) is overdetermined as the number of equations (NM) is much large than the number of variables (L) to be calculated, and is solved in the least-squared sense.

In order to apply FDA in a quantum chemical system, FM is first performed to construct a specific pairwise force field from the trajectories of ab initio MD simulations. With the pairwise force field, classical FDA can be then used to analyze the results from quantum simulations. This procedure is, therefore, named FM-FDA.

7.4.4.3 Judgment of energy distribution (JEDI)

Stauch and Dreuw proposed a method named the judgment of energy distribution analysis, which aims to analyze how the stress energy distributed in a mechanically distorted molecule based on quantum mechanical calculations [70]. The JEDI method estimates the stress energy distributed among all internal coordinates of the molecule, which includes all the bonds, bond angles, and dihedral angles.

Given a molecule that is described by a set of redundant internal coordinates $\vec{q}$, and the potential around the equilibrium geometry of the molecule $\vec{q}_0$ that can be well described by a harmonic approximation, the energy stored in each internal coordinate i can be estimated by

$$E_i = \frac{1}{2} \sum_{i}^{M} \frac{\partial^2 V(\vec{q})}{\partial q_i \partial q_j} \Big|_{\vec{q}=\vec{q}_0} \Delta q_i \Delta q_j \tag{7.9}$$

where $\frac{\partial^2 V(\vec{q})}{\partial q_i \partial q_j} \Big|_{\vec{q}=\vec{q}_0}$ is the Hessian in redundant internal coordinates $\vec{q}$ at the equilibrium structure $\vec{q}_0$, which can be estimated according to Ref. [77]. M is the number of internal coordinates. Δq_i is the change of the internal coordinate i upon transition from the relaxed structure $\vec{q}_0$ to a strained geometry in the presence of an external force.

The harmonic stress energy E_{harm} is simply the summation of the contributions E_i of each internal coordinate i and is calculated as

$$E_{\text{harm}} = \sum_i E_i = \frac{1}{2} \sum_{i,j}^{M} \frac{\partial^2 V(\vec{q})}{\partial q_i \partial q_j} \Big|_{\vec{q}=\vec{q}_0} \Delta q_i \Delta q_j \tag{7.10}$$

If the underlying assumption of the harmonic approximation is reasonable, the harmonic stress energy E_{harm} should be comparable to the energy difference E_{diff} between the relaxed and the distorted structure estimated by quantum mechanical methods. The applications of the JEDI analysis to simple molecules, such as hydrogen peroxide and ethene, demonstrated that E_{harm} is comparable to E_{diff} in a broad range of applied forces. The redundant internal coordinates were demonstrated to be a better choice among several sets of coordinates, such as normal modes, redundant internal, delocalized internal, and Z-matrix coordinates [78].

In contrast to the conventional FDA and FM-FDA method [58, 65], the mechanical work instead of the force is calculated in the JEDI analysis. Due to the harmonic approximation, the applicability of the JEDI analysis is limited to the range of low and moderate forces. Li and Ma recently proposed an energy decomposition scheme at the single coordinate level, which can be applied to the entire transition from one stable state to another [79, 80]. This energy decomposition method has been applied to systems, such as alanine dipeptide [79, 80] and rhodopsin [81], which were described by molecular mechanical force fields. However, the applicability of the method to quantum mechanically described systems has not yet been demonstrated.

7.5 Covalent mechanophores

The majority of the known covalent mechanophores are compounds with ring structures. Examples of these ring-opening mechanophores include cyclopropane-based mechanophores with a three-membered ring, cyclobutene-based mechanophores with a four-membered ring, and other ring-opening ones. The advantages of mechanophores with ring structures are that (1) the opening of a ring breaks a covalent bond and thus requires high mechanical force and (2) the ring opening does not result in a complete break of the mechanophore but a limited extension, which relaxes the external stress significantly. Thus, ring-opening mechanophores are of great potential in the design of self-healing materials.

7.5.1 Cyclopropane-based mechanophores

Gem-difluorocyclopropane (gDFC) and its derivatives are the most well-studied examples of mechanophores with a three-membered ring. gDFC (compound **1** in Fig. 7.4) has two isomers: *trans*-gDFC and *cis*-gDFC isomers. The *trans*-gDFC is thermodynamically more stable than the *cis*-gDFC, while under a transient force of extension the former can be favorably converted to the latter, which results in a net contraction of the gDFC [82]. Ab initio FCMD simulations using the complete active space self-consistent field method [92] revealed that *trans*-gDFC and *cis*-gDFC isomers under mechanical activation were converted to an s-trans/s-trans diradical

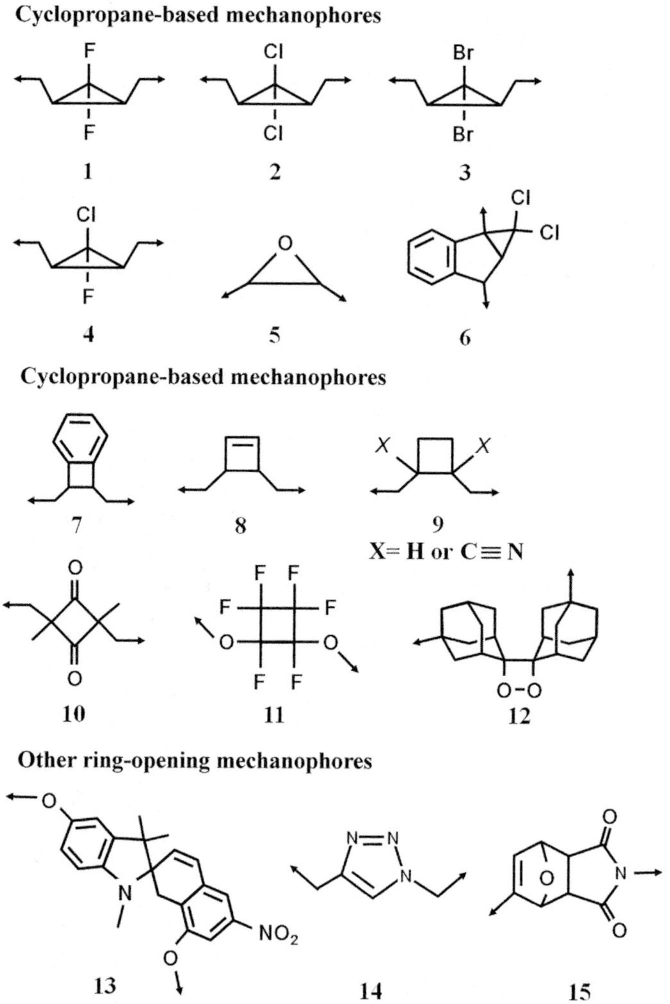

Cyclopropane-based mechanophores

Cyclopropane-based mechanophores

Other ring-opening mechanophores

Fig. 7.4 Representatives of covalent mechanophores. (**1**) *Gem*-difluorocyclopropane; [82] (**2**) *Gem*-dichlorocyclopropane; [83] (**3**) *Gem*-dibromocyclopropane; [55] (**4**) *Syn*-chloro-*gem*-chlorofluorocyclopropane [84]; (**5**) epoxidize [85]; (**6**) *Gem*-dichlorocyclopropanated indene [86]; (**7**) 1,2-disubstituted benzocyclobutene [19]; (**8**) cyclobutene; [45] (**9**) cyclobutane derivatives [87]; (**10**) cyclobutane-1,3-diones; [88] (**11**) perfluorocyclobutane; [89] (**12**) bis (adamantyl)-1,2-dioxetane [90]; (**13**) spiropyran [15]; (**14**) triazole [16]; and (**15**) furan/maleimide adducts [91]. *Arrows* indicate the PP of the applied force.

conformation via predominantly conrotatory and disrotatory ring opening, respectively [82]. This diradical is the transition state in the force-free isomerization of gDFCs. It is greatly stabilized under mechanical force with the lifetime being extended from originally less than 10^{-13} s to longer than 10^{-9} s [82]. Notably, the ring-opening reaction in the *trans*-gDFC under mechanical force proceeds through a thermally forbidden conrotatory pathway, as the stress-free *trans*-gDFC prefers a concerted, disrotatory ring opening [93]. When two of these tension-trapped diradicals are adjacent to each other, the covalent bond between them breaks via a disproportionation reaction, as confirmed by FCMD simulations on a system consisting of two neighboring gDFCs. It provided a good explanation of the observation that the ultrasound-assisted rupture probability of the gDFC containing polymers increases with the content of gDFC [94]. Craig and coworkers reported that the forbidden conrotatory and the allowed disrotatory ring-opening reaction of gDFC requires approximately 1290 and 1820 pN force, respectively, in order to induce reactions on the approximately 0.1 s timescale of the SMFS experiments [56]. In addition, under mechanochemical activation, the transition from *cis*- to *trans*-gDFC via the tension-trapped diradicals intermediate was also confirmed by SMFS experiments [95].

Another typical mechanophore of three-member ring is *gem*-dichlorocyclopropane (gDCC, compound **2** in Fig. 7.4). In contrast to gDFC, ring opening of gDCC irreversibly results in the formation of 2,3-dichloroalkene [83]. The stress-free reaction of the *cis*-gDCC is about 20 times faster than the one of the *trans*-gDCC [96]. Since the *cis*-gDCC undergoes a symmetry-allowed disrotatory ring opening, which should be coupled better to the external force than the *trans* ones, it was expected that the external force will accelerate more efficiently the reaction rate of the *cis*-gDCC and the reaction rates will differ by an even greater extent under mechanochemical activation. Surprisingly, the reaction probabilities of both isomers almost equal as indicated by NMR spectra results [83]. In an attempt to explain the counterintuitive observation, Marx and coworkers revealed the detailed reaction mechanism of both gDCC isomers with EFEI calculations and ab initio FCMD simulations. The ring opening of gDCC takes place via a two-step process: C—C bond scission in concerted with the breaking of the C—Cl bond, and then chloride migration [97]. They proposed that the equivalent rates between the mechanically assisted ring openings of *cis*- and *trans*-isomers is likely due to the sufficiently large forces generated by ultrasound, which promote both reactions to the barrierless regime [97]. Craig and coworkers reported that the force necessary for *cis*-gDCC to open its ring on the approximately 0.1 s timescale of the SMFS assays is 1300 pN and the one for *trans*-gDCC is 2300 pN [56]. Similar to the case of *trans*-gDFC, the ring of *trans*-gDCC opens via the symmetry-forbidden conrotatory pathway [56], not the symmetry-allowed disrotatory one as predicted by computational studies from Marx and coworkers [97].

Other cyclopropane derivatives under SMFS studies include *gem*-dibromocyclopropane (gDBC, compound **3** in Fig. 7.4) and *syn*-chloro-*gem*-chlorofluorocyclopropane (*syn*-Cl-gCFC, compound **4** in Fig. 7.4). Under tip velocity of 3 μm/s, the extension in the contour length of the gDBC-functionalized polymer occurs at a plateau force of roughly 1200 pN and is due to the force-induced ring-opening reaction of the gDBCs and the subsequent bromide migration into 2,3-dibromoalkenes [55]. Craig and coworkers demonstrated that under otherwise destructive shear forces gDBC-functionalized polybutadiene can be strengthened due to a mechanically assisted cross-linking, that is, mechanical activation of gDBC generates allylic bromides, which are in turn cross-linked in situ with carboxylates via nucleophilic substitution reactions [98]. As reported in the single-molecule polymer mechanochemistry experiments [84],

126 Functional Tactile Sensors

sufficient external forces induced a disrotatory ring opening of the *syn*-Cl-gCFC in an outward fashion, which violates the torquoselectivity predicted by the WH rules [17, 18] and whose transition state is of appreciable diradical feature, as in the case of the isomerization of gDFC [84]. In the absence of force, *syn*-Cl-gCFC proceeds along the disrotatory inward ring opening as allowed by the WH rules [84]. The irreversible extension of the *syn*-Cl-gCFC takes place on the time scale of the SMFS experiments at approximately 1500 pN [84].

Mechanophores based on cyclopropane have been also reported. The efforts from the group of Craig achieved the trapping of carbonyl ylides, the sonication-induced product of epoxidized (compound **5** in Fig. 7.4) polynorbornene, in the presence of small molecules [85]. Moore and coworkers [86] demonstrated that a *gem*-dichlorocyclopropanated indene (compound **6** in Fig. 7.4) when incorporated in poly(methyl acrylate) (PMA) is able to release acid under compression with high load. This acid-releasing mechanophore expanded the tool kit in the design of autonomous self-healing materials.

7.5.2 Cyclobutene-based mechanophores

1,2-Disubstituted benzocyclobutene (BCB, compound **7** in Fig. 7.4) is the first reported mechanophore that is mechanically activated to open its four-membered ring via a formal disrotatory process that is "forbidden" as predicted by the WH rules instead of the thermally allowed conrotatory pathway [19]. CoGEF studies indicated that the mechanical force performs work on the *cis*-BCB molecule to decrease the energy barrier of the disrotatory pathway while it increases the barrier of the conrotatory one [19]. The EFEI studies on the 1,2-dimethyl-BCB molecule from Marx and coworkers [99] predicted a critical force of 510 pN, the force that is sufficiently high to be able to reduce the activation energy of the disrotatory mechanism such that it becomes lower than the conrotatory pathway, while the ab initio FCMD simulations on a bare BCB molecule from Martinez and coworkers suggested a critical force of 1500 pN [20]. Marx and coworkers have also reported that under the external force of more than 750 pN, the conrotatory pathway disappears, and investigated in detail the performance of the Bell's model and the Dudko-Hummer model in the mechanochemistry of the BCB molecule [68, 99]. As suggested by MD simulations on a force-modified PES and detailed energy decomposition analysis from the group of Mosey [100], the mechanical force circumvents the WH rules via the deformation of the PES rather than the changes in the electronic structure of the system. Frank and coworkers carried out ab initio FPMD simulations and revealed that the mechanical force induces an intermediate localization of the relevant frontier orbitals, which establishes a connection between orbitals of different symmetry, and thus circumvents the conservation of orbital symmetry in the disrotatory ring opening of the *cis*-BCB molecule [101].

As a simplified system of BCB, the mechanochemical behavior of cyclobutene (CB, compound **8** in Fig. 7.4) has been also intensively investigated both experimentally and computationally [45, 58]. The group of Boulatov observed [45] that the ring-opening reaction rate of the tDCB was accelerated exponentially by the restoring forces from the photoisomerization of the stiff stilbene, a molecular force probe that forms macrocycles of varying sizes together with two linkers and tDCB (Fig. 7.2).

The distributions of the restoring forces from different macrocycles in the tDCB were revealed by the FM-FDA method. The mechanically induced rate acceleration was found to be mainly attributed to the forces on three coordinates, namely, the breaking C—C bond and two adjacent C—C—C bond angles [58]. The same three coordinates have been also demonstrated to be most relevant to the mechanical activity of the *cis*-BCB [102]. The distribution of stress energy on tDCB has recently been analyzed with the JEDI approach [70]. The mechanochemical property of CB was found to be similar to the case of BCB. The mechanical force was reported to activate a "forbidden" disrotatory opening of the *cis*-CB, which becomes the energetically favored pathway if the force exceeds 1000 pN [20].

Other cyclobutene derivatives have also received great attention. Moore and coworkers reported an ultrasound-induced cycloreversion of dicyano-substituted cyclobutanes (compound **9** in Fig. 7.4) into cyanoacrylates [103]. Subsequently, the group of Moore [87] investigated the structure-mechanochemical activity relationships of six cyclobutane derivatives, which are the *cis*- and *trans*-stereoisomers of dicyano-substituted cyclobutanes, the *cis*- and *trans*-monocyano-substituted cyclobutanes, and the *cis*- and *trans*-cyclobutanes having no cyano substituents. The CoGEF method was shown to be able to reliably predict the mechanochemical reactivity of these mechanophores. Sijbesma and coworker investigated the mechanochemical behavior of the retro [2 + 2] cycloaddition of the cyclobutane-1,3-diones (compound **10** in Fig. 7.4) via a combination of the CoGEF method and ultrasonication study [88]. Unlike the mechanochemical properties of many other mechanophores, the stability of cyclobutane-1,3-diones is strengthened under the mechanical pulling, as the chain scission of the polymer takes place somewhere other than the four-membered ring in ultrasound-induced scission experiments. DFT calculations supported an increase of the activation energy for the retro [2 + 2] cycloaddition under mechanical pulling. Craig and coworkers reported that mechanical-induced cycloreversion of the pefluorocyclobutane (compound **11** in Fig. 7.4) generated trifluorovinyl ether end groups via a stepwise mechanism with a 1,4-diradical intermediate as confirmed by radical trapping experiments. This process of pefluorocyclobutane scission is quite different to the thermally activated pathway [89]. Sijbesma and coworkers [90] realized the transduction of force into bis(adamantyl)-1,2-dioxetane (compound **12** in Fig. 7.4) in dioxetane-containing linear polymers and dioxetane-cross-linked polymer networks. Upon mechanochemical activation, the bis(adamantyl)-1,2-dioxetane unit acts as a luminescent mechanophore to emit visible light, which originates from the excited state of the adamantanone product of ring opening.

7.5.3 Other ring-opening mechanophores

Several mechanophores other than the cyclopropane- and cyclobutene-based ones were also reported to undergo mechanochemical ring opening in the presence of force. Moore and colleagues observed that the colorless spiropyran-functionalized PMA changes to a pink product via mechanically facilitated ring opening of spiropyran (compound **13** in Fig. 7.4) when subjected to pulsed ultrasound [104]. Subsequent efforts by Sottos and coworkers realized such force-induced color change in spiropyran-linked elastomeric and glassy polymers [15]. First-principle FCMD simulations indicated that the force-induced scission took place exclusively at the C—O bond of spiropyran for all applied forces, ranging from 2 to 3 nN. The CoGEF calculations at the DFT level predicted an activation barrier of 2.3 eV when the molecule is elongated to 17% and the rupture of the spiro C—O bond at an elongation of 20%. Further CoGEF studies supported the selective C—O bond rupture when compared to the results from a model with a different attachment point for the external force [15].

Bielawski and coworkers reported a ultrasound-induced cycloreversion of 1,4-disubstituted 1,2,3-triazole (compound **14** in Fig. 7.4), a kinetically stable molecule, into their azide and alkyne precursors, as confirmed by a variety of spectroscopic and chemical assays [16]. Further efforts from the same group examined regiochemical effects on molecular stability via comprehensive studies on 1,4- and 1,5-disubstituted triazoles under the scope of polymer mechanochemistry [105]. The 1,5-regioisomers were demonstrated to more susceptible to mechanical activation than the 1,4-isomer, which lead to an increase of the cycloreversion reaction rate of the 1,5-disubstituted triazole by 1.2 times as compared to the corresponding 1,4-regioisomer. The conclusion was further supported by the force dependence of the cycloreversion energy barrier as probed using computational studies [105]. In order to investigate the mechanochemical behavior of the 1,2,3-triazole at the single-molecule level using the SMFS approach. Hartke and coworkers linked a 1,2,3-triazole to an aliphatic chain to form a macrocycle, in which the aliphatic chain plays the role of a "safety line" upon rupture of the triazole [106]. The rupture events of the mechanophore can then be uniquely identified as the mechanochemical ring opening of macrocycle results in a characteristic length increase of more than 1.0 nm, which is comparable to the CoGEF calculations. A rupture force of about 1.2 nN was estimated from the force-extension curves [106].

In order to investigate how the mechanical force affects the reactivity of retro-Diels-Alder reaction of furan/maleimide adducts (compound **15** in Fig. 7.4) from the viewpoint of regiochemistry and stereochemistry, four isomers, presenting an *endo* or *exo* configuration and *proximal* or *distal* geometry, were incorporated into PMA and were subjected to pulsed ultrasound [91]. In contrast to its thermal counterpart where the stereochemical effects dominate the reactivity, the cycloreversion rate of these adducts under tension is mainly dependent on the regiochemistry, as suggested by the results from both ^{1}H NMR and the CoGEF analyses. Furthermore, mechanical force was observed to inhibit the retrocycloaddition in the *distal-exo* isomer due to poor mechanochemical coupling. For instance, the "scissile" bond in the stretched intermediate of the *distal-exo* mechanophore is almost orthogonal to the force vector [91].

7.6 Organometallic mechanophores

7.6.1 Metallocene mechanophores

In contrast to covalent mechanophores, organometallic mechanophores are the ones with coordination bonds formed between metal and organic compounds. Metallocenes are the most common organometallic compounds and their metal-cyclopentadienyl bonds are thermodynamically very stable. For example, the bond dissociation energy (BDE) of the iron-cyclopentadienyl (Fe-Cp) bond in ferrocene (compound 1 in Fig. 7.6) is estimated to be as high as 91 kcal/mol [110]. However, Tang and Craig [107] recently found that ferrocene can be mechanochemically activated from pulsed ultrasonication and the mechanical strength of the Fe-Cp bond is lower than 30 kcal/mol. DFT calculations with the CoGEF method suggested that the chain scission of ferrocene followed a heterolytic dissociation pathway, which was confirmed spectroscopic experiments. The CoGEF study unveiled also the dissociation pathway at the atomic level (see Fig. 7.5A) [107]. At the early stage of the mechanical stretching, two Cp rings were forced to rotate and transform from an eclipsed geometry to a staggered one. Further stretching distorted bonds and angles and eventually broke ferrocene into [CpFe]$^+$ and Cp$^-$. Although DFT studies significantly overestimated the energy barrier (about 100 kcal/mol, Fig. 7.5B) of the mechanical scission, the rupture force was about 3.1 nN, lower than the ones of covalent bonds. Surprisingly, the breaking point distance was as great as 4 Å, which is several times larger than the one of covalent bonds (Fig. 7.5C). The large breaking point distance could be the origins of mechanical susceptibility of ferrocene, as the external force can

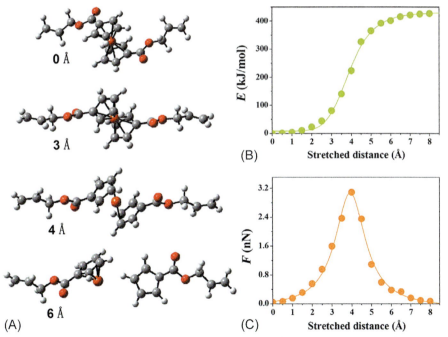

Fig. 7.5 CoGEF studies of the dissociation mechanism of ferrocene under mechanical stretching. (A) Representative structures during the dissociation pathway; (B) CoGEF potential as a function of the chain terminal distance; and (C) CoGEF force-extension curve.
Adapted with permission from Y. Sha, Y. Zhang, E. Xu, Z. Wang, T. Zhu, S. L. Craig, C. Tang, Quantitative and mechanistic mechanochemistry in ferrocene dissociation, ACS Macro Lett. 7 (V) 1174–1179. Copyright 2018 American Chemical Society.

act over a long distance and thus perform significant amount of work to lower the activation energy of the mechanical dissociation pathway. Vansco and coworkers have reported a SMFS study of ferrocene-containing polymers [111], however, they did observe the dissociation of ferrocene due to a low pulling force (<200 pN). Thus, the theoretically estimated rupture force needs further SMFS experiments to confirm.

The same groups have also studied the mechanical activation of ruthenocene (compound 2 in Fig. 7.6) in ruthenocene-containing polymers with the same techniques [108]. Similarly to ferrocene, ruthenocene is mechanically liable as well and the mechanical strength of Ru-Cp bond is comparable to the C—S bond of a thioether (BDE = 71–74 kcal/mol) [112], which is much lower than its thermodynamic stability. CoGEF studies revealed that the force-induced dissociation of ruthenocene proceeds through a two-step dissociation pathway, in contrast to a single-step dissociation in ferrocene. In the first step of ruthenocene dissociation, the structural arrangements resemble the ones in ferrocene. This two-step pathway maybe owing to the rich coordination structure in ruthenium compared to that in ferrum [108].

Giannantonio and coworkers have recently introduced ferrocene into polymethylacrylates and polyurethanes, such ferrocene-containing polymers can release iron ions under force-activated chain scission and the released ions can be detected by forming a red-colored complex with KSCN, thus ferrocene is used as an iron-releasing mechanophore here [113].

Metallocene mechanophores

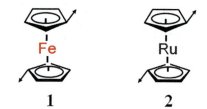

Other organometallic mechanophores

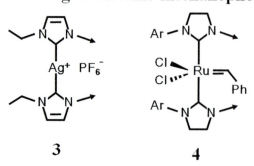

Fig. 7.6 Representatives of organometallic mechanophores: (**1**) ferrocene [107]; (**2**) ruthenocene [108]; (**3**) silver-NHC complexes [109]; and (**4**) ruthenium-NHC complexes [109]. *Arrows* indicate the pulling point of the applied force.

7.6.2 Other organometallic mechanophores

Other organometallic mechanophores, such as polymeric silver-*N*-heterocyclic carbene (NHC) complexes (compound **3** in Fig. 7.6) and ruthenium-NHC complexes (compound **4** in Fig. 7.6), are usually latent catalysts. The design concept of these mechanophores is that mechanical force is used to induce the dissociation of coordinatively saturated metal complex and force the release of catalytically active species. NHCs were known as catalysts for transesterification reactions and its catalytic activity can be suppressed when coordinatively saturated with silver(I) to form silver(I)-NHC complexes. [114] Silver(I)-NHC complexes linked to low-molecular-weight polymeric substituents (<10 kDa) were shown to give free carbenes when ultrasonicated in solution. [115] Soon after, Piermattei and coworkers reported that under mechanochemically activation polymeric silver(I)-NHC complexes with different molecular weights exhibited distinct catalytic activity for transesterification reactions between esters and alcohols [109]. It is well known that carbon-carbon double bonds can be formed under the catalysis of ruthenium-alkylidene complexes via olefin metathesis reactions [116]. Similarly, when coordinated with polymer-functionalized NHC ligands, the catalytic activity of ruthenium-alkylidene complexes can be successfully restrained and then mechanochemically activated when exposed to ultrasound [109]. However, the mechanochemical mechanisms of these metal-NHC complexes, such as rupture forces and breaking point distances, were not known and can be quantified with computational studies.

7.7 The effect of polymer chain

As indicated by the experimental and computational efforts, the mechanochemical behaviors of the same mechanophore changed significantly when subjected to mechanical pulls of different polymer chain types or attachment points [85, 91, 105, 117, 118]. The group of Craig examined how the mechanical force is delivered along the polymer backbone to the mechanophores incorporated in different polymers [117]. The SMFS experiments revealed that the critical force required to open the gDBC in polybutadiene is about 1.21 nN and is one-third times higher than the one in polynorbornene, which is around 0.74 nN. The activation length Δx for gDBC in polybutadiene is estimated to be 1.14 Å according to the Dudko-Hummer model fit, while Δx for gDBC in polynorbornene is 1.54 Å. A similar trend has been observed for the gDCC in these two polymer scaffolds. For the gDCCs, the switch from polybutadiene to polynorbornene lowers the critical force from 1.33 to 0.9 nN, and Δx changes from 1.28 to 1.64 Å [117]. For epoxidized polymers subjected to pulsed ultrasound, the generation of carbonyl ylides and the net isomerization of *cis*- to *trans*-epoxide were observed in epoxidized polynorbornenes, but not in epoxidized polybutadienes [85]. The observation by the group of Bielawshi showed that the 1,5-disubstitution rendered the cycloreversion rate of the 1,2,3-triazole 1.2 times faster than the 1,4-disubstituted one, as different substitutions applied mechanical forces to the triazole along different PPs [105]. Analogously, CoGEF calculations at the

DFT level demonstrated a dramatic difference in both the mechanically induced activation energy and the rupture force between the *proximal* and *distal* regioisomers of the furan/maleimide adducts [91]. Bailey and Mosey demonstrated the changes in the PP separations and compliances upon moving from the reactant to transition state in the force-free condition varies dramatically for different sets of PPs in the ring opening of 1,3-cyclohexadiene [118].

The force dependence can also be quite different for similar mechanophores, which differ by just one or two functional groups [101, 119]. Craig and coworkers carried out SMFS experiments and observed that α-alkene substituents on the gDCC can change its mechanochemical properties dramatically [119]. The critical force for triggering the ring-opening transition of the E-alkene-substituted gDCC is around 0.8 nN, which is 0.4 nN lower than the one for the corresponding Z-alkene isomer. The force-free activation lengths for the E and Z isomers were found to be 1.67 and 1.20 Å, respectively, as predicted by the Dudko-Hummer model fit to the experimental data, which is also in good agreement with the ones derived from the CoGEF calculations [119]. Ab initio FPMD simulations suggested that the substituents in the *cis*-BCB molecule greatly facilitated the breaking of the WH rules as compared to the unsubstituted cyclobutene counterpart [101].

The mechanophore activity can be also affected by the linker connecting the mechanophore and the polymer backbone in which it is embedded, as the linker influences the propagation of mechanical forces to mechanophores [58, 101, 102, 120]. The group of Marx has studied how the polymer chain length affects the forces transduced to the mechanophore [102, 120]. They have studied model systems of *cis*-BCB with polyethylene-like chain length ranged from 1 to 10. The dependence of the rupture force on the chain length was observed to exhibit an interesting odd-even alternation and the rupture force converges when the chain length exceeds 6 [102]. Unexpectedly, the rupture force was shown to correlate well with an out-of-plane distortion angle, which quantify the local distortion of the *cis*-BCB at its junctions with the polymer chain. They explained that the origin of the correlation is the force-dependent force constant of the C—C—C bond angle at the junction between *cis*-BCB and the polymer chain [102]. Additional efforts from the same group examined the dependence of the activation energy of the ring-opening reaction on the chain length. The activation energy showed noticeable dependence on the chain length only in the high-force regime, that is, when the force is larger than 1.5 nN, and it can vary by about 8 kcal/mol [120]. Li et al. have investigated how the linker outside of the tDCB mechanophore in a macrocycle affects its mechanochemical reactivity using the FM-FDA method [58]. They analyzed the distribution of tDCB internal forces in two different ways that the external force was applied on the tDCB, and demonstrated that the distribution on the tDCB of the force from photoisomerization of a macrocycle-containing stiff stilbene [45] is greatly different to the one when a constant pulling force is directly applied to the two end methyl groups of the tDCB. Craig and colleagues have recently examined the dependence of the mechanochemical activity of the gDBC ring-opening reaction on the length of the polymer backbone. Similar to the case of *cis*-BCB, the calculated C—C extension from the ground- to transition-state structure was reported to change in an alternating fashion when additional C—C bonds were taken into account [101].

7.8 Conclusions and perspectives

This chapter summarized the theory, experimental and computational approaches, many mechanophores with different design concepts, and the effects of polymer chain in the field of polymer mechanochemistry. Although a lot of mechanophores with different chemical geometries, elements, and functional groups have been reported, and mechanical activation of these mechanophores can produce reactive species, small molecules, active catalysts, luminescent agents, and others, we are still in the early stage of designing and discovering novel mechanophores with particular mechanochemical properties. From a chemical perspective, the chemical reactions of existing mechanophores are mainly bond scissions, electrocyclic ring openings, and retrocycloadditions. A wide open field of other chemical reactions remains unexplored. Although experimental approaches dominate in current study of mechanochemistry, computational approaches will play an increasing role in this endeavor.

For a better understanding of the mechanochemical reactivity of mechanophores, the knowledge of how forces are distributed within mechanophores and the dynamics of such force distribution is of great importance in mechanochemistry. However, the force distributions in mechanophores are largely not touched. To our best knowledge, tDCB is the only mechanophore, the distributions of internal forces [58] and stress energy [70] of which were analyzed. Thus, in the foreseen future, we will witness constantly growing publications in this field.

As the number and the information on the mechanochemical-reactivity relationship of mechanophores accumulate, public database of mechanophores should be established. More importantly, how can we use these databases to facilitate the discovery and design of novel mechanophores? At the present, the determination of whether a compound is a good mechanophore or not largely relies on chemical intuitive of experts. However, in the established field of computer-aided drug design, approaches in chemoinformatics and mathematics were applied to make better decisions in lead identification and optimization. It may be plausible to adapt similar approaches to guide the search of new mechanophores based on the existing knowledge on mechanophores.

In recent years, the potential of many mechanophores has been demonstrated in the fields such as catalysts and material science. The incorporation of mechanophores in materials that can be applied in soft electronic devices is another promising area. An encouraging example is the integration of the spiropyran-functionalized polymer in a stretchable electronic device to warn the user about the risk of device failure [121]. We expect that the combination of mechanophores and soft electronics will inspire a variety of new applications including tactile sensors.

Acknowledgments

The financial supports from the Startup Foundation for Peacock Talents, Shenzhen University, and the National Natural Science Foundation of China under grant No. 31770777 are greatly acknowledged.

References

[1] M.K. Beyer, H. Clausen-Schaumann, Mechanochemistry: the mechanical activation of covalent bonds, Chem. Rev. 105 (2005) 2921–2948.

[2] W. Kauzmann, H. Eyring, The viscous flow of large molecules, J. Am. Chem. Soc. 62 (1940) 3113–3125.

[3] F. Ritort, Single-molecule experiments in biological physics: methods and applications, J. Phys. Condens. Matter 18 (2006) R531.

[4] H. Li, B. Liu, X. Zhang, C. Gao, J. Shen, G. Zou, Single-molecule force spectroscopy on poly (acrylic acid) by AFM, Langmuir 15 (1999) 2120–2124.

[5] A.P. Wiita, S.R.K. Ainavarapu, H.H. Huang, J.M. Fernandez, Force-dependent chemical kinetics of disulfide bond reduction observed with single-molecule techniques, Proc. Natl. Acad. Sci. 103 (2006) 7222–7227.

[6] W. Li, F. Gräter, Atomistic evidence of how force dynamically regulates thiol/disulfide exchange, J. Am. Chem. Soc. 132 (2010) 16790–16795.

[7] S. Koti Ainavarapu, A. Wiita, L. Dougan, E. Uggerud, J. Fernandez, Single- molecule force spectroscopy measurements of bond elongation during a bimolecular reaction, J. Am. Chem. Soc. 130 (2008) 6479–6487.

[8] A.P. Wiita, R. Perez-Jimenez, K.A. Walther, F. Gräter, B. Berne, A. Holmgren, J.M. Sanchez-Ruiz, J.M. Fernandez, Probing the chemistry of thioredoxin catalysis with force, Nature 450 (2007) 124–127.

[9] R. Perez-Jimenez, J. Li, P. Kosuri, I. Sanchez-Romero, A.P. Wiita, D. Rodriguez-Larrea, A. Chueca, A. Holmgren, A. Miranda-Vizuete, K. Becker, et al., Diversity of chemical mechanisms in thioredoxin catalysis revealed by single-molecule force spectroscopy, Nat. Struct. Mol. Biol. 16 (2009) 890.

[10] M. Carrion-Vazquez, H. Li, H. Lu, P. Marszalek, A. Oberhauser, J. Fernandez, et al., The mechanical stability of ubiquitin is linkage dependent, Nat. Struct. Biol. 10 (2003) 738–743.

[11] P. Li, D. Makarov, Simulation of the mechanical unfolding of ubiquitin: probing different unfolding reaction coordinates by changing the pulling geometry, J. Chem. Phys. 121 (2004) 4826.

[12] R. Best, E. Paci, G. Hummer, O. Dudko, Pulling direction as a reaction coordinate for the mechanical unfolding of single molecules, J. Phys. Chem. B 112 (2008) 5968–5976.

[13] D. Brockwell, E. Paci, R. Zinober, G. Beddard, P. Olmsted, D. Smith, R. Perham, S. Radford, et al., Pulling geometry defines the mechanical resistance of a beta-sheet protein, Nat. Struct. Biol. 10 (2003) 731–737.

[14] V. Minkin, Photo-, thermo-, solvato-, and electrochromic spiroheterocyclic compounds, Chem. Rev. 104 (2004) 2751–2776.

[15] D. Davis, A. Hamilton, J. Yang, L. Cremar, D. Van Gough, S. Potisek, M. Ong, P. Braun, T. Martínez, S. White, et al., Force-induced activation of covalent bonds in mechanoresponsive polymeric materials, Nature 459 (2009) 68–72.

[16] J. Brantley, K. Wiggins, C. Bielawski, Unclicking the click: mechanically facilitated 1,3-dipolar cycloreversions, Science 333 (2011) 1606–1609.

[17] R.B. Woodward, R. Hoffmann, The conservation of orbital symmetry, Angew. Chem. Int. Ed. Engl. 8 (1969) 781–853.

[18] R.B. Woodward, R. Hoffmann, Stereochemistry of electrocyclic reactions, J. Am. Chem. Soc. 87 (1965) 395–397.

[19] C. Hickenboth, J. Moore, S. White, N. Sottos, J. Baudry, S. Wilson, Biasing reaction pathways with mechanical force, Nature 446 (2007) 423–427.

[20] M. Ong, J. Leiding, H. Tao, A. Virshup, T. Martinez, First principles dynamics and minimum energy pathways for mechanochemical ring opening of cyclobutene, J. Am. Chem. Soc. 131 (2009) 6377–6379.

[21] S. Zhurkov, B. Narzullaev, Time dependence of strength of solids, Zh. Tekh. Fiz. 23 (1953) 1677–1689.

[22] S. Zhurkov, Fracture of materials, Int. J. Fract. Mech. 1 (1965) 311–326.

[23] G.I. Bell, Models for the specific adhesion of cells to cells, Science 200 (1978) 618–627.

[24] O.K. Dudko, G. Hummer, A. Szabo, Theory, analysis, and interpretation of single-molecule force spectroscopy experiments, Proc. Natl. Acad. Sci. 105 (2008) 15755–15760.

[25] M. Schlierf, M. Rief, Single-molecule unfolding force distributions reveal a funnel-shaped energy landscape, Biophys. J. 90 (2006) L33–L35.

[26] J. Mathé, H. Visram, V. Viasnoff, Y. Rabin, A. Meller, Nanopore unzipping of individual DNA hairpin molecules, Biophys. J. 87 (2004) 3205–3212.

[27] H.A. Kramers, Brownian motion in a field of force and the diffusion model of chemical reactions, Phys. Ther. 7 (1940) 284–304.

[28] G. Hummer, A. Szabo, Kinetics from nonequilibrium single-molecule pulling experiments, Biophys. J. 85 (2003) 5–15.

[29] O.K. Dudko, G. Hummer, A. Szabo, Intrinsic rates and activation free energies from single-molecule pulling experiments, Phys. Rev. Lett. 96 (2006) 108101.

[30] R. Boulatov, Reaction dynamics in the formidable gap, Pure Appl. Chem. 83 (2010) 25–41.

[31] S.S.M. Konda, J.N. Brantley, C.W. Bielawski, D.E. Makarov, Chemical reactions modulated by mechanical stress: extended Bell theory, J. Chem. Phys. 135 (2011) 164103.

[32] G. Neuert, C.H. Albrecht, H.E. Gaub, Predicting the rupture probabilities of molecular bonds in series, Biophys. J. 93 (2007) 1215–1223.

[33] G. Schmidt-Naake, A. Frendel, M. Drache, G. Janke, Mechanochemical modification of polystyrene and polymethylmethacrylate, Chem. Eng. Technol. 24 (2001) 889–894.

[34] M. Hasegawa, M. Kimata, S.-I. Kobayashi, Mechanochemical polymerization of styrene initiated by the grinding of quartz, J. Appl. Polym. Sci. 82 (2001) 2849–2855.

[35] Y. Murata, A. Han, K. Komatsu, Mechanochemical synthesis of a novel C60 dimer connected by a germanium bridge and a single bond, Tetrahedron Lett. 44 (2003) 8199–8201.

[36] T.Q. Nguyen, H.-H. Kausch, Chain extension and degradation in convergent flow, Polymer 33 (1992) 2611–2621.

[37] T.Q. Nguyen, H.-H. Kausch, Macromolecules: Synthesis, Order and Advanced Properties, Springer, 1992, pp. 73–182.

[38] G. Cravotto, P. Cintas, Harnessing mechanochemical effects with ultrasound-induced reactions, Chem. Sci. 3 (2012) 295–307.

[39] I. Park, S.S. Sheiko, A. Nese, K. Matyjaszewski, Molecular tensile testing machines: breaking a specific covalent bond by adsorption-induced tension in brushlike macromolecules, Macromolecules 42 (2009) 1805–1807.

[40] Y. Li, A. Nese, N.V. Lebedeva, T. Davis, K. Matyjaszewski, S.S. Sheiko, Molecular tensile machines: intrinsic acceleration of disulfide reduction by dithiothreitol, J. Am. Chem. Soc. 133 (2011) 17479–17484.

[41] M. Encina, E. Lissi, M. Sarasua, L. Gargallo, D. Radic, Ultrasonic degradation of poly-vinylpyrrolidone: effect of peroxide linkages, J. Polym. Sci., Polym. Lett. Ed. 18 (1980) 757–760.

[42] K.L. Berkowski, S.L. Potisek, C.R. Hickenboth, J.S. Moore, Ultrasound-induced site-specific cleavage of azo-functionalized poly (ethylene glycol), Macromolecules 38 (2005) 8975–8978.

[43] J.M. Paulusse, R.P. Sijbesma, Reversible mechanochemistry of a PdII coordination polymer, Angew. Chem. Int. Ed. 43 (2004) 4460–4462.

[44] E.M. Puchner, H.E. Gaub, Force and function: probing proteins with AFM-based force spectroscopy, Curr. Opin. Struct. Biol. 19 (2009) 605–614.

[45] Q.-Z. Yang, Z. Huang, T.J. Kucharski, D. Khvostichenko, J. Chen, R. Boulatov, A molecular force probe, Nat. Nanotechnol. 4 (2009) 302.

[46] K.C. Neuman, A. Nagy, Single-molecule force spectroscopy: optical tweezers, magnetic tweezers and atomic force microscopy, Nat. Methods 5 (2008) 491.

[47] A.D. Mehta, M. Rief, J.A. Spudich, D.A. Smith, R.M. Simmons, Single-molecule biomechanics with optical methods, Science 283 (1999) 1689–1695.

[48] P. Kollmannsberger, B. Fabry, BaHigh-force magnetic tweezers with force feedback for biological applications, Rev. Sci. Instrum. 78 (2007) 114301.

[49] K.A. Walther, J. Brujić, H. Li, J.M. Fernández, Sub-angstrom conformational changes of a single molecule captured by AFM variance analysis, Biophys. J. 90 (2006) 3806–3812.

[50] M. Sotomayor, K. Schulten, Single-molecule experiments in vitro and in silico, Science 316 (2007) 1144–1148.

[51] A.F. Oberhauser, P.K. Hansma, M. Carrion-Vazquez, J.M. Fernández, Stepwise unfolding of titin under force-clamp atomic force microscopy, Proc. Natl. Acad. Sci. 98 (2001) 468–472.

[52] J.M. Fernández, H. Li, Force-clamp spectroscopy monitors the folding trajectory of a single protein, Science 303 (2004) 1674–1678.

[53] F.R. Kersey, W.C. Yount, S.L. Craig, Single-molecule force spectroscopy of bimolecular reactions: system homology in the mechanical activation of ligand substitution reactions, J. Am. Chem. Soc. 128 (2006) 3886–3887.

[54] F.R. Kersey, D.M. Loveless, S.L. Craig, A hybrid polymer gel with controlled rates of cross-link rupture and self-repair, J. R. Soc. Interface 4 (2007) 373–380.

[55] D. Wu, J.M. Lenhardt, A.L. Black, B.B. Akhremitchev, S.L. Craig, Molecular stress relief through a force-induced irreversible extension in polymer contour length, J. Am. Chem. Soc. 132 (2010) 15936–15938.

[56] J. Wang, T.B. Kouznetsova, Z. Niu, M.T. Ong, H.M. Klukovich, A.L. Rheingold, T.J. Martinez, S.L. Craig, Inducing and quantifying forbidden reactivity with single-molecule polymer mechanochemistry, Nat. Chem. 7 (2015) 323.

[57] Z. Huang, Q.-Z. Yang, D. Khvostichenko, T.J. Kucharski, J. Chen, R. Boulatov, Method to derive restoring forces of strained molecules from kinetic measurements, J. Am. Chem. Soc. 131 (2009) 1407–1409.

[58] W. Li, S.A. Edwards, L. Lu, T. Kubar, S.P. Patil, H. Grubmüller, G. Groenhof, F. Gräter, Force distribution analysis of mechanochemically reactive dimethylcyclobutene, ChemPhysChem 14 (2013) 2687–2697.

[59] R. Improta, F. Santoro, Excited-state behavior of trans and cis isomers of stilbene and stiff stilbene: a TD-DFT study, Chem. Eur. J. 109 (2005) 10058–10067.

[60] H. Grubmüller, B. Heymann, P. Tavan, Ligand binding: molecular mechanics calculation of the streptavidin-biotin rupture force, Science 271 (1996) 997–999.

[61] S. Izrailev, S. Stepaniants, M. Balsera, Y. Oono, K. Schulten, Molecular dynamics study of unbinding of the avidin-biotin complex, Biophys. J. 72 (1997) 1568–1581.

[62] H. Lu, B. Isralewitz, A. Krammer, V. Vogel, K. Schulten, Unfolding of titin immunoglobulin domains by steered molecular dynamics simulation, Biophys. J. 75 (1998) 662–671.

[63] F. Gräter, J. Shen, H. Jiang, M. Gautel, H. Grubmüller, Mechanically induced titin kinase activation studied by force-probe molecular dynamics simulations, Biophys. J. 88 (2005) 790–804.

[64] S. Garcia-Manyes, J. Brujić, C.L. Badilla, J.M. Fernández, Force-clamp spectroscopy of single-protein monomers reveals the individual unfolding and folding pathways of I27 and ubiquitin, Biophys. J. 93 (2007) 2436–2446.

[65] W. Stacklies, M.C. Vega, M. Wilmanns, F. Gräter, Mechanical network in titin immunoglobulin from force distribution analysis, PLoS Comput. Biol. 5 (2009) e1000306.

[66] M.K. Beyer, The mechanical strength of a covalent bond calculated by density functional theory, J. Chem. Phys. 112 (2000) 7307–7312.

[67] M.F. Iozzi, T. Helgaker, E. Uggerud, Assessment of theoretical methods for the determination of the mechanochemical strength of covalent bonds, Mol. Phys. 107 (2009) 2537–2546.

[68] J. Ribas-Arino, M. Shiga, D. Marx, Understanding covalent mechanochemistry, Angew. Chem. Int. Ed. 48 (2009) 4190–4193.

[69] K. Fukui, The path of chemical reactions-the IRC approach, Acc. Chem. Res. 14 (1981) 363–368.

[70] T. Stauch, A. Dreuw, A quantitative quantum-chemical analysis tool for the distribution of mechanical force in molecules, J. Chem. Phys. 140 (2014) 134107.

[71] W. Stacklies, C. Seifert, F. Graeter, Implementation of force distribution analysis for molecular dynamics simulations, BMC bioinform. 12 (2011) 101.

[72] B.I. Costescu, F. Gräter, Time-resolved force distribution analysis, BMC Biophys. 6 (2013) 5.

[73] F. Ercolessi, J.B. Adams, Interatomic potentials from first-principles calculations: the force-matching method, Europhys. Lett. 26 (1994) 583.

[74] S. Izvekov, M. Parrinello, C.J. Burnham, G.A. Voth, Effective force fields for condensed phase systems from ab initio molecular dynamics simulation: a new method for force-matching, J. Chem. Phys. 120 (2004) 10896–10913.

[75] S. Izvekov, G.A. Voth, A multiscale coarse-graining method for biomolecular systems, J. Phys. Chem. B 109 (2005) 2469–2473.

[76] W.G. Noid, J.-W. Chu, G.S. Ayton, V. Krishna, S. Izvekov, G.A. Voth, A. Das, H.C. Andersen, The multiscale coarse-graining method. I. A rigorous bridge between atomistic and coarse-grained models, J. Chem. Phys. 128 (2008) 244114.

[77] V. Bakken, T. Helgaker, The efficient optimization of molecular geometries using redundant internal coordinates, J. Chem. Phys. 117 (2002) 9160–9174.

[78] T. Stauch, A. Dreuw, On the use of different coordinate systems in mechanochemical force analyses, J. Chem. Phys. 143 (2015) 074118.

[79] W. Li, A. Ma, Reaction mechanism and reaction coordinates from the viewpoint of energy flow, J. Chem. Phys. 144 (2016) 114103.

[80] W. Li, A. Ma, A benchmark for reaction coordinates in the transition path ensemble, J. Chem. Phys. 144 (2016) 134104.

[81] W. Li, Residue-residue mutual work analysis of retinal-opsin interaction in rhodopsin: implications for protein-ligand binding, J. Chem. Theory Comput. 16 (2020) 1834–1842.

[82] J.M. Lenhardt, M.T. Ong, R. Choe, C.R. Evenhuis, T.J. Martinez, S.L. Craig, Trapping a diradical transition state by mechanochemical polymer extension, Science 329 (2010) 1057–1060.

[83] J.M. Lenhardt, A.L. Black, S.L. Craig, gem-Dichlorocyclopropanes as abundant and efficient mechanophores in polybutadiene copolymers under mechanical stress, J. Am. Chem. Soc. 131 (2009) 10818–10819.

[84] J. Wang, T.B. Kouznetsova, S.L. Craig, Reactivity and mechanism of a mechanically activated anti-Woodward–Hoffmann–DePuy reaction, J. Am. Chem. Soc. 137 (2015) 11554–11557.

[85] H.M. Klukovich, Z.S. Kean, A.L.B. Ramirez, J.M. Lenhardt, J. Lin, X. Hu, S.L. Craig, Tension trapping of carbonyl ylides facilitated by a change in polymer backbone, J. Am. Chem. Soc. 134 (2012) 9577–9580.

[86] C.E. Diesendruck, B.D. Steinberg, N. Sugai, M.N. Silberstein, N.R. Sottos, S.R. White, P.V. Braun, J.S. Moore, Proton-coupled mechanochemical transduction: a mechano-generated acid, J. Am. Chem. Soc. 134 (2012) 12446–12449.

[87] M.J. Kryger, A.M. Munaretto, J.S. Moore, Structure–mechanochemical activity relationships for cyclobutane mechanophores, J. Am. Chem. Soc. 133 (2011) 18992–18998.

[88] R. Groote, B.M. Szyja, F.A. Leibfarth, C.J. Hawker, N.L. Doltsinis, R.P. Sijbesma, Strain-induced strengthening of the weakest link: the importance of intermediate geometry for the outcome of mechanochemical reactions, Macromolecules 47 (2014) 1187–1192.

[89] H.M. Klukovich, Z.S. Kean, S.T. Iacono, S.L. Craig, Mechanically induced scission and subsequent thermal remending of perfluorocyclobutane polymers, J. Am. Chem. Soc. 133 (2011) 17882–17888.

[90] Y. Chen, A. Spiering, S. Karthikeyan, G.W. Peters, E. Meijer, R.P. Sijbesma, Mechanically induced chemiluminescence from polymers incorporating a 1,2-dioxetane unit in the main chain, Nat. Chem. 4 (2012) 559–562.

[91] R. Stevenson, G. De Bo, Controlling reactivity by geometry in retro-diels–alder reactions under tension, J. Am. Chem. Soc. 139 (2017) 16768–16771.

[92] B.O. Roos, The complete active space self-consistent field method and its applications in electronic structure calculations, Adv. Chem. Phys. 69 (1987) 399–445.

[93] F. Tian, S.B. Lewis, M.D. Bartberger, W.R. Dolbier, W.T. Borden, Experimental study of the stereomutation of 1,1-difluoro-2-ethyl-3-methylcyclopropane confirms the predicted preference for disrotatory ring opening and closure, J. Am. Chem. Soc. 120 (1998) 6187–6188.

[94] J.M. Lenhardt, J.W. Ogle, M.T. Ong, R. Choe, T.J. Martinez, S.L. Craig, Reactive cross-talk between adjacent tension-trapped transition states, J. Am. Chem. Soc. 133 (2011) 3222–3225.

[95] J. Wang, T.B. Kouznetsova, S.L. Craig, Single-molecule observation of a mechanically activated cis-to-trans cyclopropane isomerization, J. Am. Chem. Soc. 138 (2016) 10410–10412.

[96] W.E. Parham, K. Yong, Steric and electronic effects in the solvolysis of cis-and trans-mono-and dihalocyclopropanes, J. Org. Chem. 35 (1970) 683–685.

[97] P. Dopieralski, J. Ribas-Arino, D. Marx, Force-transformed free-energy surfaces and trajectory-shooting simulations reveal the mechano-stereochemistry of cyclopropane ring-opening reactions, Angew. Chem. Int. Ed. 50 (2011) 7105–7108.

[98] A.L.B. Ramirez, Z.S. Kean, J.A. Orlicki, M. Champhekar, S.M. Elsakr, W.E. Krause, S.L. Craig, Mechanochemical strengthening of a synthetic polymer in response to typically destructive shear forces, Nat. Chem. 5 (2013) 757.

[99] J. Ribas-Arino, M. Shiga, D. Marx, Unravelling the mechanism of force-induced ring-opening of benzocyclobutenes, Chem. Eur. J. 15 (2009) 13331–13335.

[100] G.S. Kochhar, A. Bailey, N.J. Mosey, Competition between orbitals and stress in mechanochemistry, Angew. Chem. Int. Ed. 49 (2010) 7452–7455.

[101] J. Friedrichs, M. Lüßmann, I. Frank, Conservation of orbital symmetry can be circumvented in mechanically induced reactions, ChemPhysChem 11 (2010) 3339–3342.

[102] J. Ribas-Arino, M. Shiga, D. Marx, Mechanochemical transduction of externally applied forces to mechanophores, J. Am. Chem. Soc. 132 (2010) 10609–10614.

[103] M.J. Kryger, M.T. Ong, S.A. Odom, N.R. Sottos, S.R. White, T.J. Martinez, J.S. Moore, Masked cyanoacrylates unveiled by mechanical force, J. Am. Chem. Soc. 132 (2010) 4558–4559.

[104] S.L. Potisek, D.A. Davis, N.R. Sottos, S.R. White, J.S. Moore, Mechanophore-linked addition polymers, J. Am. Chem. Soc. 129 (2007) 13808–13809.

[105] J.N. Brantley, S.S.M. Konda, D.E. Makarov, C.W. Bielawski, Regiochemical effects on molecular stability: a mechanochemical evaluation of 1,4- and 1,5-disubstituted triazoles, J. Am. Chem. Soc. 134 (2012) 9882–9885.

[106] D. Schuetze, K. Holz, J. Mueller, M.K. Beyer, U. Luening, B. Hartke, Pinpointing mechanochemical bond rupture by embedding the mechanophore into a macrocycle, Angew. Chem. Int. Ed. 54 (2015) 2556–2559.

[107] Y. Sha, Y. Zhang, E. Xu, Z. Wang, T. Zhu, S.L. Craig, C. Tang, Quantitative and mechanistic mechanochemistry in ferrocene dissociation, ACS Macro Lett. 7 (2018) 1174–1179.

[108] Y. Sha, Y. Zhang, E. Xu, C.W. McAlister, T. Zhu, S.L. Craig, C. Tang, Generalizing metallocene mechanochemistry to ruthenocene mechanophores, Chem. Sci. 10 (2019) 4959–4965.

[109] A. Piermattei, S. Karthikeyan, R.P. Sijbesma, Activating catalysts with mechanical force, Nat. Chem. 1 (2009) 133.

[110] K.E. Lewis, G.P. Smith, Bond dissociation energies in ferrocene, J. Am. Chem. Soc. 106 (1984) 4650–4651.

[111] S. Zou, M.A. Hempenius, H. Schönherr, G.J. Vancso, Force spectroscopy of individual stimulus-responsive poly (ferrocenyldimethylsilane) chains: towards a redox-driven macromolecular motor, Macromol. Rapid Commun. 27 (2006) 103–108.

[112] B. Lee, Z. Niu, J. Wang, C. Slebodnick, S.L. Craig, Relative mechanical strengths of weak bonds in sonochemical polymer mechanochemistry, J. Am. Chem. Soc. 137 (2015) 10826–10832.

[113] M. Di Giannantonio, M.A. Ayer, E. Verde-Sesto, M. Lattuada, C. Weder, K.M. Fromm, Triggered metal ion release and oxidation: ferrocene as a mechanophore in polymers, Angew. Chem. Int. Ed. 57 (2018) 11445–11450.

[114] G.A. Grasa, R.M. Kissling, S.P. Nolan, N-heterocyclic carbenes as versatile nucleophilic catalysts for transesterification/acylation reactions, Org. Lett. 4 (2002) 3583–3586.

[115] S. Karthikeyan, S.L. Potisek, A. Piermattei, R.P. Sijbesma, Highly efficient mechano-chemical scission of silver-carbene coordination polymers, J. Am. Chem. Soc. 130 (2008) 14968–14969.

[116] T.M. Trnka, R.H. Grubbs, The development of L2X2Ru CHR olefin metathesis catalysts: an organometallic success story, Acc. Chem. Res. 34 (2001) 18–29.

[117] H.M. Klukovich, T.B. Kouznetsova, Z.S. Kean, J.M. Lenhardt, S.L. Craig, A backbone lever-arm effect enhances polymer mechanochemistry, Nat. Chem. 5 (2013) 110–114.

[118] A. Bailey, N.J. Mosey, Prediction of reaction barriers and force-induced instabilities under mechanochemical conditions with an approximate model: a case study of the ring opening of 1,3-cyclohexadiene, J. Chem. Phys. 136 (2012) 044102.

[119] J. Wang, T.B. Kouznetsova, Z.S. Kean, L. Fan, B.D. Mar, T.J. Martínez, S.L. Craig, A remote stereochemical lever arm effect in polymer mechanochemistry, J. Am. Chem. Soc. 136 (2014) 15162–15165.

[120] P. Dopieralski, P. Anjukandi, M. Rückert, M. Shiga, J. Ribas-Arino, D. Marx, On the role of polymer chains in transducing external mechanical forces to benzocyclobutene mechanophores, J. Mater. Chem. 21 (2011) 8309–8316.

[121] M.H. Barbee, K. Mondal, J.Z. Deng, V. Bharambe, T.V. Neumann, J.J. Adams, N. Boechler, M.D. Dickey, S.L. Craig, Mechanochromic stretchable electronics, ACS Appl. Mater. Interfaces 10 (2018) 29918–29924.

Perovskites for tactile sensors

8

Rohit Saraf and Vivek Maheshwari[*]
Department of Chemistry, Waterloo Institute for Nanotechnology, University of Waterloo, Waterloo, ON, Canada
[*]Corresponding author: e-mail address: vmaheshw@uwaterloo.ca

8.1 Introduction

The sense of touch (specifically referring to the ability of sensing contact, force distribution, discern different kinds of forces) is pervasive in higher-level organisms, which is due to its critical role in the gathering of stimuli from surrounding. The encoded stimuli provide information that is not replicated by any other sense and hence is vital for the effective and safe functioning of the organism. In humans, the sense of touch is critical for performing many simple tasks such as the operation of tools, handling of objects, discerning texture, the stability of structures, and fluidity of surfaces. Combined with the sense of vision, the resulting combination of information encoding is critical to our functioning. Development of human-like abilities in robots or humanoids is critical for work in extreme environments such as space and progress of fields such as space travel [1, 2]. Unmanned missions which include remote exploratory vehicles to other bodies in space like the mars and moon also benefit from the sense of touch in their ability to maneuver over alien landscapes [3–5]. Tactile sensors with high sensitivity and wide range in sensing pressure magnitude are important for these applications, as it involves exploring the unknown and a single misstep can lead to critical and costly failure [6, 7]. The expense of delivering the payloads is highly prohibitive for space travel and hence any measures to lessen their weight provide critical benefit in cost [8]. At the same time, materials that pose a health challenge can be used if their benefits are significant. An example being GaAs-based solar cells which are used in solar panels for space applications. To this aspect, self-powered or light-powered tactile sensors with high performance will be highly beneficial for such applications in extreme situations since they negate the need for a continuous power source reducing payloads.

Among the strategies to develop high performance tactile sensors that can operate without a continuous power source organolead halide-based materials, specifically methylammonium lead iodide (MAPbI$_3$) offers the combination of being a semiconductor with ferroelectric/polarization effects [9, 10]. This combination of properties allows the poling of the material to generate an internal electric field, which separates the charge carriers on photo-illumination, hence allowing for simple photoferroelectric effects to generate power. The simple processing of MAPbI$_3$ into thin films with relatively low modulus (~23 GPa) [10] also allows the fabrication of flexible devices that can be mounted with tight form factors. In this chapter, we discuss a

---

Functional Tactile Sensors. https://doi.org/10.1016/B978-0-12-820633-1.00012-7
© 2021 Elsevier Ltd. All rights reserved.

simple strategy to harness this quality of MAPbI$_3$ for making a light-powered tactile sensor with high sensitivity over a broad operating pressure range. The sensor has the ability for continuous operation under applied stimuli and is self-powered under illumination. The sensing and power generation are done in the same structure. The simple fabrication of the device and its ability to harness energy over a wide range of wavelengths are added benefits.

8.2 Background on need for self-powered sensors

With the advancement and fast development of materials and technologies, the self-powered devices and sensors will become the key technologies in the near future that can drive the global industries with applications in health care, robotics, environmental and infrastructure monitoring, aerospace, and national security [11, 12]. Currently, the commercially available conventional or nonself-powered sensors usually need an external power source (such as batteries) and cannot work independently and sustainably. Such nonideal battery-based sensors are expensive and also hinder the development of next-generation smart sensor systems. Therefore, energy or light-harvesting technology that harvests energy from the environment has been regarded as a promising solution to self-power electronic devices (sensors and detectors). Besides the significant advantage of working without any power source, self-powered tactile sensors for mechanical motion and detection can achieve real-time sensing and possess fast response time as well as long cyclic lifetime, which are advantageous compared with those nonself-powered sensors that have undesirable hysteresis in response and limited cyclic lifetime [12]. Tactile (or pressure) sensors are based on different sensing mechanism including piezoresistivity [13], triboelectricity [14], capacitance [15], and piezoelectricity [16]. Among these types of tactile sensors, the triboelectric and piezoelectric nanogenerators have been actively researched for the harvesting of mechanical energy for powering devices. However, they are active only in the presence of continuously varying mechanical input and cannot continuously produce energy from static pressure; hence, their measurement is intermittent in nature.

The new paradigm of seamless interfacing of devices and sensors to monitor the human environment and act as a medium for information collection comes with the challenge of powering these devices in a continuous manner. This requires the development of appropriate power sources with a suitable form factor, energy density, and also connection to the device. An alternative to address this challenge is to develop the device and material configurations where both sensing and continuous energy harvesting from the environment to power it, can occur in the device itself, for example, solar/light-powered sensors. The challenge in such systems is (1) the decoupling of the sensing of the stimuli from the energy-harvesting mechanism is required for continuous operation of the devices irrespective of the applied stimuli and also the sensing of both static and dynamic signals in the stimuli is essential [17–19]. (2) Developing monolith structure devices to have a simple configuration that reduces the cost and the complexity of manufacturing. Further, it is highly desired to design and construct the self-powered tactile sensors that have high tactile sensitivity with a linear sensing capability over a broad dynamic operating range. This combination of characteristics is crucial for detecting diverse stimuli ranging from low-pressure (<1 kPa) to high-pressure range (>200 kPa). Moreover, if the pressure sensor operates in a large linear regime, the user can obtain accurate information from its output without the need for any additional signal processor thus meeting the increasing demand for device miniaturization and low-power consumption.

The methylammonium lead iodide ($MAPbI_3$) perovskite materials have been recently used to address the key challenge of combining the high sensitivity with a linear response over a wide dynamic pressure range in a simple monolithic tactile sensor. $MAPbI_3$ has rapidly become materials of intense research owing to its high absorption coefficients, high charge-carrier mobilities, long minority carrier diffusion lengths, low trap densities, and ease of synthesis with low material cost [20, 21]. These outstanding properties have led to their widespread application in solar cells, LEDs, energy harvesting, electro-optical devices, and sensors [20–23]. The semiconducting $MAPbI_3$ material has been reported to be ferroelectric that has a polarization effect [24–26].

8.3 Polarization effects in perovskites and their basic properties

The polarization effect in $MAPbI_3$ perovskite may arise from three major mechanisms: the rotation of dipolar MA^+ ion, the ionic polarization induced by the shift of the positive charge center of MA^+ relative to the negative charge center of the PbI_3^- cage, and the ionic polarization induced by the off-center displacement of Pb within the PbI_6 octahedra [27, 28] Theoretical calculations showed that the polarization induced by the rotation of MA^+ ions ($6–8$ μC cm^{-2}) is three times higher than that obtained by the Pb atom [28]. MA^+ ion has a large dipole moment in its gas phase of 2.29 Debye, and the permanent polarization of $MAPbI_3$ is mainly contributed by the alignment of MA^+ ion [27, 29]. Besides being a semiconducting material with a polarization effect, $MAPbI_3$ also has ion migration effects that have been proposed to lead to the formation of reversible p-n (or p-i-n) kind of structures, resulting in switchable photovoltaic effect and the anomalous photovoltaic effects (ion migration effects) [30–32]. Both these effects have been reported in $MAPbI_3$ as distinct phenomena [24, 25, 30, 31]. While distinct in origin, both these effects are the result of the material's response to applied electric fields and result in responses that critically affect the application of these materials in new electro-optical devices. In $MAPbI_3$, the possible ions which can migrate are MA^+ and I^- ions due to their low activation barrier energy. Under the applied bias (electric field), these ions can move easily and can lead to local doping effects that have been shown to cause anomalous photovoltaic effects [31–33]. The tetragonal crystal structure of $MAPbI_3$ (stable at room temperature) has been reported to be ferroelectric due to the light-induced ordering of MA^+ ions [34, 35], and results in an internal electrical field originating from electrical polarization. The internal polarization developed during poling enables the separation of the electron-hole pairs and enhances the photovoltaic efficiency of this material [35–37].

It has been shown that both these effects are present simultaneously in MAPbI$_3$ films and can lead to compensating effects. Based on the poling conditions, either effect can be made to dominate in the planar MAPbI$_3$ device (Fig. 8.1). This is also of substantial interest in fabricating hybrid energy-harvesting devices [38–40]. The MAPbI$_3$ device poled under N$_2$ atmosphere in dark conditions leads to domination of the ion migration effects, which results in an internal field (due to the p-n junctions) that will lead to separation of the charge carriers (electron-hole pairs) generated on light illumination (Fig. 8.1A) and the resulting V_{oc} (and J_{sc}) will be of the same polarity as that of the poling field (Fig. 8.1B). In contrast, the MAPbI$_3$ device poled in ambient air and under light illumination (0.1 sun) leads to polarization effect domination (Fig. 8.1A). In this case, however, due to the presence of metal electrode, a charge-compensating layer (due to the free electrons in the metal) is formed [36, 41] and the resulting V_{oc} and J_{sc} are of opposite polarity compared to the poling field direction (Fig. 8.1C). The distinction between these two effects is based on the reversal of the sign in the open-circuit voltage (V_{oc}) and the observed short-circuit current density (J_{sc}). The reversal of the V_{oc} and J_{sc} depending on the poling conditions is critical to the functioning of MAPbI$_3$ (and other similar materials)-based optoelectronic devices.

The ion migration effect in MAPbI$_3$ films has been studied by poling under N$_2$ in dark conditions (Fig. 8.2). Both V_{oc} and J_{sc} gradually decreased over the cycles (Fig. 8.2A and B) due to the rehomogenization of the ions that were diffused due to the poling fields [31]. No J_{sc} has been observed in dark (after poling; Fig. 8.2B) indicating that there was no significant ion back diffusion occurring. It has been observed that the high poling field strength and large poling time result in complete segregation of the ions, which will lead to the formation of PbI$_2$ [42] and the observed ion migration and polarization behavior disappeared. The V_{oc} increased monotonically with the poling field strength (Fig. 8.2C) due to greater ion migration, whereas J_{sc} reaches a maximum value at 2 V μm^{-1} poling fields and then reduced (Fig. 8.2C) with poling due to the higher film resistance. The dominance of each effect has been confirmed by studying the effect of polarized light and phase transition from tetragonal to cubic on the poled samples. J_{sc} does not show a dependence on the light polarization angle (Fig. 8.2D), which confirmed the nonferroelectric nature of this response. It is also known that MAPbI$_3$ undergoes a crystal structure phase transition from tetragonal (which is ferroelectric) to cubic (which is not ferroelectric) above 327 K [43, 44]. It has been shown that both V_{oc} and J_{sc} rather than disappearing increase in magnitude when the MAPbI$_3$ films heated to 330 K (Fig. 8.2E and F). This was consistent with increased ion migration (due to greater mobility) at higher temperatures [32, 42, 45]. Unpoled perovskite samples do not show any J_{sc} in the presence or the absence of light illumination (Fig. 8.2F).

The polarization effect in the MAPbI$_3$ films was investigated by poling the device in ambient air and under 0.1 sun illumination. These conditions reflect that light-induced ordering of MA$^+$ ions enhanced the polarization behavior of MAPbI$_3$ [31, 34]. Both V_{oc} and J_{sc} due to the polarization effect were reversed in the direction as compared to the ion migration effects (Fig. 8.3A and B). Further, the response

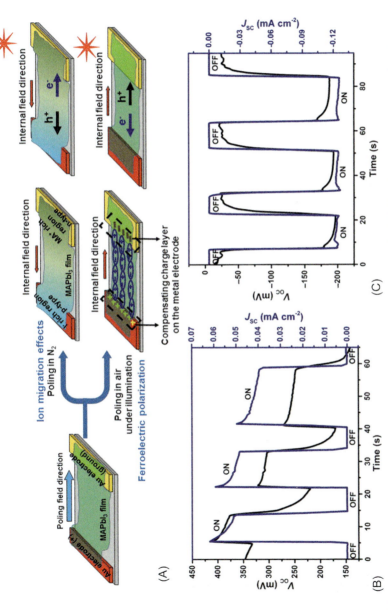

Fig. 8.1 (A) Schematic illustration showing the ion migration and polarization effects in Au/MAPbI$_3$/Au device with reversal of the internal electric field between the two cases. (B) V_{oc} and J_{sc} showing the ion migration effects in the MAPbI$_3$ films poled at 1.5 V μm^{-1} field for 5 min under N$_2$ in dark conditions. (C) V_{oc} and J_{sc} after poling at 1.5 V μm^{-1} field for 5 min in the air under light illumination (0.1 sun) results in the polarization effect. OFF refers to V_{oc} and J_{sc} measured under dark and ON refers to V_{oc} and J_{sc} measured under 0.1 sun illumination.
Reproduced from Ref. R. Saraf, L. Pu, V. Maheshwari, A light harvesting, self-powered monolith tactile sensor based on electric field induced effects in MAPbI3 perovskite, Adv. Mater. 30 (2018) 1705778.

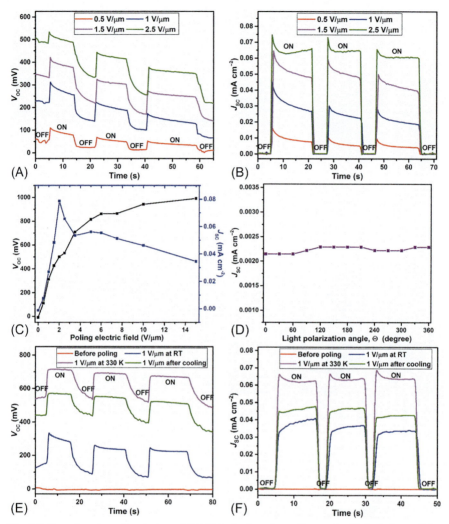

Fig. 8.2 Ion migration effect in lateral MAPbI$_3$ device after poling under N$_2$ in dark condition. (A) V_{oc} and (B) J_{sc} cycles after different poling fields. (C) V_{oc} and J_{sc} dependence on strength of the poling electric fields. (D) J_{sc} measured as a function of polarization angle of the linearly polarized light with respect to the film plane after poling the device at 1.5 V μm^{-1} for 5 min. (E) V_{oc} and (F) J_{sc} of MAPbI$_3$ device before poling at room temperature (RT) and after poling at RT, 330 K and after cooling to RT.
Reproduced from Ref. R. Saraf, L. Pu, V. Maheshwari, A light harvesting, self-powered monolith tactile sensor based on electric field induced effects in MAPbI3 perovskite, Adv. Mater. 30 (2018) 1705778.

was stable with no observable decay in their magnitude, which infers that unlike the diffusion of ions, the polarization due to the orientation of MA$^+$ ions is sustained over a longer time. It has been shown that unlike the ion migration effects, J_{sc} for polarization effect has a sinusoidal-like dependence on the angle between the electric field

Perovskites for tactile sensors 147

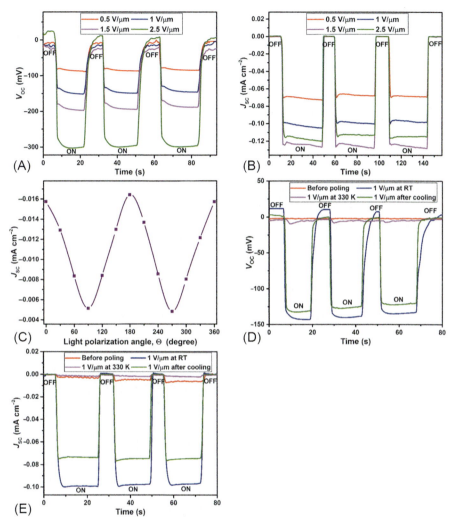

Fig. 8.3 Polarization effect in lateral MAPbI$_3$ device after poling in the air under light illumination. (A) V_{oc} and (B) J_{sc} cycles after different poling fields. (C) J_{sc} measured as a function of light polarization angle after poling the device at 1.5 V μm^{-1} for 5 min. (D) V_{oc} and (E) J_{sc} of MAPbI$_3$ device before poling at RT and after poling at RT, 330 K and after cooling to RT.
Reproduced from Ref. R. Saraf, L. Pu, V. Maheshwari, A light harvesting, self-powered monolith tactile sensor based on electric field induced effects in MAPbI3 perovskite, Adv. Mater. 30 (2018) 1705778.

of the incident polarized light and the sample plane (the in-plane ferroelectric polarization) due to second-order optical effects [46, 47], which confirmed the ferroelectric nature of this process (Fig. 8.3C). Moreover, on heating the films to 330 K which leads to a phase transition from tetragonal to cubic structure both V_{oc} and J_{sc} effects disappeared (Fig. 8.3D and E), which confirms that the polarization effects are ferroelectric in nature [25, 48]. The effects were restored on cooling the film back to room temperatures (~296 K) due to the existence of memory effect.

The synergetic integration of polymers with inorganic materials routinely occurs in natural systems such as bones and mollusk shells, leading to their outstanding properties such as high mechanical strength and toughness, high electrochemical performance, improved electronic and optical properties, and sensitivity for the detection of stimuli [49–58]. It has been shown that the integration of polystyrene (PS) chains with MAPbI$_3$ perovskite films due to specific interactions leads to modulation of both the mechanical and electro-optical properties of the perovskite films. The hydrophobic (organic) PS cross-links and strongly interacts with PbI$_2$ (due to its Lewis acid characteristics) and MA$^+$ cations (due to the π electrons of PS), resulting in the stable PS-MAPbI$_3$ films [59, 60]. The field emission scanning electron microscopy (FESEM) images of the plain MAPbI$_3$ and 1, 3, and 7 wt% PS-MAPbI$_3$ films showed uniform multicrystalline films (Fig. 8.4A–D).

It has been shown that the incorporation of the polystyrene in MAPbI$_3$ films improved the J_{sc} and V_{oc} characteristics as compared to plain MAPbI$_3$ (Fig. 8.5A and B). The comparison reveals that the 1 wt% PS-MAPbI$_3$ device shows the highest response. At the same time, the presence of PS significantly reduced the ion migration current (Fig. 8.5C), which allowed these PS films to be poled at higher field strengths compared to the plain MAPbI$_3$ films to induce stable polarization effects. The 1 wt%

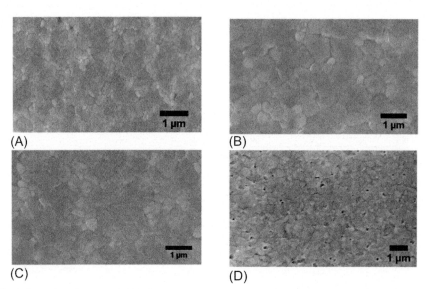

Fig. 8.4 FESEM images of (A) plain MAPbI$_3$, (B) 1 wt% PS-MAPbI$_3$, (C) 3 wt% PS-MAPbI$_3$, and (D) 7 wt% PS-MAPbI$_3$ films showing the uniform film coverage.
Reproduced from Ref. R. Saraf, T. Tsui, V. Maheshwari, Modulation of mechanical properties and stable light energy harvesting by poling in polymer integrated perovskite films: a wide range, linear and highly sensitive tactile sensor, J. Mater. Chem. A 7 (2019) 14192–14198 with permission from The Royal Society of Chemistry.

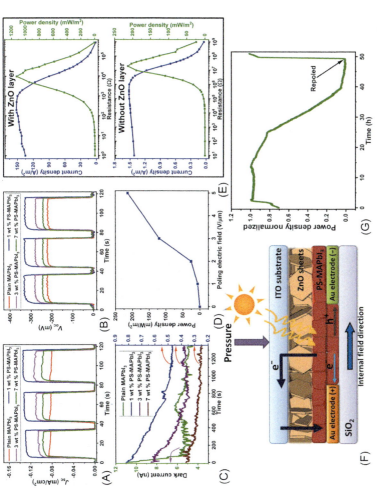

Fig. 8.5 Device performance, structure, and energy-harvesting capability. (A) J_{sc} and (B) V_{oc} cycles of devices after 2.5 V μm^{-1} poling for 5 min in the air under 0.1 sun illumination. (C) Dark current response from the devices at 3 V. (D) Power density dependence on strength of the poling electric fields for 1 wt% PS-MAPbI$_3$ device without ZnO layer. (E) Output current and power density as a function of different external resistances for 1 wt% PS-MAPbI$_3$ device (after 5 V μm^{-1} poling). (F) Schematic diagram of the self-powered PS-MAPbI$_3$ pressure sensor where ZnO nanosheets are interfaced with the PS-MAPbI$_3$ film. (G) Operational stability and continuous power generation from the 1 wt% PS-MAPbI$_3$ device (after 5 V μm^{-1} poling) interfaced with ZnO nanosheets. Reproduced from Ref. R. Saraf, T. Tsui, V. Maheshwari, Modulation of mechanical properties and stable light energy harvesting by poling in polymer integrated perovskite films: a wide range, linear and highly sensitive tactile sensor, J. Mater. Chem. A 7 (2019) 14192–14198 with permission from The Royal Society of Chemistry.

PS-MAPbI$_3$ films showed a monotonous increase in power density with poling fields (Fig. 8.5D), consistent with the expectation that the higher fields will increase internal polarization. The maximum power density of the 1 wt% PS-MAPbI$_3$ films was recorded to be 215 mW m^{-2} using an external load resistor (Fig. 8.5E), after poling at 5 V μm^{-1}.

8.4 Design of perovskite-based light-powered tactile sensors

Perovskite films are good hole conductors but their electron conductivity is limited. [61] An electron extraction layer is hence needed to further improve their energy-harvesting capability. To this effect, a top layer of ZnO nanosheets was interfaced with the PS-MAPbI$_3$ films (schematic in Fig. 8.5F, with a static load of 100 kPa). The ZnO sheets interface with the top surface of the perovskite layer and provide a pathway for efficient electron extraction across the whole device. As a result, the power density increased to 1.1 W m^{-2} (Fig. 8.5E). Continuous and stable power generation was observed in these films (with the ZnO layer) for more than 24 h (Fig. 8.5G), and subsequently, on repoling, the efficiency was recovered. Due to this, the loss was hence attributed to the depolarization of the films and not to any structural degradation. The interfacing with ZnO increased the power extraction efficiency significantly (by a factor of ~5) and continuous power generation was observed with light illumination and under constant interfacing (static load) with the ZnO layer. The static load ensures interfacing area between the two layers.

The elastic modulus of the films has been modulated in the range of ~23 GPa for pure MAPbI$_3$ to ~15 GPa with the incorporation of PS (Fig. 8.6A). The mechanical properties of these films have direct implications for their use in electromechanical and optomechanical devices and for tactile sensing. The resulting softer films were used to make tunable range tactile sensors that are combined with the light-harvesting properties of these films and hence resulted in light-powered tunable tactile sensors. The concept was based on modulating the interface between the ZnO nanosheets and the PS-MAPbI$_3$ films by an applied pressure (Fig. 8.5F). A direct correlation was observed between J_{sc} and the applied pressure for 1 wt% PS-MAPbI$_3$ film under 0.1 sun illumination after poling at 5 V μm^{-1} (Fig. 8.6B). The derivative of current density and pressure illustrated that the current accurately tracks the changes in applied pressure both in magnitude and rate (Fig. 8.6C). The observed increase in current was attributed to the increase in contact area (and not piezoelectric or triboelectric effect) as with time under a constant load the current response was maintained. The response from the sensor was correlated with the magnitude of the applied pressure as seen in cycling at different pressure loads (Fig. 8.6D). Further, the sensor was reported highly stable as there was no loss in response to more than 200 rapid loading cycles (Fig. 8.6E).

Varying the PS content in the PS-MAPbI$_3$ films will directly affect their functioning as a pressure sensor due to the change in their mechanical modulus. This was confirmed by observing the response of PS-MAPbI$_3$ films with varying amounts of PS in the precursor solution (Fig. 8.7A). Two effects were observed: first, as the PS content

Perovskites for tactile sensors 151

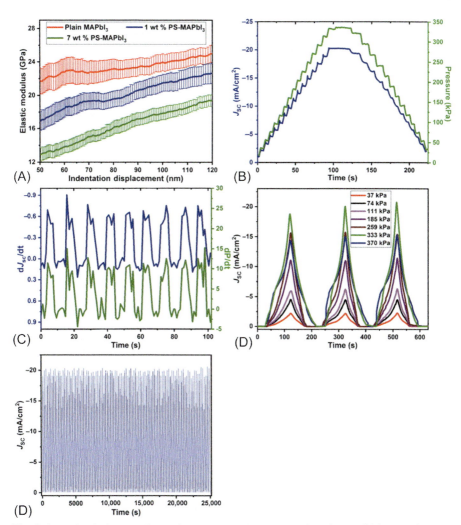

Fig. 8.6 Mechanical properties and pressure-sensing capabilities of monolithic 1 wt% PS-MAPbI$_3$ pressure sensor. (A) Elastic modulus showing the stiffness of the films. (B) J_{sc} response is in step with the dynamic and static pressure modulation. (C) Derivative of the J_{sc} response tracks that of the applied pressure with accuracy. (D) J_{sc} cycles with various applied pressure stimuli show a consistent response. (E) Cyclic stability and durability test of the pressure sensor under repetitive high-pressure loading of 333 kPa.
Reproduced from Ref. R. Saraf, T. Tsui, V. Maheshwari, Modulation of mechanical properties and stable light energy harvesting by poling in polymer integrated perovskite films: a wide range, linear and highly sensitive tactile sensor, J. Mater. Chem. A 7 (2019) 14192–14198 with permission from The Royal Society of Chemistry.

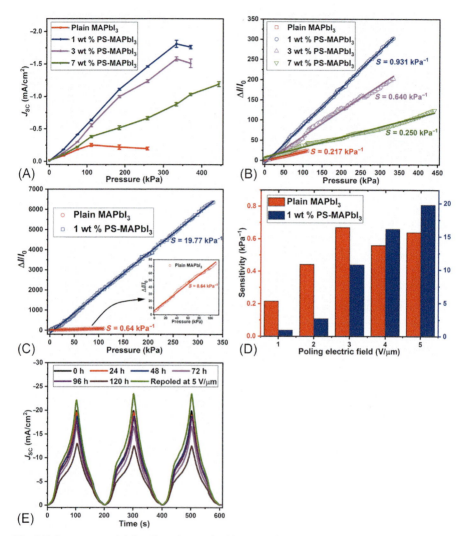

Fig. 8.7 Pressure sensitivity, linearity, and self-powered operation of the devices. (A) J_{sc} as a function of applied pressure for the devices after 1 V μm^{-1} poling for 5 min in the air under 0.1 sun illumination. (B) Relative current of the devices shows the linear response over the broad dynamic range. (C) Sensitivity and pressure range of 1 wt% PS-MAPbI$_3$ device (after 5 V μm^{-1} poling) is more than the plain MAPbI$_3$ device. (D) Sensitivity of the 1 wt% PS-MAPbI$_3$ device increases with the poling field strength. (E) J_{sc} response is maintained over 120 h to load cycles of 333 kPa for the 1 wt% PS-MAPbI$_3$ device after initial poling at 5 V μm^{-1} for 5 min. Reproduced from Ref. R. Saraf, T. Tsui, V. Maheshwari, Modulation of mechanical properties and stable light energy harvesting by poling in polymer integrated perovskite films: a wide range, linear and highly sensitive tactile sensor, J. Mater. Chem. A 7 (2019) 14192–14198 with permission from The Royal Society of Chemistry.

increased the dynamic range for pressure sensing increased, and second, however, the sensitivity does not follow a monotonic trend. Plain MAPbI$_3$ films were limited to $\sim$100 kPa pressure range before saturation in J_{sc} was observed. Introducing PS increased the dynamic range of the device progressively to more than 400 kPa with 7 wt% PS-MAPbI$_3$ films. We attribute this to the increased softness of the PS-MAPbI$_3$ films as the PS content increases. The maximum sensitivity in response was, however, observed for the 1 wt% PS-MAPbI$_3$ films (Fig. 8.7B). At a maximum poling field of 5 V μm^{-1}, the 1 wt% PS-MAPbI$_3$ devices attained a high sensitivity of 19.77 kPa^{-1} (with a linear response up to 333 kPa), which was 30 times more than the maximum sensitivity possible with the plain MAPbI$_3$ films (0.64 kPa^{-1}) (Fig. 8.7C). The sensitivity is dependent on both the increase in the interface area with applied pressure and the density of the charge carrier being generated and separated due to the internal polarization of the films. Hence, in this case, the 1 wt% PS-MAPbI$_3$ devices due to their greater polarization effects a higher sensitivity is observed compared to all other devices. The sensitivity of the 1 wt% PS-MAPbI$_3$ device increased with the strength of the poling electric field (Fig. 8.7D). This was consistent with the increased internal polarization, which will lead to more effective charge separation, thus improving both the charge collection efficiency by the ZnO films and the sensitivity to pressure modulations. However, the sensitivity of the plain MAPbI$_3$ device increased up to 3 V μm^{-1} poling field; subsequently, at higher poling fields ($>$3 V μm^{-1}), the performance decreased due to the segregation of the ions, which results in the formation of PbI$_2$ and hence the observed polarization effect decreased. The 1 wt% PS-MAPbI$_3$ device can sense pressure as low as 4 Pa (50 μL water droplet). A softer perovskite film due to the incorporated polymer was better able to dissipate the mechanical energy and hence extend the linear operating pressure range of these devices, which is similar to the observation with indentation of softer films [62]. In general, the incorporation of the polymer increased the sensitivity and linearity range of the PS-MAPbI$_3$ films compared to plain MAPbI$_3$ films. This allowed the tuning of the operating range, sensitivity, and linear range of these pressure sensors based on the polymer content. Furthermore, the 1 wt% PS-MAPbI$_3$ device once poled at 5 V μm^{-1} for 5 min can be easily operable for more than 120 h without a power source and after that, the device can be repoled to recover the performance (Fig. 8.7E), whereas the plain MAPbI$_3$ and 3 wt% and 7 wt% PS-MAPbI$_3$ devices were operable for 48–72 h. The monolithic 1 wt% PS-MAPbI$_3$ pressure sensor reported exhibits one of the best combinations of high sensitivity with a linear response over a broad dynamic pressure range, as well as the device can be self-powered and was among the best reported figure for energy-harvesting devices such as triboelectric and piezoelectric generators (which are intermittent in nature). Further, this was achieved in a simple device structure with the ability to sense both constant static stimuli and also dynamic stimuli, which is a challenge in many architectures. It has been shown that the incorporation of PS also enhanced the structural stability of MAPbI$_3$ under harsh environmental conditions, which leads to no visible degradation in performance over months; and hence allowed to make stable light-powered tactile sensors.

8.5 Future vision and challenges

Overall, perovskite-based self-powered tactile sensors have tremendous potential applications in robotics and aerospace applications; but in the meantime, they face some challenges. The lead (Pb), contained in organolead halide perovskite-based self-powered sensors, is a heavy metal and toxic element that can lead to environmental toxicological implications. However, the amount of lead present in perovskite-based self-powered sensors is below that produced annually by the coal industry when it generates an equivalent amount of electricity. The high aqueous solubility of PbI_2 poses a severe risk to human health. Future efforts may be needed to replace the toxic lead metal with other nontoxic elements [such as tin (Sn)] to develop Pb-free sensors that will broaden the applications of perovskite-based self-powered sensors in healthcare devices, human-machine interaction, wearable devices, and minimally invasive surgery (MIS). From the perspective of materials, the organolead halide perovskites should meet the long-term stability requirements under normal environmental conditions. Although it has been shown the polystyrene-perovskite composite films are stable in ambient conditions under continuous light illumination for more than 1000 h, without the use of any encapsulating layer. Future efforts may be needed to further improve the stability of perovskite-based sensors.

Efforts are required to use the polystyrene-incorporated perovskite materials in conjunction with the flexible substrates to develop the lightweight and potentially semitransparent perovskite sensors for portable optoelectronic devices. These features are particularly attractive for the automotive industry, indoor applications, and Internet of things (IoT) devices or transport. From the perspective of commercialization, new scalable processing routes must be required to achieve large-area uniform films with efficient materials utilization. Besides self-powered tactile sensors that are related to mechanical motion, the lead-free perovskite-based sensors can be explored for self-powered temperature skin sensors to continuously monitor temperature changes in the human body, which are mainly based on thermoelectric nanogenerators. This kind of skin sensors has important applications in sports and medical treatment. Triboelectric nanogenerator (TENG) as a self-powered pressure sensor has the advantages of simple structure, easy fabrication, multiple choice of materials, and low cost. Future research may be needed to develop the way of integrating TENG-based sensor with perovskite-based materials to extract more energy from the mechanical motion to power the devices.

References

[1] S. Jacobsen, E. Iversen, D. Knutti, R. Johnson, K. Biggers, Design of the Utah/M.I.T. Dextrous hand, Proceedings 1986 IEEE International Conference on Robotics and Automation, San Francisco, CA, USA, 1986, pp. 1520–1532.

[2] R.O. Ambrose, H. Aldridge, R.S. Askew, R.R. Burridge, W. Bluethmann, M. Diftler, C. Lovchik, D. Magruder, F. Rehnmark, Robonaut: NASA's Space Humanoid, 15, IEEE Intelligent Systems and their Applications, 2000, pp. 57–63.

[3] E. Tunstel, M. Maimone, A. Trebi-Ollennu, J. Yen, R. Petras, R. Willson, Mars exploration rover mobility and robotic arm operational performance, in: 2005 IEEE International Conference on Systems, Man and Cybernetics, Waikoloa, HI, 2005, pp. 1807–1814.

[4] M.A. Diftler, N.A. Radford, J.S. Mehling, M.E. Abdallah, L.B. Bridgwater, A.M. Sanders, R.S. Askew, D.M. Linn, J.D. Yamokoski, F.A. Permenter, B.K. Hargrave, Robonaut 2—the first humanoid robot in space, 2011 IEEE International Conference on Robotics and Automation, Shanghai, China, 2011, pp. 2178–2183.

[5] J. Bares, M. Hebert, T. Kanade, E. Krotkov, T. Mitchell, R. Simmons, W. Whittaker, Ambler: An Autonomous Rover for Planetary Exploration, 22, Computer, 1989, pp. 18–26.

[6] S. Parkes, M. Dunstan, I. Martin, Preparting mission operators for mars exploration, SpaceOps 2006 Conference, 2006, p. 5972.

[7] H. Kolvenbach, M. Breitenstein, C. Gehring, M. Hutter, Scalability analysis of legged robots for space exploration, Unlocking Imagination, Fostering Innovation And Strengthening Security: 68th International Astronautical Congress (IAC 2017), Curran, 2018, pp. 10399–10413.

[8] R. Bogue, Robots for space exploration, Ind. Robot 39 (2012) 323–328.

[9] R. Saraf, L. Pu, V. Maheshwari, A light harvesting, self-powered monolith tactile sensor based on electric field induced effects in MAPbI$_3$ perovskite, Adv. Mater. 30 (2018) 1705778.

[10] R. Saraf, T. Tsui, V. Maheshwari, Modulation of mechanical properties and stable light energy harvesting by poling in polymer integrated perovskite films: a wide range, linear and highly sensitive tactile sensor, J. Mater. Chem. A 7 (2019) 14192–14198.

[11] Z.L. Wang, Self-powered nanosensors and nanosystems, Adv. Mater. 24 (2012) 280–285.

[12] J. Rao, Z. Chen, D. Zhao, Y. Yin, X. Wang, F. Yi, Recent progress in self-powered skin sensors, Sensors 19 (2019) 2763.

[13] C. Pang, G.Y. Lee, T. Kim, S.M. Kim, H.N. Kim, S.H. Ahn, K.Y. Suh, A flexible and highly sensitive strain-gauge sensor using reversible interlocking of nanofibres, Nat. Mater. 11 (2012) 795–801.

[14] K.Y. Lee, H.J. Yoon, T. Jiang, X. Wen, W. Seung, S.W. Kim, Z.L. Wang, Fully packaged self-powered triboelectric pressure sensor using hemispheres-array, Adv. Energy Mater. 6 (2016) 1502566.

[15] D.J. Lipomi, M. Vosgueritchian, B.C. Tee, S.L. Hellstrom, J.A. Lee, C.H. Fox, Z. Bao, Skin-like pressure and strain sensors based on transparent elastic films of carbon nanotubes, Nat. Nanotechnol. 6 (2011) 788.

[16] W. Wu, X. Wen, Z.L. Wang, Taxel-addressable matrix of vertical-nanowire piezotronic transistors for active and adaptive tactile imaging, Science 340 (2013) 952–957.

[17] Y. Zi, L. Lin, J. Wang, S. Wang, J. Chen, X. Fan, P.K. Yang, F. Yi, Z.L. Wang, Triboelectric–pyroelectric–piezoelectric hybrid cell for high-efficiency energy-harvesting and self-powered sensing, Adv. Mater. 27 (2015) 2340–2347.

[18] Y.C. Mao, P. Zhao, G. McConohy, H. Yang, Y.X. Tong, X.D. Wang, Sponge-like piezoelectric polymer films for scalable and integratable nanogenerators and self-powered electronic systems, Adv. Energy Mater. 4 (2014) 1301624.

[19] J. Park, Y. Lee, M. Ha, S. Cho, H. Ko, Micro/nanostructured surfaces for self-powered and multifunctional electronic skins, J. Mater. Chem. B 4 (2016) 2999–3018.

[20] G. Hodes, Perovskite-based solar cells, Science 342 (2013) 317–318.

[21] M.M. Lee, J. Teuscher, T. Miyasaka, T.N. Murakami, H.J. Snaith, Efficient hybrid solar cells based on meso-superstructured organometal halide perovskites, Science 338 (2012) 643–647.

[22] N.G. Park, Organometal perovskite light absorbers toward a 20% efficiency low-cost solid-state mesoscopic solar cell, J. Phys. Chem. Lett. 4 (2013) 2423–2429.

[23] J. Burschka, N. Pellet, S.J. Moon, R. Humphry-Baker, P. Gao, M.K. Nazeeruddin, M. Grätzel, Sequential deposition as a route to high-performance perovskite-sensitized solar cells, Nature 499 (2013) 316–319.

[24] Y. Rakita, O. Bar-Elli, E. Meirzadeh, H. Kaslasi, Y. Peleg, G. Hodes, I. Lubomirsky, D. Oron, D. Ehre, D. Cahen, Tetragonal $CH_3NH_3PbI_3$ is ferroelectric, Proc. Natl. Acad. Sci. U. S. A. 114 (2017) E5504–E5512.

[25] A. Stroppa, C. Quarti, F. De Angelis, S. Picozzi, Ferroelectric polarization of $CH_3NH_3PbI_3$: a detailed study based on density functional theory and symmetry mode analysis, J. Phys. Chem. Lett. 6 (2015) 2223–2231.

[26] Y. Kutes, L. Ye, Y. Zhou, S. Pang, B.D. Huey, N.P. Padture, Direct observation of ferroelectric domains in solution-processed $CH_3NH_3PbI_3$ perovskite thin films, J. Phys. Chem. Lett. 5 (2014) 3335–3339.

[27] M. Coll, A. Gomez, E. Mas-Marza, O. Almora, G. Garcia-Belmonte, M. Campoy-Quiles, J. Bisquert, Polarization switching and light-enhanced piezoelectricity in lead halide perovskites, J. Phys. Chem. Lett. 6 (2015) 1408–1413.

[28] T. Chen, B.J. Foley, B. Ipek, M. Tyagi, J.R.D. Copley, C.M. Brown, J.J. Choi, S.H. Lee, Rotational dynamics of organic cations in the $CH_3NH_3PbI_3$ perovskite, Phys. Chem. Chem. Phys. 17 (2015) 31278–31286.

[29] J.M. Frost, K.T. Butler, A. Walsh, Molecular ferroelectric contributions to anomalous hysteresis in hybrid perovskite solar cells, APL Mater. 2 (2014), 081506.

[30] Z. Xiao, Y. Yuan, Y. Shao, Q. Wang, Q. Dong, C. Bi, P. Sharma, A. Gruverman, J. Huang, Giant switchable photovoltaic effect in organometal trihalide perovskite devices, Nat. Mater. 14 (2015) 193–198.

[31] Y. Yuan, T. Li, Q. Wang, J. Xing, A. Gruverman, J. Huang, Anomalous photovoltaic effect in organic-inorganic hybrid perovskite solar cells, Sci. Adv. 3 (2017), e1602164.

[32] Y. Yuan, J. Chae, Y. Shao, Q. Wang, Z. Xiao, A. Centrone, J. Huang, Photovoltaic switching mechanism in lateral structure hybrid perovskite solar cells, Adv. Energy Mater. 5 (2015) 1500615.

[33] Q. Wang, Y. Shao, H. Xie, L. Lyu, X. Liu, Y. Gao, J. Huang, Qualifying composition dependent p and n self-doping in $CH_3NH_3PbI_3$, Appl. Phys. Lett. 105 (2014) 163508.

[34] S. Liu, F. Zheng, I. Grinberg, A.M. Rappe, Photoferroelectric and photopiezoelectric properties of organometal halide perovskites, J. Phys. Chem. Lett. 7 (2016) 1460–1465.

[35] C. Quarti, F. Mosconi, F. De Angelis, Interplay of orientational order and electronic structure in methylammonium lead iodide: implications for solar cell operation, Chem. Mater. 26 (2014) 6557–6569.

[36] Y. Yuan, T.J. Reece, P. Sharma, S. Poddar, S. Ducharme, A. Gruverman, Y. Yang, J. Huang, Efficiency enhancement in organic solar cells with ferroelectric polymers, Nat. Mater. 10 (2011) 296–302.

[37] S.Y. Yang, J. Seidel, S.J. Byrnes, P. Shafer, C.H. Yang, M.D. Rossell, P. Yu, Y.H. Chu, J. F. Scott, J.W. Ager, L.W. Martin, R. Ramesh, Above-bandgap voltages from ferroelectric photovoltaic devices, Nat. Nanotechnol. 5 (2010) 143–147.

[38] H.W. Chen, N. Sakai, M. Ikegami, T. Miyasaka, Emergence of hysteresis and transient ferroelectric response in organo-lead halide perovskite solar cells, J. Phys. Chem. Lett. 6 (2014) 164–169.

[39] Y. Yuan, Z. Xiao, B. Yang, J. Huang, Arising applications of ferroelectric materials in photovoltaic devices, J. Mater. Chem. A 2 (2014) 6027–6041.

[40] H.S. Kim, I. Mora-Sero, V. Gonzalez-Pedro, F. Fabregat-Santiago, E.J. Juarez-Perez, N.G. Park, J. Bisquert, Mechanism of carrier accumulation in perovskite thin-absorber solar cells, Nat. Commun. 4 (2013) 2242.

[41] R.R. Mehta, B.D. Silverman, J.T. Jacobs, Depolarization fields in thin ferroelectric films, J. Appl. Phys. 44 (1973) 3379–3385.

[42] Y.C. Zhao, W.K. Zhou, X. Zhou, K.H. Liu, D.P. Yu, Q. Zhao, Quantification of light-enhanced ionic transport in lead iodide perovskite thin films and its solar cell applications, Light Sci. Appl. 6 (2017), e16243.

[43] T. Baikie, Y. Fang, J.M. Kadro, M. Schreyer, F. Wei, S.G. Mhaisalkar, M. Graetzel, T.J. White, Synthesis and crystal chemistry of the hybrid perovskite $(CH_3NH_3)PbI_3$ for solid-state sensitised solar cell applications, J. Mater. Chem. A 1 (2013) 5628–5641.

[44] C.C. Stoumpos, C.D. Malliakas, M.G. Kanatzidis, Semiconducting tin and lead iodide perovskites with organic cations: phase transitions, high mobilities, and near-infrared photoluminescent properties, Inorg. Chem. 52 (2013) 9019–9038.

[45] J. Mizusaki, K. Arai, K. Fueki, Ionic conduction of the perovskite-type halides, Solid State Ion. 11 (1983) 203–211.

[46] M. Bass, P.A. Franken, J.F. Ward, G. Weinreich, Optical rectification, Phys. Rev. Lett. 9 (1962) 446–448.

[47] A. Rice, Y. Jin, X.F. Ma, X.C. Zhang, D. Bliss, J. Larkin, M. Alexander, Terahertz optical rectification from ⟨110⟩ zinc-blende crystals, Appl. Phys. Lett. 64 (1994) 1324–1326.

[48] M.T. Weller, O.J. Weber, P.F. Henry, A.M. Di Pumpo, T.C. Hansen, Complete structure and cation orientation in the perovskite photovoltaic methylammonium lead iodide between 100 and 352 K, Chem. Commun. 51 (2015) 4180–4183.

[49] N. Reznikov, M. Bilton, L. Lari, M.M. Stevens, R. Kröger, Fractal-like hierarchical organization of bone begins at the nanoscale, Science 360 (2018), eaao2189.

[50] G. Falini, S. Albeck, S. Weiner, L. Addadi, Control of aragonite or calcite polymorphism by mollusk shell macromolecules, Science 271 (1996) 67–69.

[51] B. Wang, T.N. Sullivan, A. Pissarenko, A. Zaheri, H.D. Espinosa, M.A. Meyers, Lessons from the ocean: whale baleen fracture resistance, Adv. Mater. 31 (2019) 1804574.

[52] J.S. Peng, Q.F. Cheng, High-performance nanocomposites inspired by nature, Adv. Mater. 29 (2017) 1702959.

[53] V.K. Thakur, R.K. Gupta, Recent progress on ferroelectric polymer-based nanocomposites for high energy density capacitors: synthesis, dielectric properties, and future aspects, Chem. Rev. 116 (2016) 4260–4317.

[54] W. Liu, S.W. Lee, D.C. Lin, F.F. Shi, S. Wang, A.D. Sendek, Y. Cui, Enhancing ionic conductivity in composite polymer electrolytes with well-aligned ceramic nanowires, Nat. Energy 2 (2017) 17035.

[55] Y.Y. Kim, J.D. Carloni, B. Demarchi, D. Sparks, D.G. Reid, M.E. Kunitake, C.C. Tang, M.J. Duer, C.L. Freeman, B. Pokroy, K. Penkman, J.H. Harding, L.A. Estroff, S.P. Baker, F.C. Meldrum, Tuning hardness in calcite by incorporation of amino acids, Nat. Mater. 15 (2016) 903–910.

[56] F. Bouville, E. Maire, S. Meille, B. Van de Moortèle, A.J. Stevenson, S. Deville, Corrigendum: strong, tough and stiff bioinspired ceramics from brittle constituents, Nat. Mater. 16 (2017) 1271.

[57] J. Kang, D. Son, G.J.N. Wang, Y. Liu, J. Lopez, Y. Kim, J.Y. Oh, T. Katsumata, J. Mun, Y. Lee, L. Jin, J.B.H. Tok, Z. Bao, Tough and water-insensitive self-healing elastomer for robust electronic skin, Adv. Mater. 30 (2018) 1706846.

[58] D. Son, J. Kang, O. Vardoulis, Y. Kim, N. Matsuhisa, J.Y. Oh, J.W. To, J. Mun, T. Katsumata, Y. Liu, A.F. McGuire, M. Krason, F.M. Lopez, J. Ham, U. Kraft, Y. Lee, Y. Yun, J.B.H. Tok, Z. Bao, An integrated self-healable electronic skin system fabricated via dynamic reconstruction of a nanostructured conducting network, Nat. Nanotechnol. 13 (2018) 1057–1065.

[59] R. Saraf, V. Maheshwari, PbI_2 initiated cross-linking and integration of a polymer matrix with perovskite films: 1000 h operational devices under ambient humidity and atmosphere and with direct solar illumination, ACS Appl. Energy Mater. 2 (2019) 2214–2222.

[60] R. Saraf, V. Maheshwari, Self-powered photodetector based on electric-field-induced effects in $MAPbI_3$ perovskite with improved stability, ACS Appl. Mater. Interfaces 10 (2018) 21066–21072.

[61] E. Edri, S. Kirmayer, A. Henning, S. Mukhopadhyay, K. Gartsman, Y. Rosenwaks, G. Hodes, D. Cahen, Why lead methylammonium tri-iodide perovskite-based solar cells require a mesoporous electron transporting scaffold (but not necessarily a hole conductor), Nano Lett. 14 (2014) 1000–1004.

[62] D. Tranchida, S. Piccarolo, M. Soliman, Nanoscale mechanical characterization of polymers by AFM nanoindentations: critical approach to the elastic characterization, Macromolecules 39 (2006) 4547–4556.

Electrospun nanofibers for tactile sensors

Yichun Ding[a,b,*], Obiora Onyilagha[c], and Zhengtao Zhu[b,c,d,*]
[a]CAS Key Laboratory of Design and Assembly of Functional Nanostructures, Fujian Provincial Key Laboratory of Nanomaterials, Fujian Institute of Research on the Structure of Matter, Chinese Academy of Sciences, Fuzhou, China, [b]Biomedical Engineering Program, South Dakota School of Mines & Technology, Rapid City, SD, United States, [c]Nanoscience and Nanoengineering Program, South Dakota School of Mines & Technology, Rapid City, SD, United States, [d]Department of Chemistry, Biology, and Health Sciences, South Dakota School of Mines & Technology, Rapid City, SD, United States
*Corresponding authors: e-mail address: ycding@fjirsm.ac.cn, zhengtao.zhu@sdsmt.edu

9.1 Introduction

The fast-growing technologies of Internet of Things (IoT) and artificial intelligence (AI) are revolutionizing our life. These technologies have stimulated tremendous research and development in smart systems and devices, in which a variety of wearable sensors is utilized for real-time and continuous human health monitoring [1–3]. The tactile sensor is one of the various types of sensors that can be used in systems for human-machine interaction, human motion/movement detection, electronic skin, and material structural health monitoring [4–6]. Conventional bulk and rigid sensors can hardly fulfill the requirement of high flexibility for wearable applications; therefore, extensive research efforts are ongoing to develop new materials and/or structures for flexible tactile sensors [5]. Significant recent progresses in sensing materials, transduction methods, device configurations, and manufacturing technologies have paved the way to use flexible tactile sensors in wearable applications [7–10].

Tactile sensors are devices that can convert the external force/pressure stimuli into detectable signals; thus, these devices can recognize touch, contact, and deformation. Depending on the recognizable signals, which can be optical (e.g., color and light intensity), electrical (e.g., resistance, capacitance, voltage, and current), and magnetic, the tactile sensors are categorized into different transduction mechanisms. In particular, the electrical transducers including piezoresistive, capacitive, piezoelectric, and triboelectric ones have attracted attention due to easy material preparation and device fabrication, as well as facile signal acquisition [8]. For example, piezoresistive tactile sensors can be fabricated using porous conductive materials, and the sensors measure the resistance change in response to the stimuli, which only requires simple signal readout and low energy consumption [6]. When a tactile sensor is attached to human skin or limb joints, interaction with the surroundings or motion of the human body

Functional Tactile Sensors. https://doi.org/10.1016/B978-0-12-820633-1.00002-4
© 2021 Elsevier Ltd. All rights reserved.

deform the sensor, resulting in detectable electrical change. Such real-time detection can be applied for human health monitoring and possibly for disease diagnosis.

The wearable/flexible tactile sensor needs to not only respond to external stimuli but also have characteristics such as high sensitivity, good reproducibility and stability, and good flexibility. The active sensing material is the key component of the device. The conventional materials (e.g., silicon, metal, and metal oxide films), which are rigid, are not suitable for flexible tactile sensors. In recent years, nanomaterials and nanotechnologies have provided a promising direction to develop advanced materials for flexible tactile sensors. Nanomaterials such as metal nanowires, carbon nanotubes, graphene, MXene, and nanofibers have been widely explored for tactile sensors [11]. Among various nanomaterials, electrospun nanofibers produced by the versatile electrospinning technique have emerged as forefront active materials for tactile sensors attributed to their inherent advantages of small diameter, high aspect ratio, large surface area, high porosity, lightweight, facile fabrication, readily processing and modification, good air permeability, excellent flexibility and conformability, easy integration with wearable system, etc. [12, 13].

In this chapter, we aim to review the electrospun nanofibrous materials for flexible and wearable tactile sensors. The electrospinning technique and the resultant electrospun nanofibers are first introduced, including the electrospinning process, the features of electrospun nanofibers, and the functionalization methods. Subsequently, the transduction mechanisms, together with their specific requirements on sensing materials, are briefly discussed. Following that, tactile sensors from different electrospun nanofibers are overviewed according to the transduction mechanisms. Finally, the remaining challenges and outlooks are highlighted.

9.2 Electrospinning and electrospun nanofibers

9.2.1 Electrospinning process and setup

The electrospinning technique was first invented in 1934; the technique attracted renewed and growing attentions since Prof. Reneker's pioneer studies in the 1990s [14]. The electrospinning technique is a unique fiber-forming method that provides a straightforward while versatile process to produce fibers with nano-sized diameters (typically ranges from several to hundreds of nanometers). In electrospinning, the electrostatic force drives the process to produce the nanofibers of polymers; in combination with the various post-spinning processes, electrospinning technique is capable of producing various nanofibers including polymeric, inorganic, carbon, and composite ones [15–22]. The polymer nanofibers are typically electrospun directly from polymer solutions or melts [23, 24]. Inorganic/ceramic nanofibers are made by electrospinning the precursors followed by high-temperature pyrolysis [25–27]. Carbon nanofibers are usually made through carbonization of the polymer nanofiber precursors [22, 28–30].

A typical (needle) electrospinning setup, as shown in Fig. 9.1A, consists of four components: (1) a high-voltage electric power supply, (2) a spinning solution

Electrospun nanofibers for tactile sensors

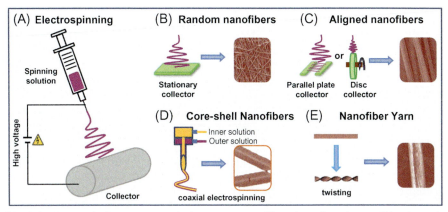

Fig. 9.1 Electrospinning process and electrospun nanofibers in various forms: (A) schematic illustration of electrospinning process and setup; (B) production of random nanofibers using a stationary collector; (C) collection of well-aligned nanofibers using a parallel-plate or high-speed rotating disc collector; (D) preparation of core-shell nanofibers by coaxial electrospinning; and (E) preparation of nanofiber yarns by twisting the as-spun nanofibers.

container (e.g., a syringe on an injection pump), (3) a spinneret (e.g., a metallic needle), and (4) a collector. During electrospinning, the spinning solution is filled in a container (e.g., syringe) with a spinneret (e.g., syringe needle), a high voltage (typically 5–40 kV, direct current) is applied through the spinneret, and a collector (electrically grounded or negative voltage applied) is placed a certain distance from the spinneret. When the electrostatic force upon applying a high voltage reaches a critical value that surpasses the surface tension and viscoelastic force, a liquid Taylor Cone is formed on the tip of the spinneret, and then a jet is ejected and travels straight for a certain distance in space. Thereafter, the jet flies toward the collector, during which the jet whips and travels in a helical-loop fashion. The phenomenon is termed as "bending instability." The bending instability results in stretching of the jet by thousands of times to form continuous and ultrafine nanofibers in a very short time period; meanwhile, the solvent evaporates during the bending instability (Fig. 9.1A) [31].

9.2.2 The features/advantages of electrospun nanofibers

The electrospun nanofibers are usually smooth and have uniform diameters under optimized electrospinning conditions; the nanofibers are deposited continuously on the collector forming a nanofibrous mat. The morphology, diameter, and structure of nanofibers can be rationally controlled by adjusting multiple factors, including the molecular structure/weight of polymer, concentration of solution and solvent, electrospinning parameters (e.g., applied voltage, feeding rate of solution, and collection distance), and the environmental conditions (e.g., temperature, humidity, and air velocity). The thickness of nanofibrous mat can be easily controlled by the amount of spinning solution, the processing time, and the dimension/size of collector [32].

Nanofibers and nanofibrous mats with different morphologies and structures can be prepared by innovative designs of the spinnerets and collectors. Nanofibers are collected as a randomly overlaid nanofibrous mat/nonwoven when using a stationary collector (e.g., flat aluminum foil), as shown in Fig. 9.1B. To prepare oriented/aligned nanofibers, several strategies have been used. The most effective and straightforward method is to adopt specifically designed collectors. As shown in Fig. 9.1C, two types of collectors are widely used for preparing the well-aligned nanofibers [33, 34]. The parallel-plate collector can obtain nanofibers with good alignment but it can just collect limited nanofibers, which may be useful for fabrication of microelectronic devices [35]. A high-speed rotating disc collector can produce on a large-scale continuously aligned nanofibers to form bundles or belts. A coaxial electrospinning method is used to prepare core-shell nanofibers, in which one component (core) is surrounded by the other (shell) coaxially; as shown in Fig. 9.1D, two different spinning solutions are fed simultaneously through a coaxial nozzle during the process [16]. By adjusting the components of solutions, hollow nanofibers can be prepared by removing the core material [36]. Similarly, multichannel nanofibers can also be prepared by designing a multiaxial nozzle [37]. Furthermore, electrospun nanofibers can be processed to a yarn/thread by twisting the as-spun nanofibrous mat (Fig. 9.1E) [38, 39].

Unlike the fibers produced by conventional spinning techniques (e.g., melt spinning, wet spinning, and dry spinning), which are micrometer sized ($c.$, 5–15 µm), the fibers prepared from electrospinning are much smaller with typical diameters down to the nanometer range ($c.$, 10–1000 nm); therefore, the fibers prepared by electrospinning are usually termed as "electrospun nanofibers." Consequently, due to the small diameters, other extraordinary properties including high aspect ratio, large surface area, high porosity, and lightweight are concomitant with electrospun nanofibers. The electrospun polymer nanofibrous mats are highly porous and flexible with good air permeability, adaptability, and conformability for integration with wearable systems. In addition, the polymer chains or the added fillers may be highly stretched during bending instability; such stretching may induce molecular orientation and increase of crystallinity, and thus improve the mechanical strength and other properties of nanofibers. For example, PVDF nanofibers contain increased content of polar β-phase and are electrically poled during the electrospinning process, leading to a high piezoelectric effect [40, 41].

9.2.3 Functionalization methods of electrospun nanofibers

Beyond the aforementioned features/advantages, another significant advantage of electrospun nanofibers is that they can be readily modified/functionalized through various strategies to acquire new properties/functionalities [42]. Fig. 9.2 lists several common methods/strategies to functionalize electrospun nanofibers. As illustrated in Fig. 9.2A, adding fillers or precursors directly to the spinning solution is the most straightforward method to prepare composite nanofibers [43]. For example, conductive composite nanofibers can be prepared by electrospinning a polymer solution containing conductive fillers (e.g., metal nanoparticles, carbon nanotubes, and graphene) or their precursors [e.g., $Ag(NO_3)$]; in the case that the precursors are used,

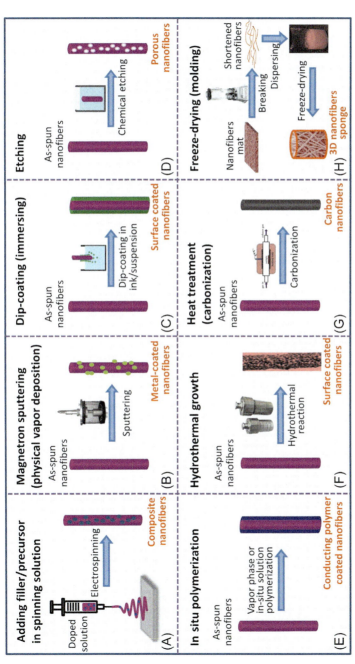

Fig. 9.2 Methods to functionalize electrospun nanofibers: (A) preparation of composite nanofibers by directly adding filler or precursor in spinning solution; (B) preparation of metal nanoparticles or film-coated nanofiber by the sputtering method (e.g., magnetron sputtering); (C) surface modification of nanofibers by dip-coating or immersing as-spun nanofibers in a solution/ink/suspension; (D) preparation of porous nanofibers through chemical etching; (E) coating conducting polymer on nanofibers through in situ polymerization; (F) surface modification of nanofibers by a hydrothermal reaction; (G) heat treatment of as-spun nanofibers (e.g., carbon nanofibers can be produced by high-temperature carbonization); and (H) assembly of electrospun nanofibers into 3D architecture by a freeze-drying molding method.

posttreatment of the obtained nanofibers (e.g., chemical reduction or heat treatment) is needed to convert the precursors to conductive fillers [16]. Physical vapor deposition (PVD) is a simple posttreatment method to deposit metal nanoparticles/film directly on polymer nanofibers to make them conductive (Fig. 9.2B). The PVD method can maintain the flexibility of the nanofibrous mat well and avoid side reactions/effects; the disadvantage is the relatively high cost. Compared to PVD, the dip-coating method by dipping or immersing nanofibers in solution/ink/suspension is more facile and simple, enabling large-scale production (Fig. 9.2C) [44]. Contrary to the coating method, if some component in nanofibers can be etched/dissolved by a chemical, then immersing nanofibers in the chemical solution can produce porous nanofibers (Fig. 9.2D), which can significantly increase the surface roughness and specific surface area.

Besides metals and carbon materials, conducting polymers are widely used for the preparation of conductive materials. For example, conducting polymers [e.g., polyaniline (PANi), polypyrrole (PPy), and poly(3,4-ethylenedioxythiophene) (PEDOT)] can be polymerized on nanofibers by in situ polymerization in solution or by vapor-phase polymerization (Fig. 9.2E) [43, 45]. The hydrothermal reaction is another facile method since the reaction takes place in aqueous solution, through which many materials such as metal oxide nanorods/nanowires can be grown on electrospun nanofibers (Fig. 9.2F). High-temperature heat treatment is another convenient method to modify/functionalize nanofibers (Fig. 9.2G). As mentioned, ceramic nanofibers are prepared by high-temperature annealing/pyrolysis of their corresponding sol-gel product and polymer nanofibers; carbon nanofibers are prepared through high-temperature carbonization of polymer nanofibers.

The as-spun nanofibrous mats are typically considered as two-dimensional (2D) materials since they have limited thickness (less than hundreds of micrometers), which may restrict their potential applications in flexible sensors. For example, three-dimensional (3D) materials (e.g., 3D conductive sponge) may be desired for wearable tactile/pressure sensors [5]. Recently, innovative methods were reported to convert the as-spun 2D nanofibrous mat into 3D monolithic sponge [46]. Fig. 9.2H shows a typical freeze-drying method. The as-electrospun nanofibers are first shortened/homogenized to short nanofibers, which are then dispersed in a solvent with added binders to form a slurry. Subsequently, the slurry is molded to a desired shape by freezing; eventually, the 3D nanofibrous sponge is obtained by freeze-drying under reduced pressure [47, 48].

9.3 Transduction mechanisms of tactile sensor

In principle, any detectable and reliable signal change in response to the action of tactile or pressure can be used as a transduction method for tactile sensors. The most widely adopted transduction methods are electrical, optical, and magnetic changes. In particular, the electrical signal changes including piezoresistive, capacitive, piezoelectric, and triboelectric are mostly used in the tactile sensor technologies. Therefore, the working principles and specific requirements on materials for these four transduction methods (Fig. 9.3) are briefly described in the following section.

Electrospun nanofibers for tactile sensors

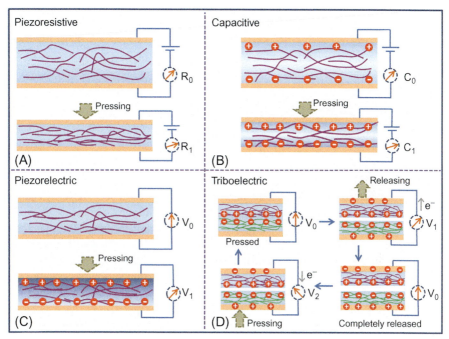

Fig. 9.3 Transduction mechanism of tactile sensors: (A) piezoresistive; (B) capacitive; (C) piezoelectric; and (D) triboelectric.

9.3.1 Piezoresistive transduction

The piezoresistive tactile sensor transduces external force/pressure into change of the resistance. Due to the simplicity of device assembly and signal readout, piezoresistive tactile sensors have recently attracted much attention. Conventionally, ductile metals and doped silicon are used as the active sensing element since their resistance (R) can change upon deformation, according to the equation:

$$R = \frac{\rho L}{A}$$

where ρ is the resistivity and L and A are the length and the cross-sectional area of a resistor, respectively. There are two drawbacks to limit these traditional materials for flexible and wearable applications: the mechanical rigidity and the limited strain range. Recently, flexible elastomeric composites and porous conductive materials have been explored for flexible piezoresistive tactile sensors, which are capable of detection of the large strain changes during human motions [5]. As shown in Fig. 9.3A, numerous temporary contact areas/points are formed in the flexible/porous materials upon external stimuli, leading to more conductive pathways and decrease of the resistance. Conductive electrospun nanofibers are intrinsically flexible and porous, and therefore can serve as good active sensing elements for piezoresistive tactile sensors.

9.3.2 Capacitive transduction

Capacitive tactile sensor is usually fabricated by sandwiching a dielectric layer between two conductive electrodes, forming a parallel-plate capacitor. The sensor measures the change of capacitance (C) of the capacitor, which is determined by the following equation:

$$C = \frac{\varepsilon_0 \varepsilon_r A}{d}$$

where ε_0 is the permittivity of vacuum, ε_r is the relative permittivity of the dielectric material, A is the overlapping area of the two electrodes, and d is the distance between the two electrodes (or thickness of the dielectric layer). Elastomeric dielectrics can be used as the active element of a capacitive tactile sensor. As shown in Fig. 9.3B, under external mechanical stimuli (e.g., compressing, bending, and shearing), the changes in A and/or d of the dielectric layer lead to capacitance change. To enhance the sensitivity of the sensor, the device can be modified by adding fillers to tune the permittivity (ε_r) of dielectric material or designing microstructures of the dielectric layer and/or electrodes. For example, microstructures such as microgroove, micro-dome, micro-pyramid, and micro-pillar can be used to optimize the contacts between the electrodes and dielectric layer and thus improve the sensitivity and response time of the tactile sensor [49–51]. In this aspect, electrospun nanofibrous mats are promising as dielectric layer materials for capacitive tactile sensors because of their inherent nanostructures, high porosity, and good flexibility.

9.3.3 Piezoelectric transduction

The piezoelectric tactile sensor is based on piezoelectric material, in which electric charges are generated upon mechanical deformation. The piezoelectric effect is closely related to the change of electric dipole moment, which either arises from the displacement of the ions' center in the material with no central symmetry [e.g., $BaTiO_3$, lead zirconate titanate (PZT), and ZnO] or from alignment of the permanent dipole moment of a material after a poling process (e.g., PVDF) [52]. The piezoelectric response is reversible and dynamic; more interestingly, the piezoelectric response does not need a power source, enabling a reliable self-powered tactile sensor or a piezoelectric nanogenerator for mechanical energy harvesting. As shown in Fig. 9.3C, when mechanical pressure is applied to the surface of a piezoelectric material, electric charges within the crystal are forced to be out of balance with excess negative and positive charges accumulated on the opposite sides of the crystal, resulting in a voltage difference at two electrodes and an electrical current through a circuit. The piezoelectric effect can be quantified using the piezoelectric coefficient (d) (unit: pC/N or m/V), which is experimentally calculated by the equation:

$$d = \frac{P}{\sigma}$$

where P is polarization (generated charges, pC/m^2) and σ is applied stress (N/m^2). Typically, the piezoelectric ceramics have much higher piezoelectric coefficient than the polymers such as PVDF, and thus the sensors based on the ceramics have higher sensitivity. However, the brittle and rigid nature makes the ceramics inferior for flexible tactile sensor. Therefore, polymer composites containing piezoelectric ceramics are used to develop highly sensitive tactile sensors. The electrospinning technique provides a simple and effective method to produce composite ceramic nanofibers by mixing fillers in the spinning solution. In addition, it has been proved that the electrospun PVDF nanofibers can be in situ electrically poled during the electrospinning process.

9.3.4 Triboelectric transduction

Since the triboelectric nanogenerator (TENG) was first invented by Prof. Wang's group in 2012 [53], the device has been investigated for self-powered sensors. The triboelectric tactile sensor operates on the triboelectric effect, which is induced from the contact or friction between two thin sheets with dissimilar triboelectric polarities. Specifically, as shown in Fig. 9.3D, when two different materials contact each other, friction between the two sheets promotes the charge transfer between the surfaces of the two materials, and equal amounts of opposite charges are accumulated at the two sides, forming a triboelectric potential layer at the interface region (step 1 in Fig. 9.3D). When the two materials are separated, compensating charges are built on the two electrodes due to electrostatic induction (step 2 in Fig. 9.3D). During the charge compensation, an electric current flows from the positive to the negative side through an external circuit until the accumulated charges are neutralized on the two electrodes (step 3 in Fig. 9.3D). Thereafter, when the two separated surfaces approach again, the current flows back from the negative to the positive side (step 4 in Fig. 9.3D). Consequently, the cyclic contact and separation of the two surfaces provide a continuous alternative current (AC) output signal between the two electrodes. Various electrospun nanofibers have been employed as triboelectric materials for wearable triboelectric tactile sensors owing to the advantages of large surface area, flexibility, lightweight, shape adaptivity, breathability, etc., of these materials.

9.4 Tactile sensors from electrospun nanofibrous materials

As described in the previous section, tactile sensors can be fabricated based on different transduction methods, and each type has specific requirements regarding the active materials. Besides the essential criteria, additional functions such as good flexibility, permeability, biocompatibility, and conformability are also desired for flexible and wearable applications. Electrospun nanofibers, which can be easily controlled and functionalized, meet these requirements; these materials are used not only as the active sensing element but also as the electrode [54, 55], spacer [56–58], packaging layer [59], etc. In this section, various electrospun nanofibers-based tactile sensors categorized by the transduction mechanism are discussed.

9.4.1 Piezoresistive tactile sensor

The sensitivity of a piezoresistive strain/tactile sensor can be assessed by the gauge factor (GF), which is calculated by the equation:

$$GF = \frac{\Delta R/R_0}{\varepsilon} \text{ or } \frac{\Delta R/R_0}{P}$$

where R_0 is the initial resistance of the conductive material, ΔR is the resistance change under external stimuli, and ε and P are the applied strain and pressure, respectively. It is intuitive to use electrospun nanofibrous materials for tactile sensors, as they are intrinsically compressible and flexible. In electrospun nanofibrous mats, numerous randomly overlaid nanofibers are interweaved and interlocked with the formation of a myriad of joints/contacts. Upon compression, more joints/contacts are temporarily built, leading to the change of resistance.

Piezoresistive tactile sensors can be fabricated by sandwiching an electrospun nanofibrous mat between two conductive electrodes. Herein, the electrospun nanofibers need to be conductive or semi-conductive to meet the demand [60–62]. Even though some conducting polymers like poly(3-hexylthiophene) (P3HT) can be directly electrospun to prepare conductive nanofibers [63, 64], the widely adopted method is to prepare polymer composite nanofibers by introducing conductive fillers, which are either added in the spinning solution prior to electrospinning [65, 66] or surface coated by post-modifications [60, 61].

Carbon materials including one-dimensional (1D) carbon nanotubes [67–69], 2D graphene [60, 61] and MXene [66] are attractive nanomaterials studied in the recent years. For example, Lou et al. reported reduced graphene oxide (rGO) encapsulated P (VDF-TrFe) nanofibers (denoted PVDF@rGO) for fabrication of an ultrasensitive piezoresistive tactile sensor [60]. The PVDF@rGO nanofibrous film was prepared by immersing an as-spun PVDF nanofibrous mat in a suspension of graphene oxide (GO), followed by adding hydrazine to reduce the GO to graphene (rGO) (Fig. 9.4A). The 2D graphene nanosheets could self-assemble to wrap on the surface of nanofibers (Fig. 9.4B). The tactile sensor was then fabricated by sandwiching the PVDF@rGO film between two PDMS films with Cu tapes as electrodes (Fig. 9.4C and D). The tactile sensor exhibited high sensitivity (15.6 kPa^{-1}), low detection limit (1.2 Pa), and long-term stability up to 100,000 cycles. The mechanism of the sensor is schematically proposed in Fig. 9.4E–G. Conductive carbonaceous nanofibers can also be prepared by high-temperature treatment on polymer nanofibers [70]; however, the carbonized nanofibers are usually brittle and less flexible.

Metal nanoparticles and conducting polymers are other commonly used conductive materials for coating on electrospun nanofibers. The metal nanoparticles can be coated on polymer nanofibers either by a physical deposition method (e.g., magnetron sputtering) [71] or through a chemical formation route. For example, silver (Ag)

Electrospun nanofibers for tactile sensors 169

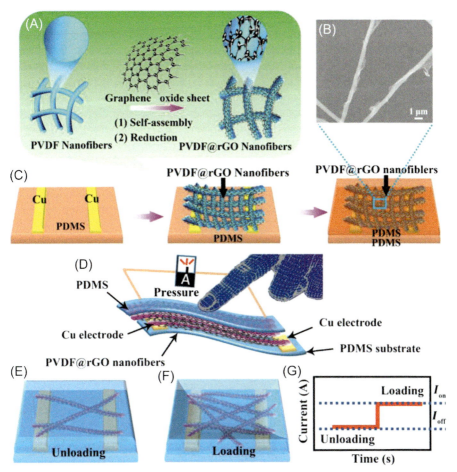

Fig. 9.4 A flexible, tactile sensor fabricated using conductive PVDF@rGO composite nanofibers as the active sensing element: (A) schematic illustration of the mechanism for the formation of PVDF fibers coated by the rGO nanosheets, followed by electrostatic interaction; (B) field emission scanning electron microscopy image of the morphology of rGO nanosheet-coated PVDF fibers; (C) schematic illustration of the fabrication of a flexible pressure sensor; (D) schematic of a typical pressure sensor; and (E–G) current changes in responses to loading and unloading.
Reprinted with permission from Z. Lou, S. Chen, L. Wang, K. Jiang, G. Shen, An ultra-sensitive and rapid response speed graphene pressure sensors for electronic skin and health monitoring, Nano Energy 23 (2016) 7–14. Copyright 2016, Elsevier.

nanoparticles can be deposited on nanofibers by treating them in the precursor $AgNO_3$ solution followed by in situ reduction using a reducing chemical such as hydrazine and borane [72–74]. Similarly, conducting polymers can be coated on nanofibers by a direct dip-coating process or in situ polymerization. Polyaniline (PANi), polypyrole (PPy), and poly(3,4-ethylenedioxythiophene) (PEDOT) are the most easily synthesized conducting polymers, and their powders/suspensions are also commercially available. To synthesize these conducting polymers, in a typical process, the catalyst/oxidant [e.g., FeCl$_3$, ammonium persulfate (APS)] is first loaded in the polymer nanofibers, followed by immersing the nanofibers in the monomer solution (e.g., aniline, pyrrole, and EDOT) (in situ solution polymerization route) or exposing in the vapor of monomers (vapor deposition polymerization route) [54, 59, 75, 76]. As shown in Fig. 9.5A, the as-spun PVDF nanofiber yarn was first dipped in the FeCl$_3$ solution, and then it was immersed in a solution of EDOT in chloroform for 20 h for in situ polymerization [75]. Thereafter, the conductive PEDOT@PVDF nanofibers were woven to a fabric as a tactile sensor. As shown in Fig. 9.5B, the oxidant FeCl$_3$ was directly mixed in the solution for electrospinning; then, the as-spun composite nanofibers were placed in a vacuum chamber injected with EDOT monomer, and PEDOT was polymerized on the surface of nanofibers by heating the reactor at 70 °C for 5 h [76].

Compared with the as-spun 2D nanofibrous mats/sheets or as-twisted 1D nanofiber yarns, the 3D monolithic sponges have much larger compression strain and thus have wider detection range. Electrospun nanofibers have been used to prepare 3D

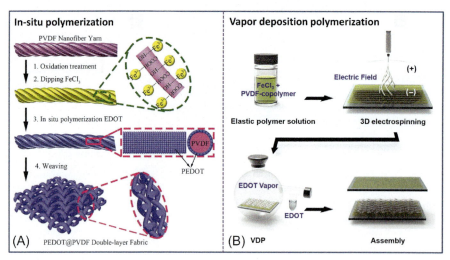

Fig. 9.5 Polymerization of PEDOT polymer on electrospun nanofibers through two different methods: (A) schematic illustration for polymerization of PEDOT on PVDF nanofiber yarn by in situ polymerization. (B) Schematic illustration for polymerization of PEDOT on electrospun nanofibers by vapor deposition polymerization (VDP).
Panel A: Reprinted with permission from Y. Zhou, J. He, H. Wang, K. Qi, N. Nan, X. You, W. Shao, L. Wang, B. Ding, S. Cui, Highly sensitive, self-powered and wearable electronic skin based on pressure-sensitive nanofiber woven fabric sensor, Sci. Rep. 7(1) (2017) 12949. Copyright 2017, Nature Publishing Group. Panel B: Reprinted with permission from O.Y. Kweon, S.J. Lee, J.H. Oh, Wearable high-performance pressure sensors based on three-dimensional electrospun conductive nanofibers, NPG Asia Mater. 10(6) (2018) 540–551. Copyright 2018, Nature Publishing Group.

conductive sponge/aerogel for tactile sensor applications [77–79]. For example, Si et al. [78] reported an ultralight and elastic conductive nanofibrous aerogel by freeze-drying of the homogenized nanofibers with addition of a biomass, konjac glucomannan (KGM), followed by high-temperature carbonization. In another example, Xu et al. [79] reported a 3D conductive sponge prepared by freeze-drying of multicomponent electrospun nanofibers. As shown in Fig. 9.6, the 3D conductive sponge was prepared by assembly of shortened nanofibers of polyacrylonitrile (PAN), polyimide (PI), and carbon (CNFs), in which the PAN and PI nanofibers served as the scaffold component and the CNFs served as the conductive component. The conductivity of the 3D sponge could be adjusted by varying the content of CNFs, and thus the sensitivity of the as-fabricated tactile sensors could be tuned correspondingly.

Besides using the resistance change from the deformation of the active materials as the transduction mechanism, piezoresistive tactile sensors can be designed by controlling the contact resistance between two electrodes [80–82]. Generally, rationally designed microstructured patterns such as pyramid arrays, microgrooves, and micro-domes are fabricated on the surface of the electrodes. When the two electrodes are packaged, the microstructures of the two electrodes are in contact, forming conductive pathways; such conductive pathways are sensitive to the external compression. However, in this type of device, the microstructures are very small and prone to permanent deformation upon continuous loading, resulting in decrease of sensitivity or loss of function. An insulating spacer (e.g., polymer bars, dots, or a mesh) between the two electrodes may help alleviate the issue. Electrospun nanofibers, due to their porous structure and readily controlled thickness and size, may be ideal to serve as the insulation spacer between two microstructured electrodes [56–58]. Gao's group [57] prepared an ultrahigh-sensitive and very easily detectable piezoresistive pressure sensor based on wrinkled PPy films as electrodes and polyvinyl alcohol (PVA) nanofibers as a spacer. Similarly, a pressure sensor consisting of a layer of PVA nanofibers sandwiched between an ultrathin, wrinkled graphene film and a pair of interdigital electrodes (IDE) was demonstrated, as shown in Fig. 9.7 [58]. The IDE served as soft substrate and provided the conductive path, the wrinkled graphene films with hierarchical microstructures served as the active sensing element, and the PVA nanofibers served as a spacer layer between graphene film and interdigital electrode. The as-prepared pressure sensor achieved a high sensitivity of $28.34 \ kPa^{-1}$ due to the synergistic effect between wrinkled graphene film and interconnected PVA nanofibers that all "point-to-point," "point-to-face," and "face-to-face" contact modes were formed.

9.4.2 Capacitive tactile sensor

Different from the piezoresistive tactile sensor that uses a conductive material, the capacitive tactile sensor uses a dielectric layer as the active sensing element, which is sandwiched between two electrodes to form a parallel-plate capacitor. As discussed in Section 9.3.2, the measured signal, capacitance, is a function of the overlapping area of the two electrodes, the thickness of the dielectric layer, and the dielectric permittivity of the dielectric material. Therefore, a deformable elastomer is usually used as

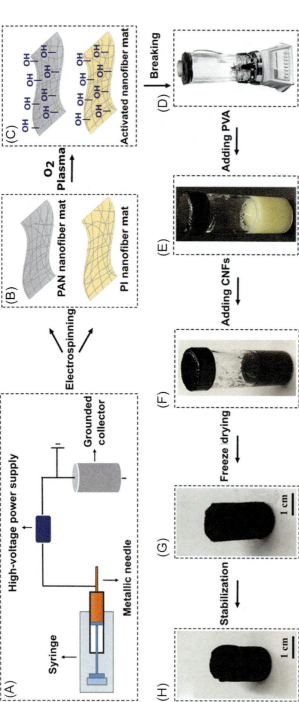

Fig. 9.6 A scheme showing the steps for the preparation of 3D conductive nanofibrous sponge. (A) Electrospinning setup; (B) the obtained PAN and PI nanofibrous mats; (C) oxygen plasma-activated nanofibrous mats; (D) PAN and PI suspension made by high-speed blender; (E) dispersion of the PAN/PI nanofibers in PVA aqueous solution; (F) mixture containing PAN/PI nanofibers, PVA, and CNFs; (G) freeze-dried 3D conductive sponge; and (H) stabilized 3D conductive sponge.

Reprinted with permission from T. Xu, Y. Ding, Z. Wang, Y. Zhao, W. Wu, H. Fong, Z. Zhu, Three-dimensional and ultralight sponges with tunable conductivity assembled from electrospun nanofibers for a highly sensitive tactile pressure sensor, J. Mater. Chem. C 5(39) (2017) 10288–10294. Copyright 2017, Royal Society of Chemistry.

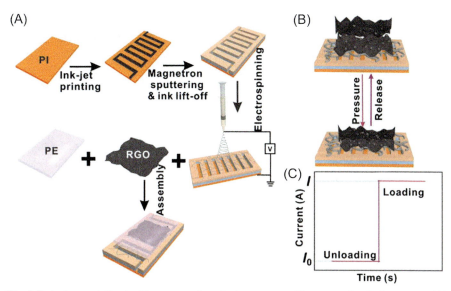

Fig. 9.7 A piezoresistive tactile sensor using electrospun nanofibers as an insulation spacer. (A) The schematic of the fabrication of flexible piezoresistive sensor; (B) the sensing mechanism of as-prepared pressure sensor; and (C) current variation in response to loading and unloading. Reprinted with permission from W. Liu, N. Liu, Y. Yue, J. Rao, F. Cheng, J. Su, Z. Liu, Y. Gao, Piezoresistive pressure sensor based on synergistical innerconnect polyvinyl alcohol nanowires/wrinkled graphene film, Small 14(15) (2018) 1704149. Copyright 2018, John Wiley & Sons, Inc.

the dielectric layer for capacitive tactile sensors. Since the dense elastomer film has very limited deformability, the sensitivity is not sufficient for wearable applications. To improve the deformability/sensitivity of the dielectric layer, air voids can be introduced into the dielectric layer, i.e., using porous materials as the dielectric layer. Electrospun polymeric nanofibrous mats, therefore, are promising dielectric layers for capacitive tactile sensors due to their high porosity.

Many polymer nanofibers, including high dielectric P(VDF-TrFE) nanofibers [83], thermoplastic polyurethane (TPU) nanofibers [84], and biodegradable polylactic-*co*-glycolic acid (PLGA) and polycaprolactone (PCL) nanofibers, have been reported as the dielectric layer of the tactile sensors [85]. Yang et al. [84] fabricated an all-nanofiber-based tactile sensor. As shown in Fig. 9.8, the tactile sensor was assembled using two PVDF nanofiber membranes (PVDFNM) screen printed with silver nanowires (Ag NWs) as electrodes and a TPU nanofiber membrane (TPUNM) as the dielectric layer. Owing to the intrinsic advantages of electrospun nanofibers, the skin-like capacitive tactile sensor showed high sensitivity, good flexibility, air permeability, high hydrophobicity and breathability, and lightweight.

While the sensitivity of capacitive tactile sensors has been significantly improved using electrospun polymer nanofibers, the sensitivity can be further enhanced by increasing the permittivity of the polymer nanofibers. The

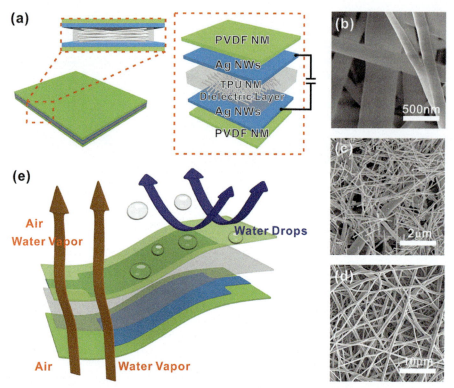

Fig. 9.8 A breathable skin-like capacitive tactile sensor fabricated using electrospun nanofibers as a dielectric layer. (A) Structure of the skin-like tactile sensor based on nanofiber membrane (NM); (B) porous surface of the PVDFNM in SEM; (C) interlaced Ag nanowires on the surface of the PVDFNM in SEM; (D) porous surface of the TPUNM; and (E) the illustration of waterproof and breathable property of the sensor.
Reprinted with permission from W. Yang, N.-W. Li, S. Zhao, Z. Yuan, J. Wang, X. Du, B. Wang, R. Cao, X. Li, W. Xu, Z.L. Wang, C. Li, A breathable and screen-printed pressure sensor based on nanofiber membranes for electronic skins, Adv. Mater. Technol. 3(2) (2018) 1700241. Copyright 2018, John Wiley & Sons, Inc.

permittivity of polymers may be enhanced by adding a small amount of conductive fillers [86–88]. For example, Yang et al. [89] studied the effect of CNT addition to the PVDF nanofibers on the performance of the capacitive tactile sensor based on the nanofiber's dielectric properties. Evidently, addition of CNTs increased the dielectric permittivity of PVDF nanofibers, which resulted in varied sensitivity of the tactile sensor by the amount of CNTs in the nanofibers. The permittivity of PVDF nanofibers increased from 3.8 to 12.6 with adding only 0.05 wt% CNTs in nanofibers; correspondingly, the sensitivity of the sensor increased from 0.09 to 0.99 kPa^{-1}. Ionic liquids (ILs) or hydrogels can also be introduced into electrospun nanofibers to improve the sensitivity of the tactile sensor due to their high ionic conductivity [90–92]. However, the water in hydrogels is easily lost under compression; to overcome the issue, core-shell nanofibers by encapsulating ionic hydrogels as the core component may prevent the evaporation of water [92].

Textile-based devices can further improve the flexibility of tactile sensors. Nanofiber yarns can be woven into designed device structures to fabricate textile-based sensors [55, 71, 93, 94]. You et al. [93] prepared core-spun yarn by coating graphene/PU composite nanofibers (as a dielectric layer) on the surface of Ni-coated cotton yarn (as an electrode); the core-spun yarn was then wound helically around the surface of an elastic thread and finally woven into a piece of fabric for tactile sensing. Wang et al. [55] used Cu/Ni nanoparticles-coated polyvinyl butyral nanofibers as electrodes for tactile sensors (Fig. 9.9A). Wu et al. [71] fabricated conductive electrodes on a fabric by transfer printing with electrospun PVA nanofibers as a template to fabricate an all-textile tactile sensor. As shown in Fig. 9.9B, Ag nanoparticles were first deposited on water-soluble PVA nanofibers by magnetron sputtering, and then the free-standing nanofibrous mat was transferred to a fabric, followed by removing the PVA nanofibers using deionized water.

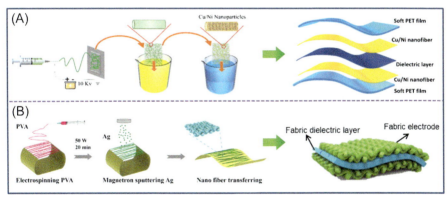

Fig. 9.9 Fabrication of conductive nanofibrous fabrics as electrodes for capacitive tactile sensors. (A) Schematic diagram of the manufacturing process of the Cu/Ni nanofiber flexible electrode of the sensor. (B) Preparation process of conductive fabric electrodes: fabricating PVA NFs by electrospinning and then magnetron sputtering of Ag on the fiber surface, followed by transferring the conductive NFs on the fabric surface.

Panel A: Reprinted with permission from J. Wang, R. Suzuki, M. Shao, F. Gillot, S. Shiratori, Capacitive pressure sensor with wide-range, bendable, and high sensitivity based on the bionic Komochi Konbu structure and Cu/Ni nanofiber network, ACS Appl. Mater. Interfaces 11(12) (2019) 11928–11935. Copyright 2019, American Chemical Society. Panel B: Reprinted with permission from R. Wu, L. Ma, A. Patil, C. Hou, S. Zhu, X. Fan, H. Lin, W. Yu, W. Guo, X.Y. Liu, All-textile electronic skin enabled by highly elastic spacer fabric and conductive fibers, ACS Appl. Mater. Interfaces 11(36) (2019) 33336–33346. Copyright 2019, American Chemical Society.

9.4.3 Piezoelectric tactile sensor

Piezoelectric materials are able to harvest energy from mechanical deformations in our daily movement, enabling the self-powered sensing of human motions. Inorganic ceramic materials such as $BaTiO_3$, PZT, and BZT-BCT are traditional piezoelectric materials with high piezoelectric coefficients. The nanofibers of these materials can be prepared by electrospinning [95, 96]; however, the nanofibrous mats are very brittle to be used for flexible devices. In contrast, piezoelectric polymeric materials are promising for flexible device applications due to their good mechanical flexibility, lightweight, and easy processing. Nanofibers of poly(vinylidene fluoride) (PVDF) and its copolymers are representative piezoelectric polymeric nanofibers that have been demonstrated for piezoelectric tactile sensor application [13]. Because the piezoelectric effect of pristine PVDF is relatively low, various methods have been applied to enhance the piezoelectric effect of PVDF nanofibers.

9.4.3.1 Electrospun nanofibers of PVDF and PVDF copolymers

PVDF and its copolymers/derivatives, poly(vinylidene fluoride-*co*-trifluoroethylene) (PVDF-TrFE) and poly(vinylidene fluoride-co-hexafluoropropene) (PVDF-HFP), are partially fluorinated semicrystalline polymers with linear molecular structures (Fig. 9.10A–C). The piezoelectric effect of PVDF is originated from the electric polarization of dipole moments in anisotropic crystalline PVDF monomers. PVDF has five crystalline phases (α, β, γ, δ, and θ) depending on the chain conformations as trans (T) or gauche (G) linkages. Among them, the α-, β-, and γ-phases are common ones in the structure; Fig. 9.10D schematically shows their representative chain conformations [13]. The symmetrically structured α-phase (TGTG') is most thermodynamically stable, but it is nonpolar; thus, it does not show a piezoelectric effect. The β (TTTT) and γ (TTTGTTTG') phases are polar, while the piezoelectric effect of the β-phase is much higher since it has higher remnant polarization and has the highest dipolar moment per unit cell (8×10^{-30}Cm) [97]. Therefore, it is essential to increase the content of the β-phase in PVDF material to have pronounced piezoelectric effects for tactile sensor applications.

The PVDF thin films prepared by conventional solution casting methods have low content of the β-phase, which can be improved by post treatments such as mechanical stretching, thermal annealing, electric poling, and adding fillers [13]. PVDF nanofibers prepared by the electrospinning technique have much higher piezoelectric effect than the film counterparts [98]. It has been confirmed that the electrospun PVDF nanofibers have increased crystallinity and content of the β-phase [99]. Because of the applied high voltage during electrospinning, the PVDF (or copolymers) solutions are subjected to high-ratio stretching, facilitating the formation of the β-phase. In addition, the high electric field provides a self-poling process during electrospinning [40, 41]. Mandal et al. [100] have experimentally verified that the electrospinning process can facilitate the preferential orientation of molecular dipoles, during which the trans-zig-zag chains of P(VDF-TrFE) orient preferentially parallel to the collector

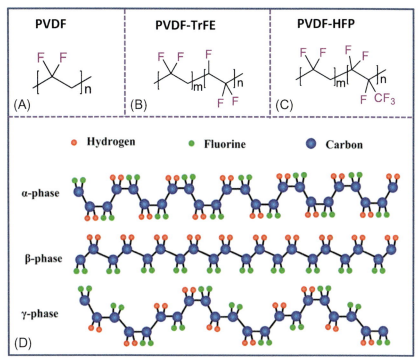

Fig. 9.10 Chemical structures of (A) PVDF, (B) PVDF-TrFE, (C) PVDF-HFP, and (D) the chain conformation for the α-, β-, and γ-phases of PVDF.

rotation direction, and the preferential orientations of the electroactive CF_2 dipoles are perpendicular to the trans-zig-zag chains (Fig. 9.11).

The β-phase content in PVDF nanofibers can be further improved by adjusting the parameters in the electrospinning process. For example, PVDF nanofibers demonstrated good piezoelectric effect with good morphology and uniform thickness, tuned by controlling the solution concentration and the collection distance [101]. The β-phase can also be increased by reducing the nanofiber diameter by adjusting the needle size and solution feeding rate [102]. Thermal treatment on the as-spun PVDF nanofibers may enhance the piezoelectric effect as well [103]. PVDF derivatives with different chemical structures can further improve the piezoelectric effects. Ren et al. [104] investigated the effect of the percentage of TrFE units in the PVDF-TrFE copolymer; they prepared three kinds of PVDF-TrFE (100/0, 55/47, and 77/23) nanofibers by electrospinning, and the results showed that the PVDF-TrFE (77/23) had the highest level of the β-phase and the highest piezoelectric effect.

Owing to the excellent piezoelectric properties, nanofibers of PVDF and its copolymers have been extensively studied for various applications such as the piezoelectric nanogenerator for energy harvesting [105, 106], self-powered sensors for human activity monitoring [107], electronic skin [108, 109], and integrated sensors for structural health monitoring [110]. Lang et al. [111] prepared a highly sensitive acoustic

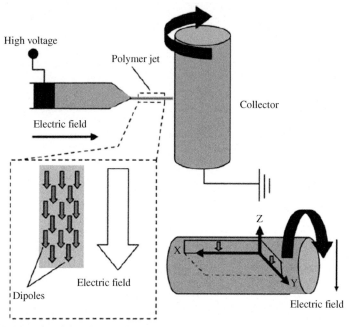

Fig. 9.11 Schematic of the electrospinning experimental setup (*top*), enlarged view of the induced dipoles in the polymer jet (*bottom left*), and the collector, showing the resultant induced dipole direction and the rotation direction (*bottom right*).
Reprinted with permission from D. Mandal, S. Yoon, K.J. Kim, Origin of piezoelectricity in an electrospun poly (vinylidene fluoride-trifluoroethylene) nanofiber web-based nanogenerator and nano-pressure sensor, Macromol. Rapid Commun. 32(11) (2011) 831–837. Copyright 2011, John Wiley & Sons, Inc.

sensor using electrospun PVDF nanofibers; the sensor could detect low-frequency (60–110 dB) sound with a sensitivity as high as 266 mV Pa^{-1}. In addition to high sensitivity, stability is important for tactile sensors to have steady applications. The contacts between the piezoelectric sensing element and electrodes, as well as the packaging of the device, are critical issues for electrospun nanofibers. To address these issues, liquid metals or microstructured materials (e.g., micro-bead arrays) have been used as the electrodes, and vacuum packaging process may be applied to pack the device [112, 113]. An all-in-one nanofibrous membrane, in which the opposite sides of a PVDF membrane are covered by conductive materials (e.g., by spraying or in situ polymerization) as electrodes, can help solve the contact issues to fabricate stable tactile sensor device [114, 115]. Twisting the electrospun nanofibers to yarns and then integrating the yarns into textile provides another effective method to fabricate flexible and robust devices [116].

9.4.3.2 Aligned electrospun PVDF nanofibers

The piezoelectric properties of electrospun PVDF nanofibers can be appreciably enhanced when the nanofibers become highly oriented [117–119], since the aligned fibers along the stress direction provide uniformly directional dipolar moments and make full use of d_{33}. Persano et al. [119] first studied the piezoelectric performance of well-aligned PVDF-TrFE nanofibers collected using a fast-rotating disc (4000 rpm), and the tactile sensor exhibited exceptional piezoelectric sensitivity (1.1 V kPa^{-1}) with the detection limit as low as 0.1 Pa. Sharma et al. [120] reported aligned electrospun core-shell PVDF-TrFE nanofibers, which showed significantly enhanced piezoelectric tactile sensor performance with sensitivity 40 times higher than that of a device based on thin-film PVDF.

Besides the parallel-plate or high-speed rotating disc as the electrospinning collector (as shown in Fig. 9.1C), various types of novel collectors have been designed to collect aligned nanofibers. For example, Hsu et al. [121] reported a collector consisting of serrated edges, which allows simultaneous collection of multiple aligned PVDF-TrFE nanofiber bundles with controlled separation. Near-field electrospinning provides a method to directly pattern aligned nanofibers [122–124]. As shown in Fig. 9.12A, well-aligned PVDF nanofibers can be arbitrarily written on a piece of paper [124]; such technique enables the design of desired patterns of nanofibers on various platforms. Mechanical stretching is a straightforward and effective method to induce the alignment of electrospun nanofibers [125]. As shown in Fig. 9.12B, the as-spun randomly distributed PVDF-TrFE nanofibrous mat was dynamically stretched up to 200%–800% strains, and the nanofibers were globally aligned after stretching. These aligned PVDF-TrFE nanofibers exhibited an enhanced piezoelectric property that the average output voltage was about 266% of the value of the original random nanofibers.

9.4.3.3 Electrospun composite/doped PVDF nanofibers

Composite nanofibers of inorganic piezoelectric material and PVDF may combine the advantages of both the good flexibility of PVDF and high piezoelectric effect of inorganic materials. Several inorganic piezoelectric materials including ZnO nanoparticles [126], ZnO nanowires/nanorods [127, 128], BaTiO$_3$ nanoparticles [129–131], BaTiO$_3$ nanowires [132], and novel lead-free ceramic nanoparticles (e.g., BNT-ST (0.78Bi$_{0.5}$Na$_{0.5}$TiO$_3$–0.22SrTiO$_3$) [133, 134], BZT-BCT (Zr$_{0.2}$Ti$_{0.8}$) O$_3$–0.5(Ba$_{0.7}$Ca$_{0.3}$)TiO$_3$) [135]) have been used to prepare PVDF composite nanofibers. Addition of the inorganic nanomaterials increases the piezoelectric effect, attributing to the intrinsic high piezoelectric coefficient of the inorganic materials. Furthermore, the added nanoparticles act as nucleating agents to induce formation of the β-phase of PVDF. Nanoclay [136] and conductive nanomaterials (e.g., Ag nanoparticles [137], CNTs [138], and graphene [139, 140]) can also work as the nucleating agents in the preparation of PVDF composite nanofibers to increase the piezoelectric effect. In this case, it should be noted that the amount of the conductive fillers needs to be below the percolation threshold to avoid the formation of conductive composite nanofibers. Synergistic enhancement can be achieved by adding multiple fillers such as a combination of ceramic nanoparticles and conductive fillers [141–144]. Zhu et al. [145] prepared core-shell piezoelectric fibers by coaxial electrospinning (Fig. 9.13A), in which the outer and inner components were PVDF-graphene oxide (PVDF-GO) nanosheets and PVDF-BaTiO$_3$ nanoparticles (BTO NPs), respectively. The core-shell nanofibers showed enhanced piezoelectric properties due to the

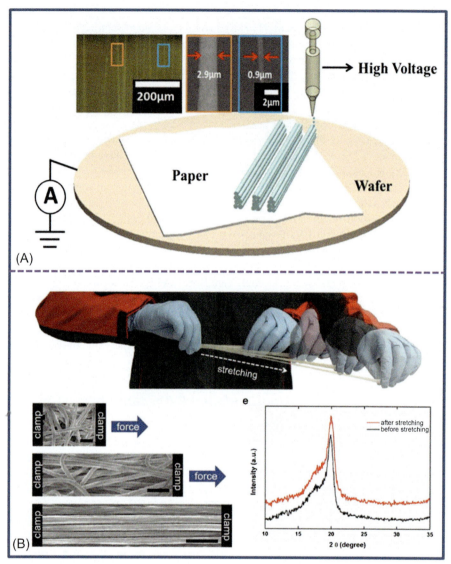

Fig. 9.12 Methods to obtain aligned electrospun nanofibers for piezoelectric tactile sensors. (A) Schematic setup of the sequentially stacked 3D fiber structure via near-field electrospinning process. (B) Images of electrospun P(VDF-TrFE) fibers during mechanical stretching process under axial stress (scale bar is 5 μm) and XRD patterns before and after the mechanical stretching process.

Panel A: Reprinted with permission from Y.K. Fuh, B.S. Wang, Near field sequentially electrospun three-dimensional piezoelectric fibers arrays for self-powered sensors of human gesture recognition, Nano Energy 30 (2016) 677–683. Copyright 2016, Elsevier. Panel B: Reprinted with permission from S. Ma, T. Ye, T. Zhang, Z. Wang, K. Li, M. Chen, J. Zhang, Z. Wang, S. Ramakrishna, L. Wei, Highly oriented electrospun P(VDF-TrFE) fibers via mechanical stretching for wearable motion sensing, Adv. Mater. Technol. 3(7) (2018) 1800033. Copyright 2018, John Wiley & Sons, Inc.

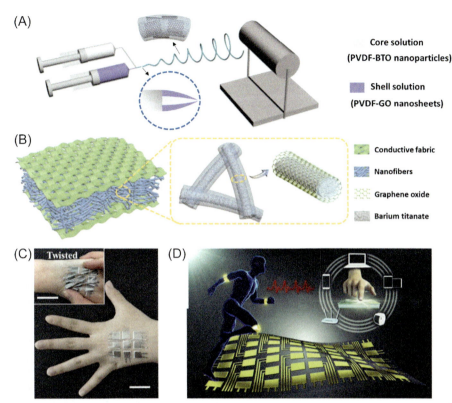

Fig. 9.13 Material fabrication and structure design of the highly shape-adaptive electronic skin. (A) Schematic illustration of the experimental setup for coaxial electrospinning process; (B) schematic illustration of a single unit of electronic skin, the magnification is partially enlarged view of the core-shell piezoelectric nanofibers; (C) optical photograph of an electronic skin (3 × 3 pixels) conformably attached on the back of the hand (scale bar is 3 mm); and (D) illustration of the proof-of-concept of fabricated electronic skin for joint motion monitoring and tactile sensing.
Reprinted with permission from M. Zhu, M. Lou, I. Abdalla, J. Yu, Z. Li, B. Ding, Highly shape adaptive fiber based electronic skin for sensitive joint motion monitoring and tactile sensing, Nano Energy 69 (2020) 104429. Copyright 2020, Elsevier.

synergistic effect of high piezoelectric inorganic BTO NPs in the core and the drastically polarized GO-doped PVDF in the shell. A shape-adaptive electronic skin was assembled by sandwiching the core-shell nanofibers between two pieces of conductive fabrics and packaged by elastic polyurethane film (Fig. 9.13B). The highly sensitive tactile sensor arrays had high mechanical deformability and could be mounted conformably onto arbitrarily curved surfaces to detect various human motions (Fig. 9.13C).

The addition of inorganic fillers in the solution of PVDF to prepare electrospun composite nanofibers suffers from several drawbacks. First, dispersion of additives in the polymer solution requires complicated processes such as sonication and surface modification, which increases the cost and limits its commercial and practical applications. Second, agglomeration and particle size effect may present challenges for manufacturing and material properties. Third, a high loading amount of the inorganic fillers would result in poor flexibility and mechanical strength. These issues may be overcome using a soluble inorganic salt [e.g., $Al(NO_3)_3$] [146] or complex (e.g., Ce^{3+} complex) [147] as precursors to form homogeneous PVDF solution. Yu et al. [148] prepared PVDF nanofibers doped with inorganic salts and investigated the effect of salt type and concentration on the piezoelectric effect of the PVDF nanofibers. The results indicated that unionized salt molecules with low dipole moment in electrospinning solvent would facilitate transformation of the amorphous polymer phase to the piezoelectric phase; however, the electrospinning process was negatively affected if a high concentration of ions existed in the spinning solution.

9.4.3.4 Electrospun nanofibers of other piezoelectric polymers

In theory, semicrystalline polymers may be piezoelectric if their molecular dipoles and/or crystalline regions can be induced to be aligned by the poling process. Polyacrylonitrile (PAN) is a semicrystalline polymer with nitrile groups, and the isotactic PAN has high alignment of dipoles. However, the large nitrile dipoles are usually difficult to be poled in the unstretched state unless the thin film is stretched many times. Street et al. [149] reported that a urea clathrate-polymerized PAN showed piezoelectric property when processed to nanofibers by electrospinning, which exhibited about 30% of the piezoelectric response of PVDF-TrFE. Biopolymers containing numerous intramolecular hydrogen bonds between amino (NH) and carboxyl (COO) groups form α-helical chain conformation easily and, therefore, may be piezoelectric. Piezoelectric biopolymer nanofibers like poly(γ-benzyl-α,L-glutamate) (PBLG) [150], gelatin [151], and synthetic spider silk [152] have been reported recently. Tactile sensors based on piezoelectric biopolymer nanofibers have the advantages of biocompatibility and biodegradability for human-monitoring applications [52].

9.4.4 Triboelectric tactile sensor

Four working modes of triboelectric nanogenerator (TENG) have been invented, i.e., the single-electrode, vertical contact, lateral sliding, and free-standing modes [153]. Among them, the single-electrode and vertical contact mode TENG can be developed for tactile sensors as they work by contact changes, whereas the lateral sliding and free-standing mode TENGs are usually designed for detection changes of bending, sliding, or moving. In the following sections, the single-electrode mode and the vertical contact mode tactile sensors made from electrospun nanofibers are discussed.

9.4.4.1 Single-electrode mode

The single-electrode TENG requires only one electrode, which is connected to the ground. The electrode serves not only as an electrode but also as a triboelectric surface in contact with another dielectric triboelectric material. When the electrode contacts the triboelectric material, charges are generated on the surface of the triboelectric material. The surface charges then pull or push electrons from the ground to the electrode depending on whether the generated charges are positive or negative, which is determined by the nature (triboelectric positive or negative) of the triboelectric material. To fabricate a nanofiber-based single-electrode triboelectric tactile sensor, one nanofibrous material is needed to serve as the triboelectric sensing element. For example, electrospun polyetherimide (PEI) [154], polyimide (PI) [155], PVDF-TrFE [156], and PVDF-HFP [157] nanofibers have been used as the triboelectric layer for the fabrication of single-electrode triboelectric tactile sensors. Nanofibrous materials as both the electrode and triboelectric layer have been used to fabricate all-nanofiber-based devices [158–160]. Li et al. [160] presented an all-fiber structured elastic and breathable electronic skin based on single-electrode mode triboelectric tactile sensor using electrospun nanofibers as both the electrode and triboelectric layer. As shown in Fig. 9.14, three layers of nanofibrous membranes of carbon nanofibers (carbon NFs), PVDF nanofibers (PVDF NFs), and polyurethane nanofibers (PU NFs) were used as the electrode, sensing layer, and substrate layer, respectively. This all-fiber single-electrode mode triboelectric tactile sensor demonstrated an excellent sensitivity of 0.18 V kPa^{-1} in the pressure range of 0–175 kPa.

9.4.4.2 Vertical contact mode

The vertical contact mode of TENG requires two different layers of triboelectric materials with different triboelectric polarity in the triboelectric series. The detailed working mechanism is described in Fig. 9.3D. Typically, the higher the difference between the triboelectric positive (+) and triboelectric negative (−) materials (i.e., the higher the rank difference in the triboelectric series), the higher the output signal of the as-fabricated device. To fabricate a vertical contact mode tactile sensor, an interlayer spacer is usually required at the edges of the device in order to create a gap between the two layers of triboelectric materials for reversible contact and separation. Nanofibers coupling with polymer membrane/film or two electrospun nanofibers with different ranks in the triboelectric series, such as polyamide 6 (PA6) membrane (+)/PTFE nanofibers (−) [161], PA nanofibers (+)/PI aerogel (−) [162], PEI aerogel (+)/PVDF nanofibers (−) [163], TPU nanofibers (+)/PVDF nanofibers (−) [164], poly(3-hydroxybutyrate-*co*-3-hydroxyvalerate) (PHBV) nanofibers (+)/PVDF nanofibers (−) [165], PLA nanofibers (+)/PVDF-HFP nanofibers [45], and PVP fibers (+)/PVDF nanofibers [166], have been assembled into vertical contact mode triboelectric tactile sensors. As shown in Fig. 9.15, Li et al. [167] prepared two electrospun nanofibers, PVDF and PAN nanofibers, which were further modified by coating with PDMS or PA6 to tailor their triboelectric polarity, mechanical strength, and hydrophobic properties. A hydrophobic, flexible, and wearable TENG tactile sensor was then fabricated by packaging the two nanofibrous mats. The triboelectric tactile sensors are promising, next-generation, self-powered wearable electronics for various applications [168].

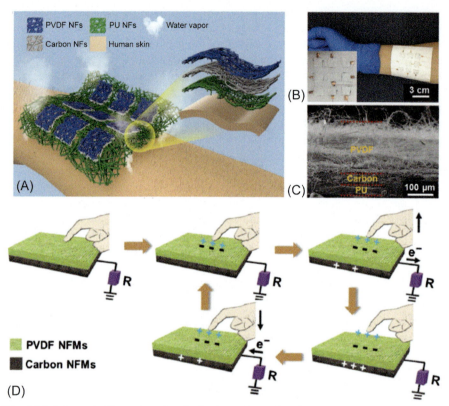

Fig. 9.14 Schematic illustration of a single-electrode mode triboelectric tactile sensor. (A) Schematic illustration of a 3 × 3 pixel electronic skin. Insets: partially enlarged view of a single electronic skin pixel; (B) optical photograph of a 3 × 3 pixel electronic skin conformably attached on the arm, the inset is an electronic skin optical picture; (C) SEM image of the cross-sectional view of the electronic skin; and (D) sensing mechanism of the electronic skin as a pressure sensor.

Reprinted with permission from Z. Li, M. Zhu, J. Shen, Q. Qiu, J. Yu, B. Ding, All-fiber structured electronic skin with high elasticity and breathability, Adv. Funct. Mater. 30(6) (2020) 1908411. Copyright 2020, John Wiley & Sons, Inc.

9.5 Conclusions and perspective

Many types of tactile sensors have been developed based on different transduction mechanisms including piezoresistivity, capacitance, piezoelectricity, and triboelectricity. The newly developed versatile electrospinning technique provides an effective tool to produce various nanofibers with advantages of ultrasmall diameter, large specific area, high porosity, good flexibility, etc. These electrospun nanofibers can be directly used or further modified/functionalized for the fabrication of tactile sensors. This chapter presents an overview of the electrospinning technique, the features of

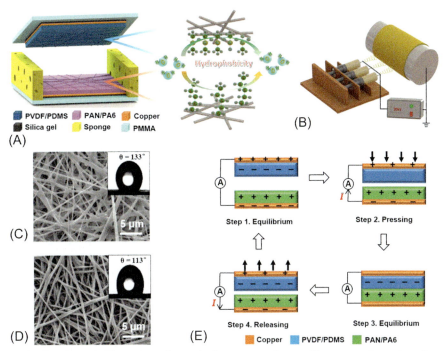

Fig. 9.15 Schematic illustration of a vertical contact mode nanofibrous membrane-based triboelectric nanogenerator (NM-TENG). (A) Structure design of the NM-TENG; (B) schematic illustration of the electrospinning for nanofibers' synthesis; (C) SEM image of PVDF/PDMS nanofibrous membrane etched by NaOH; (D) SEM image of the PAN/PA6 nanofibrous membrane etched by HCl; and (E) electricity generation mechanism of the NM-TENG.
Reprinted with permission from Z. Li, J. Shen, I. Abdalla, J. Yu, B. Ding, Nanofibrous membrane constructed wearable triboelectric nanogenerator for high performance biomechanical energy harvesting, Nano Energy 36 (2017) 341–348. Copyright 2017, Elsevier.

nanofibers, various methods to modify as-spun nanofibers, and the recent advances in various types of tactile sensors based on electrospun nanofibrous materials. These electrospun nanofibers-based tactile sensors showed excellent performance including high sensitivity, low detection limit, and more importantly, good flexibility and conformability; therefore, they have been demonstrated in various wearable applications such as human motion detection, electronic skin, and soft robots. It may be envisioned that the electrospun nanofibrous mat will have broad applications in the coming era of intelligent and smart devices and systems.

Several imminent challenges and issues need to be further studied for the practical applications of the nanofiber-based tactile sensors. First, even though various nanofibers can be produced by the electrospinning process, the process needs to be further optimized to control the quality of the produced nanofibers, including uniform fiber diameter, size and thickness, morphology, rationally designed shapes, and deposition on various platforms such as wearable textiles or the human body. Second, the product throughput of the electrospinning technique is rather low, particularly in preparation of the well-aligned nanofibers; such low throughput largely limits the scalable application of the nanofibers. Third, the material requirements for tactile sensors based on different working mechanisms are very different. For the piezoresistive tactile sensor, the nanofibers need to be conductive, but if the conductivity is too high, the sensitivity becomes low. Therefore, the conductivity and structure of the nanofibrous mat must be well controlled and optimized. For the other three types of tactile sensors, the nanofibers cannot be conductive and require high dielectric permittivity, high piezoelectric coefficient, or high triboelectric effect. These properties are intrinsically poor for polymeric materials. To improve these desired properties, composite nanofibers of polymers and fillers are needed, in which trade-off between flexibility and electric performance needs to be considered. Fourth, even though the electrospun nanofibers are flexible, the other components including electrode and package materials may not be flexible; novel flexible electrode and package materials are needed to fabricate "true" flexible devices. Fabrication of all-nanofiber-based devices may be a viable approach. In addition, since the electrospun nanofibers are highly porous, ultrathin, and have relatively poor mechanical strength, the contacts between the nanofibers and electrodes and/or package processes need to be carefully optimized. Last but not least, the results reported in the literature are often for single device and simple demonstration of specific applications in a research lab setting, and the practical applications of these devices need real-world validation. Additionally, integration of tactile sensors with other components (e.g., other sensors, power supply devices, and data collection, and processing module), as well as the related cross talk effect and stability issues, challenges the future development of the technology.

Acknowledgments

The authors would like to acknowledge the funding support from the National Aeronautics and Space Administration (NASA Cooperative Agreement: 80NSSC18M0022), and Y. Ding would like to acknowledge the funding support from the National Natural Science Foundation of China (Project Number: 51903235). The authors would also like to acknowledge the Biomedical Engineering Program at the South Dakota School of Mines & Technology.

References

[1] T.R. Ray, J. Choi, A.J. Bandodkar, S. Krishnan, P. Gutruf, L. Tian, R. Ghaffari, J.A. Rogers, Bio-integrated wearable systems: a comprehensive review, Chem. Rev. 119 (8) (2019) 5461–5533.

[2] Y. Ding, J. Yang, C.R. Tolle, Z. Zhu, A highly stretchable strain sensor based on electrospun carbon nanofibers for human motion monitoring, RSC Adv. 6 (82) (2016) 79114–79120.

[3] O. Onyilagha, Y. Ding, Z. Zhu, Freestanding electrospun nanofibrous materials embedded in elastomers for stretchable strain sensors, in: Micro-and Nanotechnology Sensors,

Systems, and Applications XI, International Society for Optics and Photonics, 2019, p. 1098216.

[4] C. Wang, L. Dong, D. Peng, C. Pan, Tactile sensors for advanced intelligent systems, Adv. Intell. Syst. 1 (8) (2019) 1900090.

[5] Y. Ding, T. Xu, O. Onyilagha, H. Fong, Z. Zhu, Recent advances in flexible and wearable pressure sensors based on piezoresistive 3D monolithic conductive sponges, ACS Appl. Mater. Interfaces 11 (7) (2019) 6685–6704.

[6] Y. Ding, J. Yang, C.R. Tolle, Z. Zhu, Flexible and compressible PEDOT: PSS@melamine conductive sponge prepared via one-sep dip coating as piezoresistive pressure sensor for human motion detection, ACS Appl. Mater. Interfaces 10 (18) (2018) 16077–16086.

[7] S. Chen, K. Jiang, Z. Lou, D. Chen, G. Shen, Recent developments in graphene-based tactile sensors and E-skins, Adv. Mater. Technol. 3 (2) (2018) 1700248.

[8] Y. Zang, F. Zhang, C.-a. Di, D. Zhu, Advances of flexible pressure sensors toward artificial intelligence and health care applications, Mater. Horiz. 2 (2) (2015) 140–156.

[9] Kenry, J.C. Yeo, C.T. Lim, Emerging flexible and wearable physical sensing platforms for healthcare and biomedical applications, Microsyst. Nanoeng. 2 (1) (2016) 16043.

[10] G. Wang, Y. Zhao, Y. Ding, J. Yang, Z. Zhu, Design of wireless body area network with motion sensors using new materials, in: R. Wang, Z. Chen, W. Zhang, Q. Zhu (Eds.), Proceedings of the 11th International Conference on Modelling, Identification and Control (ICMIC2019), Springer Singapore, Singapore, 2020, pp. 707–717.

[11] Y. Huang, X. Fan, S.-C. Chen, N. Zhao, Emerging technologies of flexible pressure sensors: materials, modeling, devices, and manufacturing, Adv. Funct. Mater. 29 (12) (2019) 1808509.

[12] W. Han, Y. Wang, J. Su, X. Xin, Y. Guo, Y.-Z. Long, S. Ramakrishna, Fabrication of nanofibrous sensors by electrospinning, Sci. China Technol. Sci. 62 (6) (2019) 886–894.

[13] X. Wang, F. Sun, G. Yin, Y. Wang, B. Liu, M. Dong, Tactile-sensing based on flexible PVDF nanofibers via electrospinning: a review, Sensors 18 (2) (2018) 330.

[14] D.H. Reneker, I. Chun, Nanometre diameter fibres of polymer, produced by electrospinning, Nanotechnology 7 (1996) 216–223.

[15] W. Xu, Y. Ding, S. Jiang, W. Ye, X. Liao, H. Hou, High permittivity nanocomposites fabricated from electrospun polyimide/BaTiO$_3$ hybrid nanofibers, Polym. Compos. 37 (3) (2016) 794–801.

[16] Y. Ding, H. Hou, Y. Zhao, Z. Zhu, H. Fong, Electrospun polyimide nanofibers and their applications, Prog. Polym. Sci. 61 (2016) 67–103.

[17] W. Xu, Y. Ding, S. Jiang, L. Chen, X. Liao, H. Hou, Polyimide/BaTiO$_3$/MWCNTs three-phase nanocomposites fabricated by electrospinning with enhanced dielectric properties, Mater. Lett. 135 (2014) 158–161.

[18] Y. Ding, Q. Wu, D. Zhao, W. Ye, M. Hanif, H. Hou, Flexible PI/BaTiO$_3$ dielectric nanocomposite fabricated by combining electrospinning and electrospraying, Eur. Polym. J. 49 (9) (2013) 2567–2571.

[19] Y. He, D. Han, J. Chen, Y. Ding, S. Jiang, C. Hu, S. Chen, H. Hou, Highly strong and highly tough electrospun polyimide/polyimide composite nanofibers from binary blend of polyamic acids, RSC Adv. 4 (104) (2014) 59936–59942.

[20] Z. Zhou, X.-F. Wu, Y. Ding, M. Yu, Y. Zhao, L. Jiang, C. Xuan, C. Sun, Needleless emulsion electrospinning for scalable fabrication of core–shell nanofibers, J. Appl. Polym. Sci. 131 (20) (2014).

[21] J. Zhu, Y. Ding, S. Agarwal, A. Greiner, H. Zhang, H. Hou, Nanofibre preparation of non-processable polymers by solid-state polymerization of molecularly self-assembled monomers, Nanoscale 9 (46) (2017) 18169–18174.

[22] X. Yang, Y. Ding, Z. Shen, Q. Sun, F. Zheng, H. Fong, Z. Zhu, J. Liu, J. Liang, X. Wang, High-strength electrospun carbon nanofibrous mats prepared via rapid stabilization as frameworks for Li-ion battery electrodes, J. Mater. Sci. 54 (17) (2019) 11574–11584.

[23] Y. Shen, L. Chen, S. Jiang, Y. Ding, W. Xu, H. Hou, Electrospun nanofiber reinforced all-organic PVDF/PI tough composites and their dielectric permittivity, Mater. Lett. 160 (2015) 515–517.

[24] W. Xu, Y. Ding, R. Huang, Z. Zhu, H. Fong, H. Hou, High-performance polyimide nanofibers reinforced polyimide nanocomposite films fabricated by co-electrospinning followed by hot-pressing, J. Appl. Polym. Sci. 135 (47) (2018) 46849.

[25] X. Wang, S. Karanjit, L. Zhang, H. Fong, Q. Qiao, Z. Zhu, Transient photocurrent and photovoltage studies on charge transport in dye sensitized solar cells made from the composites of TiO_2 nanofibers and nanoparticles, Appl. Phys. Lett. 98 (8) (2011), 082114.

[26] Z. Zhu, L. Zhang, J.Y. Howe, Y. Liao, J.T. Speidel, S. Smith, H. Fong, Aligned electrospun ZnO nanofibers for simple and sensitive ultraviolet nanosensors, Chem. Commun. (18) (2009) 2568–2570.

[27] G. He, Y. Cai, Y. Zhao, X. Wang, C. Lai, M. Xi, Z. Zhu, H. Fong, Electrospun anatase-phase TiO_2 nanofibers with different morphological structures and specific surface areas, J. Colloid Interface Sci. 398 (2013) 103–111.

[28] W. Nan, Y. Zhao, Y. Ding, A.R. Shende, H. Fong, R.V. Shende, Mechanically flexible electrospun carbon nanofiber mats derived from biochar and polyacrylonitrile, Mater. Lett. 205 (2017) 206–210.

[29] X. Peng, W. Ye, Y. Ding, S. Jiang, M. Hanif, X. Liao, H. Hou, Facile synthesis, characterization and application of highly active palladium nano-network structures supported on electrospun carbon nanofibers, RSC Adv. 4 (80) (2014) 42732–42736.

[30] J. Zhu, Y. Ding, X. Liao, W. Xu, H. Zhang, H. Hou, Highly flexible electrospun carbon/graphite nanofibers from a non-processable heterocyclic rigid-rod polymer of polybisbenzimidazobenzophenanthroline-dione (BBB), J. Mater. Sci. 53 (12) (2018) 9002–9012.

[31] Y. Ding, W. Xu, T. Xu, Z. Zhu, H. Fong, Theories and principles behind electrospinning, in: Y. Liu, C. Wang (Eds.), Advanced Nanofibrous Materials Manufacture Technology Based on Electrospinning, CRC Press, 2019, pp. 22–51.

[32] J. Xue, T. Wu, Y. Dai, Y. Xia, Electrospinning and electrospun nanofibers: methods, materials, and applications, Chem. Rev. 119 (8) (2019) 5298–5415.

[33] S. Jiang, G. Duan, L. Chen, X. Hu, Y. Ding, C. Jiang, H. Hou, Thermal, mechanical and thermomechanical properties of tough electrospun poly(imide-co-benzoxazole) nanofiber belts, New J. Chem. 39 (10) (2015) 7797–7804.

[34] W. Xu, Y. Ding, T. Yang, Y. Yu, R. Huang, Z. Zhu, H. Fong, H. Hou, An innovative approach for the preparation of high-performance electrospun poly(p-phenylene)-based polymer nanofiber belts, Macromolecules 50 (24) (2017) 9760–9772.

[35] D. Li, Y. Wang, Y. Xia, Electrospinning of polymeric and ceramic nanofibers as uniaxially aligned arrays, Nano Lett. 3 (8) (2003) 1167–1171.

[36] J.T. McCann, D. Li, Y. Xia, Electrospinning of nanofibers with core-sheath, hollow, or porous structures, J. Mater. Chem. 15 (7) (2005) 735–738.

[37] Y. Zhao, X. Cao, L. Jiang, Bio-mimic multichannel microtubes by a facile method, J. Am. Chem. Soc. 129 (4) (2007) 764–765.

[38] W. Xu, Y. Ding, Making polymer fibers strong and tough simultaneously, Sci. China Mater. 63 (3) (2020) 481–482.

[39] X. Liao, M. Dulle, J.M. de Souza e Silva, R.B. Wehrspohn, S. Agarwal, S. Förster, H. Hou, P. Smith, A. Greiner, High strength in combination with high toughness in robust and sustainable polymeric materials, Science 366 (6471) (2019) 1376–1379.

[40] J. Joseph, M. Kumar, S. Tripathy, G.D.V.S. Kumar, S.G. Singh, S.R.K. Vaniari, A highly flexible tactile sensor with self-poled electrospun PVDF nanofiber, in: IEEE Sensors, 2018, pp. 1–4.

[41] E. Ghafari, N. Lu, Self-polarized electrospun polyvinylidene fluoride (PVDF) nanofiber for sensing applications, Compos. Part B Eng. 160 (2019) 1–9.

[42] W. Ma, Y. Ding, M. Zhang, S. Gao, Y. Li, C. Huang, G. Fu, Nature-inspired chemistry toward hierarchical superhydrophobic, antibacterial and biocompatible nanofibrous membranes for effective UV-shielding, self-cleaning and oil-water separation, J. Hazard. Mater. 384 (2020) 121476.

[43] S. Liu, W. Xu, C. Ding, J. Yu, D. Fang, Y. Ding, H. Hou, Boosting electrochemical performance of electrospun silicon-based anode materials for lithium-ion battery by surface coating a second layer of carbon, Appl. Surf. Sci. 494 (2019) 94–100.

[44] Y. Ding, W. Xu, W. Wang, H. Fong, Z. Zhu, Scalable and facile preparation of highly stretchable electrospun PEDOT:PSS@PU fibrous nonwovens toward wearable conductive textile applications, ACS Appl. Mater. Interfaces 9 (35) (2017) 30014–30023.

[45] Z. Qin, X. Chen, Y. Yin, G. Ma, Y. Jia, J. Deng, K. Pan, Flexible janus electrospun nanofiber films for wearable triboelectric nanogenerator, Adv. Mater. Technol. 5 (2) (2020) 1900859.

[46] T. Xu, Y. Ding, Z. Liang, H. Sun, F. Zheng, Z. Zhu, Y. Zhao, H. Fong, Three-dimensional monolithic porous structures assembled from fragmented electrospun nanofiber mats/membranes: methods, properties, and applications, Prog. Mater. Sci. 112 (2020) 100656.

[47] T. Xu, F. Zheng, Z. Chen, Y. Ding, Z. Liang, Y. Liu, Z. Zhu, H. Fong, Halloysite nanotubes sponges with skeletons made of electrospun nanofibers as innovative dye adsorbent and catalyst support, Chem. Eng. J. 360 (2019) 280–288.

[48] T. Xu, Z. Wang, Y. Ding, W. Xu, W. Wu, Z. Zhu, H. Fong, Ultralight electrospun cellulose sponge with super-high capacity on absorption of organic compounds, Carbohydr. Polym. 179 (2018) 164–172.

[49] Y. Wan, Z. Qiu, Y. Hong, Y. Wang, J. Zhang, Q. Liu, Z. Wu, C.F. Guo, A highly sensitive flexible capacitive tactile sensor with sparse and high-aspect-ratio microstructures, Adv. Electron. Mater. 4 (4) (2018) 1700586.

[50] T.D. Nguyen, H.S. Han, H.-Y. Shin, C.T. Nguyen, H. Phung, H.V. Hoang, H.R. Choi, Highly sensitive flexible proximity tactile array sensor by using carbon micro coils, Sensors Actuators A Phys. 266 (2017) 166–177.

[51] D. Pyo, S. Ryu, K.-U. Kyung, S. Yun, D.-S. Kwon, High-pressure endurable flexible tactile actuator based on microstructured dielectric elastomer, Appl. Phys. Lett. 112 (6) (2018), 061902.

[52] M.T. Chorsi, E.J. Curry, H.T. Chorsi, R. Das, J. Baroody, P.K. Purohit, H. Ilies, T.D. Nguyen, Piezoelectric biomaterials for sensors and actuators, Adv. Mater. 31 (1) (2019) 1802084.

[53] F.-R. Fan, Z.-Q. Tian, Z.L. Wang, Flexible triboelectric generator, Nano Energy 1 (2) (2012) 328–334.

[54] H.H. Shi, N. Khalili, T. Morrison, H.E. Naguib, Self-assembled nanorod structures on nanofibers for textile electrochemical capacitor electrodes with intrinsic tactile sensing capabilities, ACS Appl. Mater. Interfaces 10 (22) (2018) 19037–19046.

[55] J. Wang, R. Suzuki, M. Shao, F. Gillot, S. Shiratori, Capacitive pressure sensor with wide-range, bendable, and high sensitivity based on the bionic Komochi Konbu structure and Cu/Ni nanofiber network, ACS Appl. Mater. Interfaces 11 (12) (2019) 11928–11935.

[56] H. Ren, L. Zheng, G. Wang, X. Gao, Z. Tan, J. Shan, L. Cui, K. Li, M. Jian, L. Zhu, Y. Zhang, H. Peng, D. Wei, Z. Liu, Transfer-medium-free nanofiber-reinforced graphene film and applications in wearable transparent pressure sensors, ACS Nano 13 (5) (2019) 5541–5548.

[57] C. Luo, N. Liu, H. Zhang, W. Liu, Y. Yue, S. Wang, J. Rao, C. Yang, J. Su, X. Jiang, Y. Gao, A new approach for ultrahigh-performance piezoresistive sensor based on wrinkled PPy film with electrospun PVA nanowires as spacer, Nano Energy 41 (2017) 527–534.

[58] W. Liu, N. Liu, Y. Yue, J. Rao, F. Cheng, J. Su, Z. Liu, Y. Gao, Piezoresistive pressure sensor based on synergistical innerconnect polyvinyl alcohol nanowires/wrinkled graphene film, Small 14 (15) (2018) 1704149.

[59] Z. Zhao, B. Li, L. Xu, Y. Qiao, F. Wang, Q. Xia, Z. Lu, A sandwich-structured piezoresistive sensor with electrospun nanofiber mats as supporting, sensing, and packaging layers, Polymers 10 (6) (2018) 575.

[60] Z. Lou, S. Chen, L. Wang, K. Jiang, G. Shen, An ultra-sensitive and rapid response speed graphene pressure sensors for electronic skin and health monitoring, Nano Energy 23 (2016) 7–14.

[61] Y. Ai, Z. Lou, S. Chen, D. Chen, Z.M. Wang, K. Jiang, G. Shen, All rGO-on-PVDF-nanofibers based self-powered electronic skins, Nano Energy 35 (2017) 121–127.

[62] H.-D. Zhang, Y.-J. Liu, J. Zhang, J.-W. Zhu, Q.-H. Qin, C.-Z. Zhao, X. Li, J.-C. Zhang, Y.-Z. Long, Electrospun ZnO/SiO_2 hybrid nanofibers for flexible pressure sensor, J. Phys. D Appl. Phys. 51 (8) (2018), 085102.

[63] Q. Gao, H. Meguro, S. Okamoto, M. Kimura, Flexible tactile sensor using the reversible deformation of poly(3-hexylthiophene) nanofiber assemblies, Langmuir 28 (51) (2012) 17593–17596.

[64] K. Yin, L. Zhang, C. Lai, L. Zhong, S. Smith, H. Fong, Z. Zhu, Photoluminescence anisotropy of uni-axially aligned electrospun conjugated polymer nanofibers of MEH-PPV and P3HT, J. Mater. Chem. 21 (2) (2011) 444–448.

[65] K. Qi, K. Wang, X. You, X. Tao, M. Li, Y. Zhou, Y. Zhang, J. He, W. Shao, S. Cui, Core-sheath nanofiber yarn for textile pressure sensor with high pressure sensitivity and spatial tactile acuity, J. Colloid Interface Sci. 561 (2020) 93–103.

[66] P. Sobolčiak, A. Tanvir, K.K. Sadasivuni, I. Krupa, Piezoresistive sensors based on electrospun mats modified by 2D $Ti_3C_2T_x$ MXene, Sensors 19 (20) (2019) 4589.

[67] X. Guan, G. Zheng, K. Dai, C. Liu, X. Yan, C. Shen, Z. Guo, Carbon nanotubes-adsorbed electrospun PA66 nanofiber bundles with improved conductivity and robust flexibility, ACS Appl. Mater. Interfaces 8 (22) (2016) 14150–14159.

[68] N. Wang, Z. Xu, P. Zhan, K. Dai, G. Zheng, C. Liu, C. Shen, A tunable strain sensor based on a carbon nanotubes/electrospun polyamide 6 conductive nanofibrous network embedded into poly(vinyl alcohol) with self-diagnosis capabilities, J. Mater. Chem. C 5 (18) (2017) 4408–4418.

[69] L. Wang, Y. Chen, L. Lin, H. Wang, X. Huang, H. Xue, J. Gao, Highly stretchable, anticorrosive and wearable strain sensors based on the PDMS/CNTs decorated elastomer nanofiber composite, Chem. Eng. J. 362 (2019) 89–98.

[70] Y. Liu, F. Meng, Y. Zhou, S.M. Mugo, Q. Zhang, Graphene oxide films prepared using gelatin nanofibers as wearable sensors for monitoring cardiovascular health, Adv. Mater. Technol. 4 (11) (2019) 1900540.

[71] R. Wu, L. Ma, A. Patil, C. Hou, S. Zhu, X. Fan, H. Lin, W. Yu, W. Guo, X.Y. Liu, All-textile electronic skin enabled by highly elastic spacer fabric and conductive fibers, ACS Appl. Mater. Interfaces 11 (36) (2019) 33336–33346.

[72] Z. Wang, L. Zhang, J. Liu, C. Li, A flexible bimodal sensor based on an electrospun nanofibrous structure for simultaneous pressure–temperature detection, Nanoscale 11 (30) (2019) 14242–14249.

[73] W.-P. Hu, B. Zhang, J. Zhang, W.-L. Luo, Y. Guo, S.-J. Chen, M.-J. Yun, S. Ramakrishna, Y.-Z. Long, Ag/alginate nanofiber membrane for flexible electronic skin, Nanotechnology 28 (44) (2017) 445502.

[74] P. Bi, X. Liu, Y. Yang, Z. Wang, J. Shi, G. Liu, F. Kong, B. Zhu, R. Xiong, Silver-nanoparticle-modified polyimide for multiple artificial skin-sensing applications, Adv. Mater. Technol. 4 (10) (2019) 1900426.

[75] Y. Zhou, J. He, H. Wang, K. Qi, N. Nan, X. You, W. Shao, L. Wang, B. Ding, S. Cui, Highly sensitive, self-powered and wearable electronic skin based on pressure-sensitive nanofiber woven fabric sensor, Sci. Rep. 7 (1) (2017) 12949.

[76] O.Y. Kweon, S.J. Lee, J.H. Oh, Wearable high-performance pressure sensors based on three-dimensional electrospun conductive nanofibers, NPG Asia Mater. 10 (6) (2018) 540–551.

[77] Z. Han, Z. Cheng, Y. Chen, B. Li, Z. Liang, H. Li, Y. Ma, X. Feng, Fabrication of highly pressure-sensitive, hydrophobic, and flexible 3D carbon nanofiber networks by electrospinning for human physiological signal monitoring, Nanoscale 11 (13) (2019) 5942–5950.

[78] Y. Si, X. Wang, C. Yan, L. Yang, J. Yu, B. Ding, Ultralight biomass-derived carbonaceous nanofibrous aerogels with superelasticity and high pressure-sensitivity, Adv. Mater. 28 (43) (2016) 9512–9518.

[79] T. Xu, Y. Ding, Z. Wang, Y. Zhao, W. Wu, H. Fong, Z. Zhu, Three-dimensional and ultralight sponges with tunable conductivity assembled from electrospun nanofibers for a highly sensitive tactile pressure sensor, J. Mater. Chem. C 5 (39) (2017) 10288–10294.

[80] H. Park, Y.R. Jeong, J. Yun, S.Y. Hong, S. Jin, S.-J. Lee, G. Zi, J.S. Ha, Stretchable array of highly sensitive pressure sensors consisting of polyaniline nanofibers and Au-coated polydimethylsiloxane micropillars, ACS Nano 9 (10) (2015) 9974–9985.

[81] S.-J. Park, J. Kim, M. Chu, M. Khine, Flexible piezoresistive pressure sensor using wrinkled carbon nanotube thin films for human physiological signals, Adv. Mater. Technol. 3 (1) (2018) 1700158.

[82] W. Zhong, Q. Liu, Y. Wu, Y. Wang, X. Qing, M. Li, K. Liu, W. Wang, D. Wang, A nanofiber based artificial electronic skin with high pressure sensitivity and 3D conformability, Nanoscale 8 (24) (2016) 12105–12112.

[83] Y. Kim, S. Jang, B.J. Kang, J.H. Oh, Fabrication of highly sensitive capacitive pressure sensors with electrospun polymer nanofibers, Appl. Phys. Lett. 111 (7) (2017), 073502.

[84] W. Yang, N.-W. Li, S. Zhao, Z. Yuan, J. Wang, X. Du, B. Wang, R. Cao, X. Li, W. Xu, Z. L. Wang, C. Li, A breathable and screen-printed pressure sensor based on nanofiber membranes for electronic skins, Adv. Mater. Technol. 3 (2) (2018) 1700241.

[85] M.A.U. Khalid, M. Ali, A.M. Soomro, S.W. Kim, H.B. Kim, B.-G. Lee, K.H. Choi, A highly sensitive biodegradable pressure sensor based on nanofibrous dielectric, Sensors Actuators A Phys. 294 (2019) 140–147.

[86] W. Xu, Y. Ding, S. Jiang, J. Zhu, W. Ye, Y. Shen, H. Hou, Mechanical flexible PI/MWCNTs nanocomposites with high dielectric permittivity by electrospinning, Eur. Polym. J. 59 (2014) 129–135.

[87] W. Xu, Y. Feng, Y. Ding, S. Jiang, H. Fang, H. Hou, Short electrospun carbon nanofiber reinforced polyimide composite with high dielectric permittivity, Mater. Lett. 161 (2015) 431–434.

[88] W. Xu, Y. Ding, Y. Yu, S. Jiang, L. Chen, H. Hou, Highly foldable PANi@CNTs/PU dielectric composites toward thin-film capacitor application, Mater. Lett. 192 (2017) 25–28.

[89] X. Yang, Y. Wang, X. Qing, A flexible capacitive sensor based on the electrospun PVDF nanofiber membrane with carbon nanotubes, Sensors Actuators A Phys. 299 (2019) 111579.

[90] M.S. Reza, K.R. Ayag, M.K. Yoo, K.J. Kim, H. Kim, Electrospun spandex nanofiber webs with ionic liquid for highly sensitive, low hysteresis piezocapacitive sensor, Fibers Polym. 20 (2) (2019) 337–347.

[91] X. Yang, Y. Wang, X. Qing, Electrospun ionic nanofiber membrane-based fast and highly sensitive capacitive pressure sensor, IEEE Access 7 (2019) 139984–139993.

[92] M.-F. Lin, J. Xiong, J. Wang, K. Parida, P.S. Lee, Core-shell nanofiber mats for tactile pressure sensor and nanogenerator applications, Nano Energy 44 (2018) 248–255.

[93] X. You, J. He, N. Nan, X. Sun, K. Qi, Y. Zhou, W. Shao, F. Liu, S. Cui, Stretchable capacitive fabric electronic skin woven by electrospun nanofiber coated yarns for detecting tactile and multimodal mechanical stimuli, J. Mater. Chem. C 6 (47) (2018) 12981–12991.

[94] C. Hou, Z. Xu, W. Qiu, R. Wu, Y. Wang, Q. Xu, X.Y. Liu, W. Guo, A biodegradable and stretchable protein-based sensor as artificial electronic skin for human motion detection, Small 15 (11) (2019) 1805084.

[95] J. He, X. Guo, J. Yu, S. Qian, X. Hou, M. Cui, Y. Yang, J. Mu, W. Geng, X. Chou, A high-resolution flexible sensor array based on PZT nanofibers, Nanotechnology 31 (15) (2020) 155503.

[96] N. Cui, X. Jia, A. Lin, J. Liu, S. Bai, L. Zhang, Y. Qin, R. Yang, F. Zhou, Y. Li, Piezo-electric nanofiber/polymer composite membrane for noise harvesting and active acoustic wave detection, Nanoscale Adv. 1 (12) (2019) 4909–4914.

[97] B. Bera, M. Das Sarkar, Piezoelectricity in PVDF and PVDF based piezoelectric nanogenerator: a concept, IOSR J. Appl. Phys. 9 (3) (2017) 95–99.

[98] X. Liu, S. Xu, S. Kuang, D. Tan, X. Wang, Nanoscale investigations on β-phase orientation, piezoelectric response, and polarization direction of electrospun PVDF nanofibers, RSC Adv. 6 (110) (2016) 109061–109066.

[99] N. Manikandan, S. Muruganand, K. Sriram, P. Balakrishnan, A. Suresh Kumar, Fabrication of piezoelectric polyvinylidene fluoride (PVDF) polymer-based tactile sensor using electrospinning method, Nano Hybrids Composites 12 (2017) 42–50.

[100] D. Mandal, S. Yoon, K.J. Kim, Origin of piezoelectricity in an electrospun poly (vinylidene fluoride-trifluoroethylene) nanofiber web-based nanogenerator and nano-pressure sensor, Macromol. Rapid Commun. 32 (11) (2011) 831–837.

[101] G. Wang, T. Liu, X.-C. Sun, P. Li, Y.-S. Xu, J.-G. Hua, Y.-H. Yu, S.-X. Li, Y.-Z. Dai, X.-Y. Song, C. Lv, H. Xia, Flexible pressure sensor based on PVDF nanofiber, Sensors Actuators A Phys. 280 (2018) 319–325.

[102] R.K. Singh, S.W. Lye, J. Miao, PVDF nanofiber sensor for vibration measurement in a string, Sensors 19 (17) (2019) 3739.

[103] G. Ico, A. Myung, B.S. Kim, N.V. Myung, J. Nam, Transformative piezoelectric enhancement of P(VDF-TrFE) synergistically driven by nanoscale dimensional reduction and thermal treatment, Nanoscale 10 (6) (2018) 2894–2901.

[104] G. Ren, F. Cai, B. Li, J. Zheng, C. Xu, Flexible pressure sensor based on a poly(VDF-TrFE) nanofiber web, Macromol. Mater. Eng. 298 (5) (2013) 541–546.

[105] J. Chang, M. Dommer, C. Chang, L. Lin, Piezoelectric nanofibers for energy scavenging applications, Nano Energy 1 (3) (2012) 356–371.

[106] B. Azimi, M. Milazzo, A. Lazzeri, S. Berrettini, M.J. Uddin, Z. Qin, M.J. Buehler, S. Danti, Electrospinning piezoelectric fibers for biocompatible devices, Adv. Healthc. Mater. 9 (1) (2020) 1901287.

[107] Y. Xin, J. Zhu, H. Sun, Y. Xu, T. Liu, C. Qian, A brief review on piezoelectric PVDF nanofibers prepared by electrospinning, Ferroelectrics 526 (1) (2018) 140–151.

[108] A. Hanif, T.Q. Trung, S. Siddiqui, P.T. Toi, N.-E. Lee, Stretchable, transparent, tough, ultrathin, and self-limiting skin-like substrate for stretchable electronics, ACS Appl. Mater. Interfaces 10 (32) (2018) 27297–27307.

[109] X. Wang, W.-Z. Song, M.-H. You, J. Zhang, M. Yu, Z. Fan, S. Ramakrishna, Y.-Z. Long, Bionic single-electrode electronic skin unit based on piezoelectric nanogenerator, ACS Nano 12 (8) (2018) 8588–8596.

[110] S. Lotfian, C. Giraudmaillet, A. Yoosefinejad, V.K. Thakur, H.Y. Nezhad, Electrospun piezoelectric polymer nanofiber layers for enabling in situ measurement in high-performance composite laminates, ACS Omega 3 (8) (2018) 8891–8902.

[111] C. Lang, J. Fang, H. Shao, X. Ding, T. Lin, High-sensitivity acoustic sensors from nanofibre webs, Nat. Commun. 7 (1) (2016) 11108.

[112] L. Yu, L. Wang, D. Wu, Y. Zhao, D. Sun, Enhanced piezoelectric performance of electrospun PVDF nanofibers with liquid metal electrodes, ECS J. Solid State Sci. Technol. 7 (9) (2018) N128–N131.

[113] Y.W. Kim, H.B. Lee, S.M. Yeon, J. Park, H.J. Lee, J. Yoon, S.H. Park, Enhanced piezoelectricity in a robust and harmonious multilayer assembly of electrospun nanofiber mats and microbead-based electrodes, ACS Appl. Mater. Interfaces 10 (6) (2018) 5723–5730.

[114] K. Maity, D. Mandal, All-organic high-performance piezoelectric nanogenerator with multilayer assembled electrospun nanofiber mats for self-powered multifunctional sensors, ACS Appl. Mater. Interfaces 10 (21) (2018) 18257–18269.

[115] Y.J. Hwang, S. Choi, H.S. Kim, Highly flexible all-nonwoven piezoelectric generators based on electrospun poly(vinylidene fluoride), Sensors Actuators A Phys. 300 (2019) 111672.

[116] E. Yang, Z. Xu, L.K. Chur, A. Behroozfar, M. Baniasadi, S. Moreno, J. Huang, J. Gilligan, M. Minary-Jolandan, Nanofibrous smart fabrics from twisted yarns of electrospun piezopolymer, ACS Appl. Mater. Interfaces 9 (28) (2017) 24220–24229.

[117] M. Asadnia, A. Kottapalli, J. Miao, M. Triantafyllou, Ultra-sensitive and stretchable strain sensor based on piezoelectric polymeric nanofibers, in: 28th IEEE International Conference on Micro Electro Mechanical Systems (MEMS), IEEE, 2015, pp. 678–681.

[118] S. You, L. Zhang, J. Gui, H. Cui, S. Guo, A flexible piezoelectric nanogenerator based on aligned P(VDF-TrFE) nanofibers, Micromachines 10 (5) (2019) 302.

[119] L. Persano, C. Dagdeviren, Y. Su, Y. Zhang, S. Girardo, D. Pisignano, Y. Huang, J.A. Rogers, High performance piezoelectric devices based on aligned arrays of nanofibers of poly(vinylidenefluoride-co-trifluoroethylene), Nat. Commun. 4 (1) (2013) 1633.

[120] T. Sharma, S. Naik, J. Langevine, B. Gill, J.X.J. Zhang, Aligned PVDF-TrFE nanofibers with high-density PVDF nanofibers and PVDF core–shell structures for endovascular pressure sensing, IEEE Trans. Biomed. Eng. 62 (1) (2015) 188–195.

[121] Y.-H. Hsu, C.-H. Chan, W.C. Tang, Alignment of multiple electrospun piezoelectric fiber bundles across serrated gaps at an incline: a method to generate textile strain sensors, Sci. Rep. 7 (1) (2017) 15436.

[122] Y.K. Fuh, B.S. Wang, C.-Y. Tsai, Self-powered pressure sensor with fully encapsulated 3D printed wavy substrate and highly-aligned piezoelectric fibers array, Sci. Rep. 7 (1) (2017) 6759.

[123] Y.K. Fuh, S.C. Lee, C.Y. Tsai, Application of highly flexible self-powered sensors via sequentially deposited piezoelectric fibers on printed circuit board for wearable electronics devices, Sensors Actuators A Phys. 268 (2017) 148–154.

[124] Y.K. Fuh, B.S. Wang, Near field sequentially electrospun three-dimensional piezoelectric fibers arrays for self-powered sensors of human gesture recognition, Nano Energy 30 (2016) 677–683.

[125] S. Ma, T. Ye, T. Zhang, Z. Wang, K. Li, M. Chen, J. Zhang, Z. Wang, S. Ramakrishna, L. Wei, Highly oriented electrospun P(VDF-TrFE) fibers via mechanical stretching for wearable motion sensing, Adv. Mater. Technol. 3 (7) (2018) 1800033.

[126] W. Deng, T. Yang, L. Jin, C. Yan, H. Huang, X. Chu, Z. Wang, D. Xiong, G. Tian, Y. Gao, H. Zhang, W. Yang, Cowpea-structured PVDF/ZnO nanofibers based flexible self-powered piezoelectric bending motion sensor towards remote control of gestures, Nano Energy 55 (2019) 516–525.

[127] H. Parangusan, D. Ponnamma, M.A.A. Al-Maadeed, Stretchable electrospun PVDF-HFP/Co-ZnO nanofibers as piezoelectric nanogenerators, Sci. Rep. 8 (1) (2018) 754.

[128] P. Fakhri, B. Amini, R. Bagherzadeh, M. Kashfi, M. Latifi, N. Yavari, S. Asadi Kani, L. Kong, Flexible hybrid structure piezoelectric nanogenerator based on ZnO nanorod/PVDF nanofibers with improved output, RSC Adv. 9 (18) (2019) 10117–10123.

[129] S. Siddiqui, H.B. Lee, D.-I. Kim, L.T. Duy, A. Hanif, N.-E. Lee, An omnidirectionally stretchable piezoelectric nanogenerator based on hybrid nanofibers and carbon electrodes for multimodal straining and human kinematics energy harvesting, Adv. Energy Mater. 8 (2) (2018) 1701520.

[130] X. Hu, X. Yan, L. Gong, F. Wang, Y. Xu, L. Feng, D. Zhang, Y. Jiang, Improved piezoelectric sensing performance of P(VDF–TrFE) nanofibers by utilizing BTO nanoparticles and penetrated electrodes, ACS Appl. Mater. Interfaces 11 (7) (2019) 7379–7386.

[131] X. Guan, B. Xu, J. Gong, Hierarchically architected polydopamine modified BaTiO$_3$@P (VDF-TrFE) nanocomposite fiber mats for flexible piezoelectric nanogenerators and self-powered sensors, Nano Energy 70 (2020) 104516.

[132] W. Guo, C. Tan, K. Shi, J. Li, X.-X. Wang, B. Sun, X. Huang, Y.-Z. Long, P. Jiang, Wireless piezoelectric devices based on electrospun PVDF/BaTiO$_3$ NW nanocomposite fibers for human motion monitoring, Nanoscale 10 (37) (2018) 17751–17760.

[133] S.H. Ji, J.H. Cho, Y.H. Jeong, J.-H. Paik, J.D. Yun, J.S. Yun, Flexible lead-free piezoelectric nanofiber composites based on BNT-ST and PVDF for frequency sensor applications, Sensors Actuators A Phys. 247 (2016) 316–322.

[134] S.H. Ji, J.H. Cho, J.-H. Paik, J. Yun, J.S. Yun, Poling effects on the performance of a lead-free piezoelectric nanofiber in a structural health monitoring sensor, Sensors Actuators A Phys. 263 (2017) 633–638.

[135] J. Liu, B. Yang, L. Lu, X. Wang, X. Li, X. Chen, J. Liu, Flexible and lead-free piezoelectric nanogenerator as self-powered sensor based on electrospinning BZT-BCT/P (VDF-TrFE) nanofibers, Sensors Actuators A Phys. 303 (2020) 111796.

[136] Y. Xin, X. Qi, H. Tian, C. Guo, X. Li, J. Lin, C. Wang, Full-fiber piezoelectric sensor by straight PVDF/nanoclay nanofibers, Mater. Lett. 164 (2016) 136–139.

[137] L. Jin, Y. Zheng, Z. Liu, J. Li, H. Zhai, Z. Chen, Y. Li, Design of an ultrasensitive flexible bend sensor using a silver-doped oriented poly(vinylidene fluoride) nanofiber web for respiratory monitoring, ACS Appl. Mater. Interfaces 12 (1) (2020) 1359–1367.

[138] H. Li, W. Zhang, Q. Ding, X. Jin, Q. Ke, Z. Li, D. Wang, C. Huang, Facile strategy for fabrication of flexible, breathable, and washable piezoelectric sensors via welding of nanofibers with multiwalled carbon nanotubes (MWCNTs), ACS Appl. Mater. Interfaces 11 (41) (2019) 38023–38030.

[139] K. Roy, S.K. Ghosh, A. Sultana, S. Garain, M. Xie, C.R. Bowen, K. Henkel, D. Schmeißer, D. Mandal, A self-powered wearable pressure sensor and pyroelectric breathing sensor based on GO interfaced PVDF nanofibers, ACS Appl. Nano Mater. 2 (4) (2019) 2013–2025.

[140] P. Li, L. Zhao, Z. Jiang, M. Yu, Z. Li, X. Li, Self-powered flexible sensor based on the graphene modified P(VDF-TrFE) electrospun fibers for pressure detection, Macromol. Mater. Eng. 304 (12) (2019) 1900504.

[141] S.M. Hosseini, A.A. Yousefi, Piezoelectric sensor based on electrospun PVDF-MWCNT-Cloisite 30B hybrid nanocomposites, Org. Electron. 50 (2017) 121–129.

[142] K. Shi, B. Sun, X. Huang, P. Jiang, Synergistic effect of graphene nanosheet and BaTiO$_3$ nanoparticles on performance enhancement of electrospun PVDF nanofiber mat for flexible piezoelectric nanogenerators, Nano Energy 52 (2018) 153–162.

[143] A. Ahmed, Y. Jia, Y. Huang, N.A. Khoso, H. Deb, Q. Fan, J. Shao, Preparation of PVDF-TrFE based electrospun nanofibers decorated with PEDOT-CNT/rGO composites for piezo-electric pressure sensor, J. Mater. Sci. Mater. Electron. 30 (15) (2019) 14007–14021.

[144] S. Cherumannil Karumuthil, S. Prabha Rajeev, U. Valiyaneerilakkal, S. Athiyanathil, S. Varghese, Electrospun poly(vinylidene fluoride-trifluoroethylene)-based polymer nanocomposite fibers for piezoelectric nanogenerators, ACS Appl. Mater. Interfaces 11 (43) (2019) 40180–40188.

[145] M. Zhu, M. Lou, I. Abdalla, J. Yu, Z. Li, B. Ding, Highly shape adaptive fiber based electronic skin for sensitive joint motion monitoring and tactile sensing, Nano Energy 69 (2020) 104429.

[146] Y.M. Yousry, K. Yao, S. Chen, W.H. Liew, S. Ramakrishna, Mechanisms for enhancing polarization orientation and piezoelectric parameters of PVDF nanofibers, Adv. Electron. Mater. 4 (6) (2018) 1700562.

[147] S. Garain, S. Jana, T.K. Sinha, D. Mandal, Design of in situ poled Ce^{3+}-doped electrospun PVDF/graphene composite nanofibers for fabrication of nanopressure sensor and ultrasensitive acoustic nanogenerator, ACS Appl. Mater. Interfaces 8 (7) (2016) 4532–4540.

[148] B. Yu, M. Mao, H. Yu, T. Huang, W. Zuo, H. Wang, M. Zhu, Enhanced piezoelectric performance of electrospun polyvinylidene fluoride doped with inorganic salts, Macromol. Mater. Eng. 302 (11) (2017) 1700214.

[149] R.M. Strect, M. Minagawa, A. Vengrenyuk, C.L. Schauer, Piezoelectric electrospun polyacrylonitrile with various tacticities, J. Appl. Polym. Sci. 136 (20) (2019) 47530.

[150] D.-N. Nguyen, W. Moon, Piezoelectric polymer microfiber-based composite for the flexible ultra-sensitive pressure sensor, J. Appl. Polym. Sci. 137 (2020) 48884.

[151] S.K. Ghosh, P. Adhikary, S. Jana, A. Biswas, V. Sencadas, S.D. Gupta, B. Tudu, D. Mandal, Electrospun gelatin nanofiber based self-powered bio-e-skin for health care monitoring, Nano Energy 36 (2017) 166–175.

[152] N. Shehata, I. Kandas, I. Hassounah, P. Sobolčiak, I. Krupa, M. Mrlik, A. Popelka, J. Steadman, R. Lewis, Piezoresponse, mechanical, and electrical characteristics of synthetic spider silk nanofibers, Nanomaterials 8 (8) (2018) 585.

[153] S.S. Kwak, H.-J. Yoon, S.-W. Kim, Textile-based triboelectric nanogenerators for self-powered wearable electronics, Adv. Funct. Mater. 29 (2) (2019) 1804533.

[154] Y. Cheng, C. Wang, J. Zhong, S. Lin, Y. Xiao, Q. Zhong, H. Jiang, N. Wu, W. Li, S. Chen, B. Wang, Y. Zhang, J. Zhou, Electrospun polyetherimide electret nonwoven for bi-functional smart face mask, Nano Energy 34 (2017) 562–569.

[155] Y. Kim, X. Wu, J.H. Oh, Fabrication of triboelectric nanogenerators based on electrospun polyimide nanofibers membrane, Sci. Rep. 10 (1) (2020) 2742.

[156] S.-R. Kim, J.-H. Yoo, J.-W. Park, Using electrospun AgNW/P(VDF-TrFE) composite nanofibers to create transparent and wearable single-electrode triboelectric nano-generators for self-powered touch panels, ACS Appl. Mater. Interfaces 11 (16) (2019) 15088–15096.

[157] Z. Qin, Y. Yin, W. Zhang, C. Li, K. Pan, Wearable and stretchable triboelectric nan-ogenerator based on crumpled nanofibrous membranes, ACS Appl. Mater. Interfaces 11 (13) (2019) 12452–12459.

[158] X. Wang, Y. Zhang, X. Zhang, Z. Huo, X. Li, M. Que, Z. Peng, H. Wang, C. Pan, A highly stretchable transparent self-powered triboelectric tactile sensor with metallized nanofibers for wearable electronics, Adv. Mater. 30 (12) (2018) 1706738.

[159] S. Zhao, J. Wang, X. Du, J. Wang, R. Cao, Y. Yin, X. Zhang, Z. Yuan, Y. Xing, D.Y.H. Pui, C. Li, All-nanofiber-based ultralight stretchable triboelectric nanogenerator for self-powered wearable electronics, ACS Appl. Energy Mater. 1 (5) (2018) 2326–2332.

[160] Z. Li, M. Zhu, J. Shen, Q. Qiu, J. Yu, B. Ding, All-fiber structured electronic skin with high elasticity and breathability, Adv. Funct. Mater. 30 (6) (2020) 1908411.

[161] P. Zhao, N. Soin, K. Prashanthi, J. Chen, S. Dong, E. Zhou, Z. Zhu, A.A. Narasimulu, C. D. Montemagno, L. Yu, J. Luo, Emulsion electrospinning of polytetrafluoroethylene (PTFE) nanofibrous membranes for high-performance triboelectric nanogenerators, ACS Appl. Mater. Interfaces 10 (6) (2018) 5880–5891.

[162] H.-Y. Mi, X. Jing, M.A.B. Meador, H. Guo, L.-S. Turng, S. Gong, Triboelectric nano-generators made of porous polyamide nanofiber mats and polyimide aerogel film: output optimization and performance in circuits, ACS Appl. Mater. Interfaces 10 (36) (2018) 30596–30606.

[163] H.-Y. Mi, X. Jing, Q. Zheng, L. Fang, H.-X. Huang, L.-S. Turng, S. Gong, High-performance flexible triboelectric nanogenerator based on porous aerogels and electrospun nanofibers for energy harvesting and sensitive self-powered sensing, Nano Energy 48 (2018) 327–336.

[164] Y. Yin, J. Wang, S. Zhao, W. Fan, X. Zhang, C. Zhang, Y. Xing, C. Li, Stretchable and tailorable triboelectric nanogenerator constructed by nanofibrous membrane for energy harvesting and self-powered biomechanical monitoring, Adv. Mater. Technol. 3 (5) (2018) 1700370.

[165] B. Yu, H. Yu, T. Huang, H. Wang, M. Zhu, A biomimetic nanofiber-based triboelectric nan-ogenerator with an ultrahigh transfer charge density, Nano Energy 48 (2018) 464–470.

[166] C. Garcia, I. Trendafilova, R. Guzman de Villoria, J. Sanchez del Rio, Self-powered pres-sure sensor based on the triboelectric effect and its analysis using dynamic mechanical analysis, Nano Energy 50 (2018) 401–409.

[167] Z. Li, J. Shen, I. Abdalla, J. Yu, B. Ding, Nanofibrous membrane constructed wearable triboelectric nanogenerator for high performance biomechanical energy harvesting, Nano Energy 36 (2017) 341–348.

[168] B. Zhang, L. Zhang, W. Deng, L. Jin, F. Chun, H. Pan, B. Gu, H. Zhang, Z. Lv, W. Yang, Z.L. Wang, Self-powered aceleration sensor based on liquid metal triboelectric nan-ogenerator for vibration monitoring, ACS Nano 11 (7) (2017) 7440–7446.

Tactile sensors based on buckle structure

Yuhuan Lv, Mingti Wang, Lizhen Min, and Kai Pan*
Beijing Key Laboratory of Advanced Functional Polymer Composites, State Key Laboratory of Organic-Inorganic Composites, College of Materials Science and Engineering, Beijing University of Chemical Technology, Beijing, China
*Corresponding author: e-mail address: pankai@mail.buct.edu.cn

10.1 Introduction

Tactile sensors have advanced rapidly in the past few decades on account of the progress in the manufacturing of advanced electronic materials and structures. With many examples ranging from experimental demonstrations to commercially available products, tactile sensors have aroused our interests and contributed to developments in various areas, such as flexible/stretchable electronics, safety monitoring system, personalized health monitoring, and smart clothing. As research continues, we consider that the structural design and material selection are the key factors, which affect the performance of tactile sensors all the time. So, if you were to access excellent tactile sensors, you not only attach importance to structural design, but also think conductive material over seriously. As one of the most widely used geometric structures, the buckle structures exist far and wide in nature. The research of buckled tactile sensors has developed a research boom due to the advantages in high stretchability, high mechanical extensibility, and comfortable contact with human–computer interaction through wearable flexible devices.

This chapter provides fundamental insights into the buckle structure preparation, and the content mainly revolves around several common methods for the preparation of buckle structures and conductive materials. Readers who are interested in buckled tactile sensors are expected to find it helpful.

10.2 Buckle structure in tactile sensor

Industries such as "Big Data" and "Internet of things (IOT)" are in the ascendant for the 5G era in recent years. The research into flexible electronic devices has gradually received widespread attention [1–3]. As the core of flexible electronic devices, high performance is an eternal theme in the tactile sensors' design and preparation.

During the fabrication of flexible tactile sensors, conductive materials often directly affect the sensor's performance, such as sensitivity, response time, and strain coefficient. Therefore, investigating conductive materials, including material selection and structural design, is the main way to improve the performance of flexible tactile sensors. In terms of the selection of conductive materials, traditional conductive materials, such as metal or semiconductor conductive materials [4, 5], because of the high electrical conductivity, have higher advantages in improving sensitivity and reducing energy consumption. However, these traditional conductive materials are all inorganic rigid materials, which are brittle. In addition, the preparation process of such inorganic conductive material devices is also cumbersome; so, they have greater limitations in the application of flexible sensors. With the research breakthrough of two-dimensional (2D) materials and the development of conductive polymer materials, the selection of conductive materials has gradually entered the second stage, such as the study of 2D materials graphene (GR) and its derivatives, conductive polymer polyaniline (PANI), polythiophene, polypyrrole (PPy), and other new conductive materials [6–8]. In addition to stable physical and chemical characteristics, these materials also have a variety of optical, electrical, magnetic, and thermal properties. Furthermore, it is easy to modify and compound that has a great advantage in the fabrication of the conductive layer of a flexible tactile sensor. By gradually optimizing the selection of conductive materials, flexible tactile sensors have achieved rapid development, but the sensor still has a large contradiction between device flexibility and high sensitivity at the same time.

To ensure flexibility and high sensitivity synthetically, the microstructural construction and microstructure design by combining the characteristics and flexibility of the conductive material is the main way to resolve the contradiction between device flexibility and sensitivity, which is also one of the main scientific problems in this area. Among the many geometric structures, the buckle is one of the rich and fascinating structures in nature; it can be said that the buckles are all over nature [9–13] (Fig. 10.1). Therefore, more and more researchers have focused their attention on the research of buckle structure sensors. The buckle structure, which is usually caused by the bending deformation of the material due to uneven force, is widely existing in nature. For example, there is a large number of buckle structures on the surface of the human skin (Fig. 10.1G). The structural design of the sensor with a pleated structure has many advantages [14–17]: (1) realizing the transformation of materials from 2D structure to three-dimensional (3D) structure, and improving the specific surface area of materials; (2) controlling the prestretching ratio can provide reserved stretching space to make the material both more flexible and stretchable; (3) the buckle structure can grasp and fix the surface material to improve the stability of the device during long-term use; and (4) the method of fabricating buckles is simple and easy, and it has the potential for practical development. Different from the traditional 3D structure, the structural characteristics and advantages of the buckle structure sensor need to be further developed. By selecting new conductive materials and combining the innovative structure design, it is expected to obtain an unprecedented new flexible sensor.

10.3 Methods of buckle structures

From magnificent mountains in the macro to skin surface in the micro, buckle structure is greatly common in the natural word. On account of the multiple existence patterns, there is difference in the specific morphology and modalities of buckle structures [9]. Hence,

Tactile sensors based on buckle structure 199

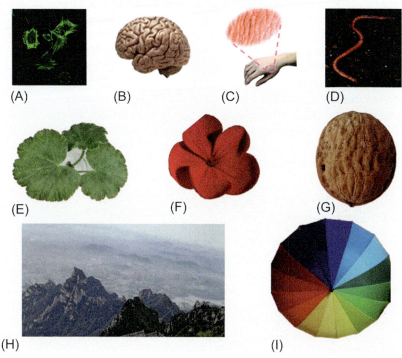

Fig. 10.1 Buckle structure in nature: (A) confocal laser scanning microscopy (CLSM) images of DPCs' (dental papilla cells) cell cytoskeleton with buckle surfaces (scale bar: 2 μm). (B) Highly buckled brain cortex with bumps and grooves (scale bar: 1 cm). (C) Human skin with buckled structures. (D) Earthworm with buckles. (E) Leaf with microscopic buckle structure. (F) Flower with numerous buckle structures. (G) Buckled walnut. (H) Majestic Mount Tai. (I) Umbrella with buckles. (E–I) Photos taken by the authors.
Panel (A) Reproduced with permission M. Bai, J. Xie, X. Liu, et al., Microenvironmental stiffness regulates dental papilla cell differentiation: implications for the importance of fibronectin-paxillin-beta-catenin axis, ACS Appl. Mater. Interfaces 10 (32) (2018) 26917–27. Copyright 2018, American Chemical Society, Panel (B) Reproduced with permission Z.M. Al-Majdoub, H. Al Feteisi, B. Achour, et al., Proteomic quantification of human blood-brain barrier SLC and ABC transporters in healthy individuals and dementia patients, Mol. Pharm. 16 (3) (2019) 1220–33. Copyright 2019, American Chemical Society, Panel (C) Reproduced with permission J. Jia, G. Huang, J. Deng, et al., Skin-inspired flexible and high-sensitivity pressure sensors based on rGO films with continuous-gradient wrinkles, Nanoscale 11 (10) (2019) 4258–66. Copyright 2019, American Chemical Society, Panel (D) Reproduced with permission E.J. Petersen, R.A. Pinto, L. Zhang, et al., Effects of polyethyleneimine-mediated functionalization of multi-walled carbon nanotubes on earthworm bioaccumulation and sorption by soils, Environ. Sci. Technol. 45 (8) (2011) 3718–24. Copyright 2011, American Chemical Society.

the fabrication methods of buckle are diverse, which conclude the microstructural construction and microstructure design, such as stretch–release, thermal contraction, and so on. However, it is found that each method has its advantages and deficiencies along with the fabrication process. In consequence, except in one way, combining two or multiple methods is also efficacious and frequently selected to prepare buckle structure, which utilize their advantages and features synthetically.

In consideration of its excellent performances of large relative surface area, obtaining stretchability, and nonuniform surfaces, the buckle structures are always introduced to the piezoresistive sensors, which have good prospects in outstanding properties. Herein, according to the scientific researchers previous reports, the frequently used fabrication methods of buckle structures applied in the piezoresistive sensors are summarized and generalized below.

10.3.1 Stretch–release

The stretch–release method refers to prestretching the elastic substrate and releasing it after coating the selected materials or relative material processing, which utilizes the mismatch of the mechanical properties in essence. The elastic substrates are usually required for stretch–release method and there are diverse kinds of frequently used materials, such as polydimethylsiloxane (PDMS) [18], thermoplastic polyurethane (TPU), acrylic ester, and other stretchable materials [19]. As for the stretch–release method, the stretching direction, stretching ratio, and coating thickness of selected materials are critical factors for the buckle structures' preparation. With regard to the stretching direction, it can be summarized as uniaxial stretching, biaxial stretching, and multiaxial stretching approximately [20, 21]. Uniaxial stretching means that the strain occurs in one direction along the long side intensively and the other short side provides the constraint [22]. The formative buckle structures are regular and arranged orderly on the whole. However, the biaxial or multiaxial stretching can construct relatively complex and neoteric buckle structures in contrast to the uniaxial stretching [23]. Furthermore, the stretching ratio is pivotal to the wavelength, which is an effective measure to regulate the buckles. The larger strain ratio is, the denser the wavelength [24]. In particular, the coating thickness is the principal element for the structure, which also determines the wavelength. In conclusion, it's important to consider the combination of the three factors discussed above to design and fabricate the ideal buckle structures [25, 26]. Typically, as shown in Fig. 10.2, Pan's group proposed the buckle GR films for the piezoresistive sensors with the different gradient direction, which expounds the correlation of stretching direction, stretching ratio, and coating thickness better. And, two modes of gradient buckle structures were presented [11, 27]. The first mode signifies that the stretching direction is paralleled with the gradient direction, and another mode is perpendicular. The gradient structure was realized by the nonuniform coating thickness. Along the gradient direction, the coating thickness increased gradually, and the formative wavelength became larger synchronously. Furthermore, it is found that these gradient buckle structures are sequential and well organized, which provide potential for pressure-sensitive properties. This category of GR film sensor with gradient buckle structures could not only exhibit the excellent properties, like high sensitivity and wide detection range, but also has the

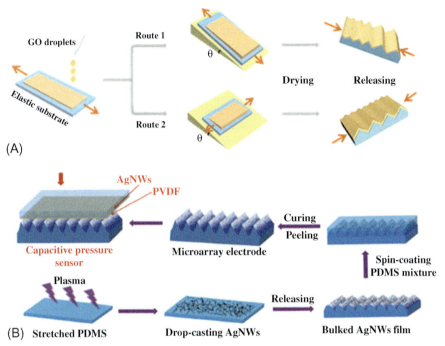

Fig. 10.2 Schematic diagram of stretch approaches for buckle structures. (A) Gradient wrinkle structure with different gradient direction based on GR materials. Picture made by the authors. (B) The buckle PDMS film prepared by stretched.
Reproduced with permission X. Shuai, P. Zhu, W. Zeng, et al., Highly sensitive flexible pressure sensor based on silver nanowires-embedded polydimethylsiloxane electrode with microarray structure, ACS Appl. Mater. Interfaces 9 (31) (2017) 26314–24. Copyright 2017, American Chemical Society.

function of integrating and amplifying output signals. Those superior performances lay the foundation for motion signal detection, visual programming, and other application. Furthermore, Wong's group presented the buckle silver nanowires (AgNWs) films on the PDMS substrate through the stretch method [28]. And, the capacitive pressure sensor with buckle was developed with high sensitivity and flexibility, which can be applied in electronic skins, wearable health-care monitors, and noncontact detection.

To sum up, the stretch–release method is simple and easy to implement. The conditions of stretching direction, ratio, and coating thickness are important for regulating the buckle structures to acquire a targetable structure.

10.3.2 Mold

For the sake of simplicity and convenience, the buckle structure could be obtained by the mold of substrates. There are plenty of materials to be selected according to the requirements and demands, such as polymer materials and metal materials [29]. With the development of technologies and science, the fabrication methods of molds are multitudinous, which include printing, etching, sculpturing, and so on. The mold foundation is to be coated with the conductive materials to form the buckle structure. Hence, the coating methods should also be taken into account for the match of the conductive materials and mold substrates to acquire the ideal buckle morphology, like electroplating, spin coating, and so on. For instance, Bao's group selected the wafer mold to prepare the elastic conducting polymer of PDMS and PPy with replicated buckle microstructures [8]. And Cho's research group dropped the AgNW solutions on the wrinkle template to obtain the self-aligned wrinkled AgNWs simply [30]. Inspired by the *Epipremnum aureum* leaves, Zhan's group proposed a PDMS/CNT pressure sensor with unique microstructures through molding [31] (Fig. 10.3). Furthermore, the designed sensors exhibit the excellent performances of ultrahigh sensitivity, a wide dynamic range, and rapid response times, and have great potential in human–computer user interfaces, robotics, and industrial monitoring.

In a word, the key point of constructing structure using molds is mold design and manufacture. The micro- or macrostructures of mold show great impact on the properties of sensors, and it is the foremost step to select the mold materials appropriately and design the mold structure reasonably.

10.3.3 Thermal

The thermal method refers to utilizing the expansion and shrinkage of the substrate by heating and cooling, which is a little similar to the omnibearing stretch release. The conductive materials are coated on the expanded substrate and the buckle structures are obtained after cooling, on account of the mismatching of the flexibility and ductility in the conductive materials and substrates [32]. There are many kinds of substrate

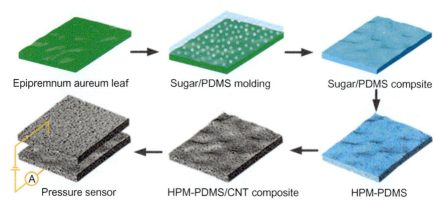

Fig. 10.3 Schematic diagram of mold approaches for wrinkle structures constructed by PDMS and CNT.
Reproduced with permission T. Zhao, T. Li, L. Chen, et al., Highly sensitive flexible piezoresistive pressure sensor developed using biomimetically textured porous materials, ACS Appl. Mater. Interfaces 11 (32) (2019) 29466–73. Copyright 2019, American Chemical Society.

materials expended by heating, like polyethylene (PE) and polyvinyl chloride (PVC), which should be selected according to the characteristic of the conductive materials [33]. Furthermore, the heating time and thermal temperature are primary for the final formative buckle morphology. Michelle's group proposed highly flexible buckled carbon nanotube film by a thermal method [15]. Polystyrene (PS) was chosen as the substrate and shrunk by heating after coating the carbon nanotubes (CNTs) to fabricate the buckle microstructure. The application of the sensors for detecting human motion by mounting the sensors onto the joint areas of the finger, elbow, and knee was demonstrated with the high sensitivity. Analogously, the pressure sensor based on MXene and PS substrate through a thermal method was prepared by Chen (Fig. 10.4), which was flexible and exhibited high sensitivity under the high-precision pressure [34] (Fig. 10.4).

The thermal method is frequently used and the different buckle structures can be acquired through selecting the substrate and adjusting the time and temperature properly.

10.3.4 Swell

The swelling method is analogous to the thermal method and the distinction lies at the different external stimuli, which is utilizing the expansion and shrink of the substrate by solvent stimulus. Furthermore, the different substrate materials match the diverse solvents for stimulus. It's nonnegligible to select the conductive material, substrate materials, and solvents legitimately and properly. And, the performances of the pressure sensors are correlative with the deposition time and swelling time. On account of

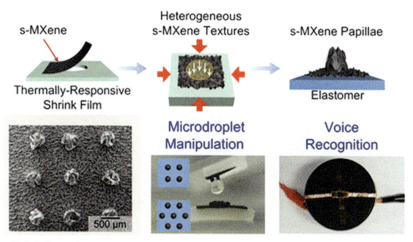

Fig. 10.4 Schematic diagram of thermal approaches for wrinkle structures. The wrinkle structures were prepared by MXene and PS substrate.
Reproduced with permission Y. Zhang, T.H. Chang, L. Jing, et al., Heterogeneous, 3D architecturing of 2D titanium carbide (MXene) for microdroplet manipulation and voice recognition, ACS Appl. Mater. Interfaces 12 (7) (2020) 8392–402. Copyright 2020, American Chemical Society.

the omnibearing swell and shrink, the buckle structures often look anomalous and irregular, but look uniform on the whole. Wang's research group has presented a buckled pressure sensor induced by polymer swelling [35]. The conductive layer of Ag materials was attached to the PDMS by electroless deposition and the Ag buckles were formed via polymer swelling (Fig. 10.5). The pressure sensor, combined the buckle structure with a patterned or uneven elastic substrate, offers great promise for improving the testing sensitivity with high reliability and repeatability.

As for the swelling, it mainly refers to the difference swelling ratio between the conductive material layers and the substrates in essence. It is a versatile strategy for depositing conductive buckles on elastic substrates, which can resolve the conglutination and attachment of the conductive layers and substrates to a certain extent.

10.3.5 Composite methods

It is conspicuous that each method has its advantages and characteristics. How to utilize the advantages reasonably is a problem to be solved urgently. The research results demonstrate that combining two or multiple methods to fabricate the sensors is an effective approach to exploit the advantages to the full. At the same time, the associations of multiple methods provide the conveniences and approaches to construct the hierarchical structure easily, which have potential for developing the prominent performances of sensors. The selection of the preparation method is according to the premeditated structures and functions of pressure sensors. For example, the thermal and stretch–release method is put into use to stretch GR pressure sensors with hierarchical

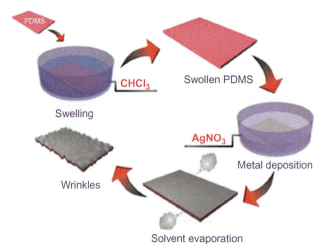

Fig. 10.5 Schematic diagram of swell approaches for wrinkle structures prepared via Ag deposition on swollen PDMS.
Reproduced with permission N. Gao, X. Zhang, S. Liao, et al., Polymer swelling induced conductive wrinkles for an ultrasensitive pressure sensor, ACS Macro Lett. 5 (7) (2016) 823–7. Copyright 2016, American Chemical Society.

Tactile sensors based on buckle structure 205

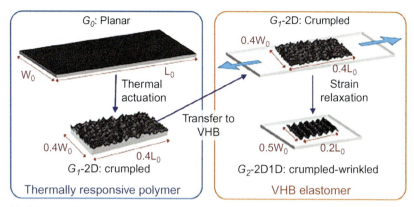

Fig. 10.6 Schematic diagram of typical composite methods for wrinkle structures. The composite methods combined the stretch and thermal shrink.
Reproduced with permission T.H. Chang, Y. Tian, C. Li, et al., Stretchable graphene pressure sensors with Shar-Pei-like hierarchical wrinkles for collision-aware surgical robotics, ACS Appl. Mater. Interfaces 11 (10) (2019) 10226–36. Copyright 2019, American Chemical Society.

buckles, which is inspired by the Shar Pei skin [36] (Fig. 10.6). The conductive material was deposited on the PS substrate and formed the crump microstructures by the thermal stimulus. Then, the conductive material layer was transferred to the prestretch substrate and the buckles were constructed by the release of substrate. The successful implementation of hierarchical crump-buckle structures prevented the films from the development of pores and cracks under deformation and provided the proposal for the further applications of soft robots. The fabrication process is so complex that the interrelated effects of multiple methods should be taken into account synthetically. The GR-based piezoresistive pressure sensor was proposed through the stretch and laser method [37]. The first step is prestretching the substrate to get buckles; second, precise laser was applied to construct microstructures on the buckles. The two phases accomplished the manufacture of hierarchical buckle with different specifications and scales. And, the precise and rigid performances of laser lay the foundation for the design of sophisticated structures and targeted applications [38]. Furthermore, the composite method combines and highlights the advantages of both buckled structure on flexibility and laser on efficient reduction, which can be applied in various fields.

Its obvious that the composite methods can combine the peculiarities and features felicitously to obtain the predicted microstructures and accomplish the better performances, which provides possibility to design and fabricate smart sensors.

Above all, there are multitudinous methods for constructing the buckle structures, such as mold, stretch, thermal, laser, and so on. To select the suitable fabrication method, the conductive materials, presupposed morphology, application area, and other impact factors should be considered from all angles. It's essential to take full advantages of the properties of the fabrication method, which shows the potential and promising future in the piezoresistive pressure sensors for smart robots, motion detection, and other intelligent application fields.

10.4 Conductive materials for buckled tactile sensor

The selection of conductive material is another factor that affects the tactile sensors' characteristics. With the emergence of nanomaterials, until now, the development of nanomaterials has gradually matured and is divided into zero-dimension (0D) nanomaterials, one-dimension (1D) nanomaterials, two-dimension (2D) nanomaterials, etc. by scale, which are different from conventional materials, even when made of the same substance. Many physical and chemical properties, such as melting point, vapor pressure, optical properties, chemical reactivity, magnetism, superconductivity, and plastic deformation of nanomaterials, show special properties due to the small size effect, surface effect, quantum size effect, macroscopic quantum tunneling effect, and dielectric limit region of nanomaterials [39, 40].

Nowadays, more people focus on improving the performance of electronic devices particularly in the strain sensor with good flexibility, wide sensing range, and high sensitivity. However, conventional electronic devices based on metal or semiconductor materials, such as sensors, usually show low sensitivity and a narrow sensing range. In order to solve these problems, nanoconductive materials are selected, such as 0D nanoparticles, 1D CNTs, nanofibers, metal nanowires, 2D GR and GR derivatives, GR-like materials such as MXenes, etc., and some conductive polymers such as PPy, poly(3,4-ethylenedioxythiophene) (PEDOT), etc. Nanomaterial strain sensors have higher controllable sensitivity, but these strain sensors cannot detect a large range due to the fact that the single material can only form an inherent conductive network under stress, and that the limited connection between sensitive materials caused by the small strain results in the low sensitivity. Therefore, it cannot be used to detect the larger movement from small deformation to substantial deformation. The hybrid composite of various conductive materials is an effective way to generate new materials and meet the demand. The synergistic effect of the properties of various materials can improve the sensitivity and stability of the sensor itself [41]. Through the selection of conductive materials and the structural design, the conductive layer and base layer is beneficial to improve the working performance and stability of the sensor itself. The curved structure design is often used to fabricate the sensitive layer of the tactile sensor to improve the performance of the sensor. In this section, we summarize the conductive materials selected by researchers in recent years for the preparation of flexible sensors and the structural design of different materials, which will be summarized one by one in the following narration.

10.4.1 0D materials

Zero-dimension (0D) nanomaterials such as metal nanoparticles and nanocrystals are widely used in the flexible electronic devices due to the plasticity of their morphological structure (Fig. 10.7). The possibility of reconstructing the conductive network by means of dynamic mechanical deformation and high surface area is the reason why 0D nanomaterials can be used in flexible devices. Considering that real-time, continuous health monitoring and disease management are essential for wearable biosensors, the

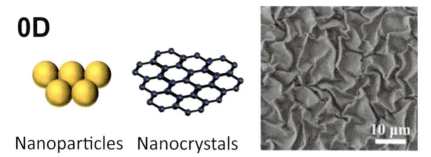

Fig. 10.7 The 0D material morphology and a scanning electron microscope (SEM) image of the 2D Ag wrinkled film on flat PDMS surface.
Reproduced with permission N. Gao, X. Zhang, S. Liao, et al., Polymer swelling induced conductive wrinkles for an ultrasensitive pressure sensor, ACS Macro Lett. 5 (7) (2016) 823–7. Copyright 2016, American Chemical Society.

conductors that remain conductive under strain and the strategies based on stretchable materials or structures, which allow the electrodes to withstand significant strains before they lose conductivity, are the basic building blocks of these systems.

The sensitivity of the conductive layer with nonplanar morphology is much higher than that of the flat film. A continuous layer of Ag nanoparticles was deposited on the surface of the swelling PDMS by chemical deposition. After solvent shrinkage, wrinkled silver film was formed on the PDMS soft substrate. Silver nanoparticles are used to ensure the conductivity of the whole sensor and provide a certain tolerance for the existence of fold structure, which provide the basic guarantee for the output signal of the whole stretchable sensor. By adjusting the deposition time of the polymer substrates, the amplitude and wavelength of wrinkling can be adjusted to meet the requirements of ultrasensitive pressure sensors [35].

In addition to metal-based nanoparticles, carbon-based 0D materials also play an important role in flexible sensors with buckle structures. Compared with metal nanoparticles, carbon-based nanoparticles have better mechanical properties, stronger surface activity, and easier surface recombination with other materials.

Flexible strain sensors have high requirements for detecting various movements of the human body, but how to integrate high sensitivity and high stretchability into a flexible strain sensor at the same time is still a challenge. High stretchability can ensure a large working range and improve the durability of the strain sensor, while high sensitivity can make the strain sensor capture slight changes in motion. Equipped with high sensitivity and high tensile performance on a flexible strain sensor at the same time can effectively broaden the application range. However, there is usually a trade-off between the high stretchability and high sensitivity.

Graphene nanocrystalline carbon (GNC) film is a carbon film in which GR nanocrystals are embedded in an amorphous matrix, trap excess electrons, and act as a good channel for electron conduction. The dense distribution of graphene nanocrystals also indicates the tunneling effect of adjacent graphene nanocrystals in GNC films. Therefore, a GNC film based on the special nanoscale structure has good potential to obtain high gauge factor (GF). On the other hand, the use of buckles or cracks on the micro-scale structure design can effectively improve the tensile performance of the strain sensor [42, 43]. Based on the multiscale structure design idea, Diao's group prepared a supersensitive, high-tensile strain sensor using a GR nanocrystalline film with the buckle structure [43].

From the perspective of structure, compared to 1D materials stacked grid, 0D materials stacked structure often have a very high density. Compared with 2D materials, 0D nanomaterials cannot directly prepare the buckle structure, so 0D materials are often used as fillers or additional parts in shaping conductive materials to give higher conductive properties to conductive sensing layers and the possibility of shaping tiny multilevel structures.

10.4.2 1D materials

One-dimension (1D) nanomaterials such as nanofibers, nanorods, carbon nanotubes, and nanowires, due to their strong plasticity, are easy to adhere to 0D particles, and have strong interface bonding (Fig. 10.8). The surface of 1D nanomaterials is prone to generate tiny buckled structures due to the interlattice interaction or chemical treatment. The continuous development of electrospinning technology and new 1D nanomaterials affect and promote each other. And, the plasticity of 1D materials makes it extremely easy to form films, and build other multidimensional structures, which provides conditions to fabricate the buckle structure and then prepare the stretchable flexible device. Moreover, 1D nanomaterials can easily form a loose mesh structure; so, the structure constructed by 1D materials can be either isotropic or anisotropic. It provides the possibility for the preparation of flexible devices with high expansion and contraction characteristics [9].

Metal-based materials are also widely used as conductive layers in flexible electronic devices due to their excellent electrical properties [44]. Among them, AgNWs can be used to make transparent conductors due to the loose structure of the nanowire grid that are most used in the sensors' sensitive layers. Moreover, the high electrical conductivity of AgNWs provides a lower resistance for the conductor, and the adhesion of AgNWs on the flexible substrate provides good cycle stability for the electronic devices [45].

With the continuous development and application of carbon-based materials, the most used 1D materials for preparing stretchable flexible devices, carbon-based 1D materials: CNTs. There are certain functional groups bound to the surface, and the CNTs obtained by different preparation methods have different surface structures due to the different preparation methods and the posttreatment processes. CNTs are not always straight, but there are bumps and depressions in local areas [40, 46–48].

The working principle of a piezoresistive sensor is to detect changes in resistance during driving. This is a typical structure coupling two conductive rough surfaces together. By doing so, the number of electrical contacts can be changed by applying mechanical pressure; thereby, the resistivity between the electrodes effectively increases or decreases. Each individual carbon nanotube is entangled and adhered to each other by van der Waals force to form a fused film [46]. The entanglement

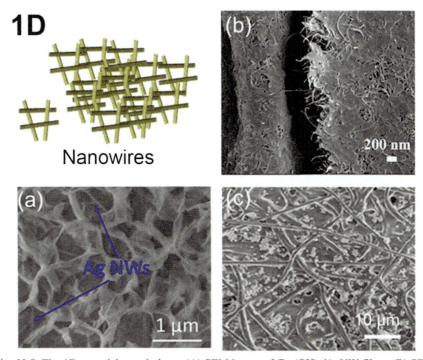

Fig. 10.8 The 1D material morphology: (A) SEM image of Co (OH)$_2$/AgNW films. (B) SEM image of the CNTs/PU yarn. (C) SEM image of the elastic conductor reinforced by PVDF nanofibers.
Panel (A) Reproduced with permission H. Sheng, X. Zhang, Y. Ma, et al., Ultrathin, wrinkled, vertically aligned Co(OH)2 nanosheets/ag nanowires hybrid network for flexible transparent supercapacitor with high performance, ACS Appl. Mater. Interfaces 11 (9) (2019) 8992–9001. Copyright 2019, American Chemical Society, Panel (B) Reproduced with permission H.L. Sun, K. Dai, W. Zhai, et al., A highly sensitive and stretchable yarn strain sensor for human motion tracking utilizing a wrinkle-assisted crack structure, ACS Appl. Mater. Interfaces 11 (39) (2019) 36052–62. Copyright 2019, American Chemical Society, Panel (C) Reproduced with permission H. Jin, M.O.G. Nayeem, S. Lee, et al., Highly durable nanofiber-reinforced elastic conductors for skin-tight electronic textiles, ACS Nano 13 (7) (2019) 7905–12. Copyright 2019, American Chemical Society.

of the percolation network makes the buckle CNTs film more elastic, which provides deformation possibilities for the electronic devices. The uniaxial or biaxial winkles are obtained through different shrinking methods, which will not be repeated here [49]. When buckle CNTs are assembled on the PDMS-patterned film, the strain sensor can have sensitive resistance responses to various deformations, and produce different resistance responses under bending, compression, and strain. In addition, through electrodeposition of PANI, the carbon nanotube film has developed into a flexible supercapacitor, which has super flexibility. After different deformations and different cycles, the capacitance retention rate can be stabilized at a certain value, which indicates that the use of buckle CNT-based electrodes with fixed microstructures is conducive to obtaining an excellent electrochemical performance [50].

With the upgrading of the electrospinning technology, the different structure nanofiber materials have emerged as the times require. The polymer nanofiber network in the pressure sensor provides conditions for improving the sensitivity of the sensor and expanding the detection range. The nonconductive nanofiber structure realizes the internal deformation transition of the stress sensor sensitive layer after being subjected to stress. In addition, a frame structure is provided for the conductive layer, which improves the stability of the conductive layer and further improves the strain response stability of the overall sensor [51–53].

10.4.3 2D materials

Two-dimensional (2D) materials such as GR, GR derivatives, and MXenes are prone to buckles and surface defects on their 2D structures (Fig. 10.9). After passing through other processing methods, larger-sized buckles will be formed. For the 2D material itself it is easy to build a conductive network, and provides performance support for the subsequent preparation of piezoresistive sensors. The mechanical properties of 2D materials are good. For example, Young's modulus has good advantages in terms of volume, compressive strength, and impact strength. In the process of preparing the sensor, the operation is simple and the plasticity is good. Hence, there are many types of 2D materials that can be selected according to different preparation targets to achieve the desired product and assembly effect, and then improve the performance of the prepared products [54–56].

Among them, GR is a 2D single-atomic layer thick honeycomb planar carbon atom layer structure [55], and the bond energy is very strong and cannot be destroyed because the carbon atoms in GR are linked by three δ covalent bonds [55]. Hence, the GR lattice structure is very stable. In addition, GR also has ultralight and ultrathin characteristics, and its elastic modulus is extremely large. GO is an oxide of GR; while maintaining the structure of GR itself, through chemical methods, a large number of oxygen-containing groups are introduced into the GR molecular structure [57]. Therefore, in terms of thermal, optical, electrical, and biocompatibility, GO has a wide range of development prospects, but the conductivity is seriously affected. Reduced graphene oxide (rGO) removes the oxygen-containing functional groups to achieve chemical and physical reduction of GO. Due to the removal of polar functional groups, the hydrophilicity of rGO is weakened, and the conductive sites are reduced [58] but it can still show excellent thermal and optical properties [59]; and due to the loss of polar functional groups, the π-bond connection between rGO layers has excellent electrical conductivity. Due to its excellent properties, rGO has become a popular material for making buckled structures and even stretchable flexible devices. rGO itself is a 2D nanostructure, which can be stacked according to the expected structure to equip different flexible structures [36].

MXene is a type of transition metal carbide, nitride, or carbonitride [60], which has plentiful structures, excellent conductivity close to metal, good hydrophilicity, and magnetic properties. In addition, MXene also has good UV absorption properties.

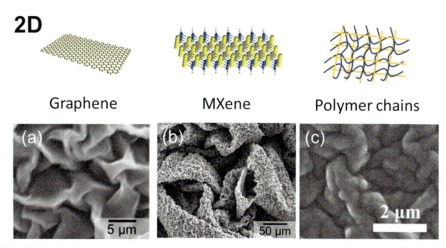

Fig. 10.9 The 2D material morphology: (A) SEM images of a rGO buckled film. (B) SEM images of MXene-CB (carbon black nanoparticles) film structures. (C) SEM image of the microstructured PPy films with the three-scale nested surface buckles.
Panel (A) Reproduced with permission T.H. Chang, Y. Tian, C. Li, et al., Stretchable graphene pressure sensors with Shar-Pei-like hierarchical wrinkles for collision-aware surgical robotics, ACS Appl. Mater. Interfaces 11 (10) (2019) 10226–36. Copyright 2019, American Chemical Society, Panel (B) Reproduced with permission T.H. Chang, T.R. Zhang, H.T. Yang, et al., Controlled crumpling of two-dimensional titanium carbide (MXene) for highly stretchable, bendable, efficient supercapacitors, ACS Nano 12 (8) (2018) 8048–59. Copyright 2018, American Chemical Society, Panel (C) Reproduced with permission C.F. Yang, L.L. Li, J.X. Zhao, et al., Highly sensitive wearable pressure sensors based on three-scale nested wrinkling microstructures of polypyrrole films, ACS Appl. Mater. Interfaces 10 (30) (2018) 25811–8. Copyright 2018, American Chemical Society.

Thanks to the hexagonal lattice, the presence of a metal layer makes it rigid and show higher Young's modulus [57].

Since MXene has higher mechanical stiffness and weaker interlayer interactions, increasing its stretchability is still a big challenge. Therefore, it is necessary to further develop the assembly technology of mechanically stable MXene architecture. Utilizing the instability of the interface, the regular buckle structure of the high-dimensional MXene nanocoating can be generated. These textures can be used to make controlled wetting surfaces, high-area capacitive electrodes, and transfer the curled MXene nanocoating to the elastomer to make MXene/elastomer with high tensile properties [61].

In addition to GR-based materials, GR-like materials, and 2D structural materials constructed from 1D nanomaterials, there is also a class of polymer-based materials that can self-buckle to shape 2D film materials with surface microstructures used to prepare flexible electronic devices. Compared with metal electrodes, all-organic electrodes have the advantages of simple fabrication, low cost, and large flexibility.

Poly(3,4-ethylenedioxythiophene) (PEDOT) is one of the most used polymer-based materials. For example, the surface of the poly(3,4-ethylenedioxythiophene): polystyrene sulfonate (PEDOT: PSS) conductive film was attached to a PDMS substrate, and the conductive film was contracted by the substrate's relaxation effect. Then, the high-performance skin-like electronics were obtained. By grafting poly (methacrylamide-co-acrylic acid) (PMAAc) on the surface of PDMS to improve the stretchability and stability of the electrode, the PMAAc forms hydrogen bonds with the PEDOT: PSS film, and introduces deep buckles into the conductive layer to form a strong interaction with it [62].

Except the PEDOT, the PPy is also widely investigated by researchers because it can produce self-wrinkling. To give an example, in a mixed acid solution, when the PPy film was oxidized and polymerized on an elastic polysiloxane substrate, it buckled in situ to produce two-scale nested buckles. Subsequent heating/cooling treatment can induce the surface buckle, thereby controlling the formation of a three-scale nested buckled microstructure. The multiscale nested microstructure combines the characteristics of buckled morphology and adaptive ability of the PPy film to provide the piezoresistive pressure sensor with high sensitivity, low detection stress, ultrafast response, excellent durability, and stability [63]. In addition, the PPy film can also be used directly to construct the conductive sensitive layer of the sensor, and it can also cooperate with other nanomaterials to improve the sensitivity and stability of the sensor [52].

10.4.4 Hybrid composite materials

The hybrid composite between 0D, 1D, and 2D materials can achieve the synergistic effect of "1 + 1 > 2," which greatly improves the performance of the raw materials, and may even have new properties different from the raw materials. According to the desired performance, the expectant hybrid composite materials are obtained by combining multiple dimension materials with a suitable coupling method, with which clever composite structures such as bionic structures, layer-by-layer structures, island structures, and buckle structures are designed. You can find the same and different characteristics between different materials during the experiment, and extract the target characteristics from them.

Hybrid composite materials usually include but are not limited to the composite of different materials in the same dimensions, such as nanowires and CNTs, and similar materials in different dimensions, such as the composite of carbon nanotubes and GR. Through the hybridization of different materials, targeted enhanced function can be given to the device, and then electronic devices with multifunctional sensing can be obtained.

Composites of nanomaterials in the same dimension, such as the self-assembled multiwalled carbon nanotubes (MWCNTs) networks and the conductive polymer PANI array on the surface of a glassy carbon electrode can be synthesized by electrochemistry in situ. The ordered PANI nanowire array and MWCNT network was integrated to form the 3D-nanostructured PANI/MWCNT composite material. The PANI/MWCNT nanocomposites have excellent electrical properties, high specific capacitance, and good cycle stability due to their unique microstructure and conductivity [64].

Although the recombination of nanomaterials with the same dimension can provide support for the new material in terms of strain stability and electrical properties, it cannot greatly change the percolation network structure inside the overall material.

Tactile sensors based on buckle structure 213

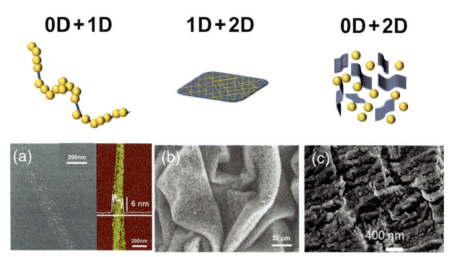

Fig. 10.10 Hybrid composite materials morphology: (A) 0D + 1D: SEM image and AFM image of Ag nanoparticles grown on a GO nanoribbon. (B) 1D + 2D: SEM image of the crumpled nanofiber membrane. (C) 0D + 2D: field-emission scanning electron microscopy (FE-SEM) images of wrinkle, stretchable, nanohybrid fiber (WSNF) of rGO/PU/Au.
Panel (A) X. Zhou, G. Lu, X. Qi, et al., A method for fabrication of graphene oxide nanoribbons from graphene oxide wrinkles, J. Phys. Chem. C. 113 (44) (2009) 19119–19122. Copyright 2009, American Chemical Society, Panel B Reproduced with permission Z. Qin, Y.Y. Yin, W. Z. Zhang, et al., Wearable and stretchable triboelectric nanogenerator based on crumpled nanofibrous membranes, ACS Appl. Mater. Interfaces 11 (13) (2019) 12452–9. Copyright 2019, American Chemical Society, Panel (C) Reproduced with permission P.T. Toi, T.Q. Trung, T.M. L. Dang, et al., Highly electrocatalytic, durable, and stretchable nanohybrid fiber for on-body sweat glucose detection, ACS Appl. Mater. Interfaces 11 (11) (2019) 10707–17. Copyright 2019, American Chemical Society.

Therefore, researchers are more inclined to recombine nanomaterials with different dimensions (Fig. 10.10), such as 0D + 1D, 1D + 2D, and 0D + 2D binary composites, and even 0D + 1D + 2D ternary composites, but currently ternary material composites rarely use and introduce buckle structures to the stretchable electronic devices.

Except as shown in the picture, hydroxyl-functionalized multiwalled carbon nanotubes (OH-fMWCNTs) coated with silver nanoparticles can provide highly sensitive sensing performance. Through the stretching and deposition methods, Ag@OH-fMWCNTs and PDMS were fabricated into a flexible strain sensor with buckle sandwich structure. And, the structure prepared by covering 0D materials with 1D materials can provide highly sensitive sensing performance [65, 66].

The GO was deposited on the prestretched PDMS, then reduced, and fixed on an electrospinning device. Polyvinylidene fluoride-hexafluoropropylene (PVDF-HFP) was spun on it, which can convert the unstretched flexible nanofiber film into stretchable material. The PVDF-HFP nanofibers have been used as charged materials due to their excellent negative frictional polarity and high specific surface area. The introduction of the pleated structure has achieved excellent flexibility and ductility.

The layer-by-layer stacking of the 0D nanofiber grid and the 2D rGO film combines the excellent properties of the two [67].

A buckled, stretchable nanohybrid fiber has the advantages of stretchability, high sensitivity, low detection limit, high selectivity to interferences, and high environmental stability. It combines 2D materials with 0D materials such as nanohybrid fibers with Au nanoparticles to obtain a new material through electrospinning. The method can realize multiple functions of sensing detection, multilevel structure nesting design, and the synergy of multiple materials, which makes the volume of the sensor reach a smaller range and improves the stability of the sensor. In daily life, it is more conducive for people to continuously monitor human physiological characteristics in real-time [68].

In addition, there is another issue worthy of attention in flexible electronic devices: the selection of substrate materials. From the method of forming the material buckle structure, it is not difficult to find that, due to the rigidity of the conductive material itself, the buckles of the whole material are mainly caused by the deformation of the substrate to drive the conductive material. For example, the conductive groups on the surface of the VHB 4910 (acrylic adhesive tape, 3 M Corporation) and a certain elasticity make it possible to tightly combine with conductive materials to ensure the stability of the sensor [11, 17, 27]. Due to the plasticity of PDMS during the polymerization process, it can be reengraved with a template [50, 69–71]. The PS substrates can use simple thermal shrinkage to obtain the desired buckle structure. The rough fabric and fiber substrates offer the possibility of multilevel nesting designs [15, 49, 61]. Therefore, the high affinity between the conductive material and the substrate is also an important factor to ensure the formation of the target structure and the sensitivity and stability of the overall device.

The development of materials can promote the development of related fields, and usually the newer and higher requirements for materials appearing in the development process of various fields will stimulate the birth of new materials. The pursuits of higher sensitivity performance, faster response time, wider detection range, better antiinterference performance, longer service life, and better mechanical properties have all proposed higher demend for the development and preparation of materials. How to make full use of the excellent performance of a material and avoid its shortcomings, how to find a balance among various performance requirements and develop new materials based on the existing materials, and how to design a new type of composite structure to make materials work together without the problem of performance inhibition between materials are still problems that we need to think about and explore.

10.5 Overview

Tactile sensors have developed rapidly in the past few decades due to the demand for excellent advanced electronics industries and structures. Some potential applications in tactile sensors such as flexible electronic devices, personal health monitoring, intelligent skin, and security detection systems have attracted many researchers' attention.

Usually, tactile sensors are composed of a rigid substrate and flexible conductive materials. There is a trade-off effect between these, which severely limits the development needs of tactile sensors. How to improve the relationship between them to obtain high-performance, high-sensitivity, and high-flexibility tactile sensors is the main direction of future research. Structure and conductive materials are the main factors determining sensor performance. Tactile sensors based on the buckle structure show excellent performance advantages. It is foreseeable that the innovative design of sensor composite structures, including the regulation of macro- and microstructures, combined with the characteristics of conductive materials, will further promote the development of high-performance tactile sensors.

References

[1] Z. Liu, D. Qi, G. Hu, et al., Surface strain redistribution on structured microfibers to enhance sensitivity of fiber-shaped stretchable strain sensors, Adv. Mater. 30 (5) (2018), 1704229.

[2] S. Wang, J. Xu, W. Wang, et al., Skin electronics from scalable fabrication of an intrinsically stretchable transistor array, Nature 555 (7694) (2018) 83–88.

[3] I. You, S.-E. Choi, H. Hwang, et al., E-skin tactile sensor matrix pixelated by position-registered conductive microparticles creating pressure-sensitive selectors, Adv. Funct. Mater. 28 (31) (2018), 1801858.

[4] S. Gong, W. Schwalb, Y. Wang, et al., A wearable and highly sensitive pressure sensor with ultrathin gold nanowires, Nat. Commun. 5 (2014) 3132.

[5] J. Lee, S. Kim, J. Lee, et al., A stretchable strain sensor based on a metal nanoparticle thin film for human motion detection, Nanoscale 6 (20) (2014) 11932–11939.

[6] Y. Ma, N. Liu, L. Li, et al., A highly flexible and sensitive piezoresistive sensor based on MXene with greatly changed interlayer distances, Nat. Commun. 8 (1) (2017) 1207.

[7] Y. Jiang, C. Hu, H. Cheng, et al., Spontaneous, straightforward fabrication of partially reduced graphene oxide-polypyrrole composite films for versatile actuators, ACS Nano 10 (4) (2016) 4735–4741.

[8] L. Pan, A. Chortos, G. Yu, et al., An ultra-sensitive resistive pressure sensor based on hollow-sphere microstructure induced elasticity in conducting polymer film, Nat. Commun. 5 (1) (2014), 3002.

[9] X. Hu, Y. Dou, J. Li, et al., Buckled structures: fabrication and applications in wearable electronics, Small 15 (32) (2019), e1804805.

[10] B. Gao, A. Elbaz, Z. He, et al., Bioinspired Kirigami fish-based highly stretched wearable biosensor for human biochemical-physiological hybrid monitoring, Adv. Mater. Technol. 3 (4) (2018), 1700308.

[11] J. Jia, G. Huang, J. Deng, et al., Skin-inspired flexible and high-sensitivity pressure sensors based on rGO films with continuous-gradient wrinkles, Nanoscale 11 (10) (2019) 4258–4266.

[12] R. Wang, Z. Liu, G. Wan, et al., Controllable preparation of ordered and hierarchically buckled structures for inflatable tumor ablation, volumetric strain sensor, and communication via inflatable antenna, ACS Appl. Mater. Interfaces 11 (11) (2019) 10862–10873.

[13] T.S.Z. Aida, Y. Yamauchi, An anisotropic hydrogel actuator enabling earthworm-like directed peristaltic crawling, Angew. Chem. Int. Ed. 57 (2018) 15772–15776.

[14] S.J.K.J. Park, M. Chu, et al., Flexible piezoresistive pressure sensor using wrinkled carbon nanotube thin films for human physiological signals, Adv. Mater. Technol. 3 (2017), 1700158.

[15] S.-J. Park, J. Kim, M. Chu, et al., Highly flexible wrinkled carbon nanotube thin film strain sensor to monitor human movement, Adv. Mater. Technol. 1 (5) (2016), 1600053.

[16] W. Chen, X. Gui, L. Yang, et al., Wrinkling of two-dimensional materials: methods, properties and applications, Nanoscale Horiz. 4 (2) (2019) 291–320.

[17] Y. Qiu, M. Wang, W. Zhang, et al., An asymmetric graphene oxide film for developing moisture actuators, Nanoscale 10 (29) (2018) 14060–14066.

[18] J. Zang, S. Ryu, N. Pugno, et al., Multifunctionality and control of the crumpling and unfolding of large-area graphene, Nat. Mater. 12 (4) (2013) 321–325.

[19] J. Xu, J. Chen, M. Zhang, et al., Highly conductive stretchable electrodes prepared by in situ reduction of wavy graphene oxide films coated on elastic tapes, Adv. Electron. Mater. 2 (6) (2016), 1600022.

[20] J.W. Durham 3rd, Y. Zhu, Fabrication of functional nanowire devices on unconventional substrates using strain-release assembly, ACS Appl. Mater. Interfaces 5 (2) (2013) 256–261.

[21] K.I. Jang, K. Li, H.U. Chung, et al., Self-assembled three dimensional network designs for soft electronics, Nat. Commun. 8 (2017) 15894.

[22] Q. Liu, J. Chen, Y. Li, et al., High-performance strain sensors with fish-scale-like graphene-sensing layers for full-range detection of human motions, ACS Nano 10 (8) (2016) 7901–7906.

[23] Y. Tan, Z. Chu, Z. Jiang, et al., Gyrification-inspired highly convoluted graphene oxide patterns for ultralarge deforming actuators, ACS Nano 11 (7) (2017) 6843–6852.

[24] Z. Niu, H. Dong, B. Zhu, et al., Highly stretchable, integrated supercapacitors based on single-walled carbon nanotube films with continuous reticulate architecture, Adv. Mater. 25 (7) (2013) 1058–1064.

[25] X. Wang, H. Hu, Y. Shen, et al., Stretchable conductors with ultrahigh tensile strain and stable metallic conductance enabled by prestrained polyelectrolyte nanoplatforms, Adv. Mater. 23 (27) (2011) 3090–3094.

[26] J. Zang, C. Cao, Y. Feng, et al., Stretchable and high-performance supercapacitors with crumpled graphene papers, Sci. Rep. 4 (2014) 6492.

[27] M. Wang, Y. Qiu, J. Jia, et al., Wavelength-gradient graphene films for pressure-sensitive sensors, Adv. Mater. Technol. 4 (1) (2019), 1800363.

[28] X. Shuai, P. Zhu, W. Zeng, et al., Highly sensitive flexible pressure sensor based on silver nanowires-embedded polydimethylsiloxane electrode with microarray structure, ACS Appl. Mater. Interfaces 9 (31) (2017) 26314–26324.

[29] M. Jian, K. Xia, Q. Wang, et al., Flexible and highly sensitive pressure sensors based on bionic hierarchical structures, Adv. Funct. Mater. 27 (9) (2017), 1606066.

[30] G. Lee, S.G. Lee, Y. Chung, et al., Omnidirectionally and highly stretchable conductive electrodes based on noncoplanar zigzag mesh silver nanowire arrays, Adv. Electron. Mater. 2 (8) (2016), 1600158.

[31] T. Zhao, T. Li, L. Chen, et al., Highly sensitive flexible piezoresistive pressure sensor developed using biomimetically textured porous materials, ACS Appl. Mater. Interfaces 11 (32) (2019) 29466–29473.

[32] J. Mu, C. Hou, G. Wang, et al., An elastic transparent conductor based on hierarchically wrinkled reduced graphene oxide for artificial muscles and sensors, Adv. Mater. 28 (43) (2016) 9491–9497.

[33] P.Y. Chen, J. Sodhi, Y. Qiu, et al., Multiscale graphene topographies programmed by sequential mechanical deformation, Adv. Mater. 28 (18) (2016) 3564–3571.

[34] Y. Zhang, T.H. Chang, L. Jing, et al., Heterogeneous, 3D architecturing of 2D titanium carbide (MXene) for microdroplet manipulation and voice recognition, ACS Appl. Mater. Interfaces 12 (7) (2020) 8392–8402.

[35] N. Gao, X. Zhang, S. Liao, et al., Polymer swelling induced conductive wrinkles for an ultrasensitive pressure sensor, ACS Macro Lett. 5 (7) (2016) 823–827.

[36] T.H. Chang, Y. Tian, C. Li, et al., Stretchable graphene pressure sensors with Shar-Pei-like hierarchical wrinkles for collision-aware surgical robotics, ACS Appl. Mater. Interfaces 11 (10) (2019) 10226–10236.

[37] J. Jia, G. Huang, M. Wang, et al., Multi-functional stretchable sensors based on a 3D-rGO wrinkled microarchitecture, Nanoscale Adv. 1 (11) (2019) 4406–4414.

[38] L. Li, J. Zhang, Z. Peng, et al., High-performance pseudocapacitive microsupercapacitors from laser-induced graphene, Adv. Mater. 28 (5) (2016) 838–845.

[39] S.J. Klaine, P.J.J. Alvarez, G.E. Batley, et al., Nanomaterials in the environment: behavior, fate, bioavailability, and effects, Environ. Toxicol. Chem. 27 (9) (2008) 1825–1851.

[40] A.S. Arico, P. Bruce, B. Scrosati, et al., Nanostructured materials for advanced energy conversion and storage devices, Nat. Mater. 4 (5) (2005) 366–377.

[41] X.L. Shi, S.R. Liu, Y. Sun, et al., Lowering internal friction of 0D-1D-2D ternary nanocomposite-based strain sensor by fullerene to boost the sensing performance, Adv. Funct. Mater. 28 (22) (2018) 10.

[42] M.A. Meyers, A. Mishra, D.J. Benson, Mechanical properties of nanocrystalline materials, Prog. Mater. Sci. 51 (4) (2006) 427–556.

[43] P.D. Xue, C. Chen, D.F. Diao, Ultra-sensitive flexible strain sensor based on graphene nanocrystallite carbon film with wrinkle structures, Carbon 147 (2019) 227–235.

[44] H. Sheng, X. Zhang, Y. Ma, et al., Ultrathin, wrinkled, vertically aligned Co(OH)2 nanosheets/ag nanowires hybrid network for flexible transparent supercapacitor with high performance, ACS Appl. Mater. Interfaces 11 (9) (2019) 8992–9001.

[45] H.W. Fan, K.R. Li, Q. Li, et al., Prepolymerization-assisted fabrication of an ultrathin immobilized layer to realize a semi-embedded wrinkled AgNW network for a smart electrothermal chromatic display and actuator, J. Mater. Chem. C 5 (37) (2017) 9778–9785.

[46] H.L. Sun, K. Dai, W. Zhai, et al., A highly sensitive and stretchable yarn strain sensor for human motion tracking utilizing a wrinkle-assisted crack structure, ACS Appl. Mater. Interfaces 11 (39) (2019) 36052–36062.

[47] Y.P. Zhai, Y.Q. Dou, D.Y. Zhao, et al., Carbon materials for chemical capacitive energy storage, Adv. Mater. 23 (42) (2011) 4828–4850.

[48] Z.D. Han, A. Fina, Thermal conductivity of carbon nanotubes and their polymer nanocomposites: a review, Prog. Polym. Sci. 36 (7) (2011) 914–944.

[49] S.J. Park, J. Kim, M. Chu, et al., Flexible piezoresistive pressure sensor using wrinkled carbon nanotube thin films for human physiological signals, Adv. Mater. Technol. 3 (1) (2018) 7.

[50] C.J. Zhang, H. Li, A.M. Huang, et al., Rational Design of a Flexible CNTs@PDMS film patterned by bio-inspired templates as a strain sensor and supercapacitor, Small 15 (18) (2019) 8.

[51] H. Jin, M.O.G. Nayeem, S. Lee, et al., Highly durable nanofiber-reinforced elastic conductors for skin-tight electronic textiles, ACS Nano 13 (7) (2019) 7905–7912.

[52] C. Luo, N.S. Liu, H. Zhang, et al., A new approach for ultrahigh-performance piezoresistive sensor based on wrinkled PPy film with electrospun PVA nanowires as spacer, Nano Energy 41 (2017) 527–534.

[53] W. Liu, N. Liu, Y. Yue, et al., Piezoresistive pressure sensor based on synergistical innerconnect polyvinyl alcohol nanowires/wrinkled graphene film, Small 14 (15) (2018), 1704149.

[54] F. Bonaccorso, L. Colombo, G.H. Yu, et al., Graphene, related two-dimensional crystals, and hybrid systems for energy conversion and storage, Science 347 (6217) (2015) 10.

[55] A.H. Castro Neto, F. Guinea, N.M.R. Peres, et al., The electronic properties of graphene, Rev. Mod. Phys. 81 (1) (2009) 109–162.

[56] G. Fiori, F. Bonaccorso, G. Iannaccone, et al., Electronics based on two-dimensional materials, Nat. Nanotechnol. 9 (10) (2014) 768–779.

[57] M. Naguib, O. Mashtalir, J. Carle, et al., Two-dimensional transition metal carbides, ACS Nano 6 (2) (2012) 1322–1331.

[58] S. Pei, H.-M. Cheng, The reduction of graphene oxide, Carbon 50 (9) (2012) 3210–3228.

[59] Y. Huang, J.J. Liang, Y.S. Chen, An overview of the applications of graphene-based materials in supercapacitors, Small 8 (12) (2012) 1805–1834.

[60] M. Naguib, V.N. Mochalin, M.W. Barsoum, et al., 25th anniversary article: MXenes: a new family of two-dimensional materials, Adv. Mater. 26 (7) (2014) 992–1005.

[61] T.H. Chang, T.R. Zhang, H.T. Yang, et al., Controlled crumpling of two-dimensional titanium carbide (MXene) for highly stretchable, bendable, efficient supercapacitors, ACS Nano 12 (8) (2018) 8048–8059.

[62] G. Li, Z.G. Qiu, Y. Wang, et al., PEDOT:PSS/grafted-PDMS electrodes for fully organic and intrinsically stretchable skin-like electronics, ACS Appl. Mater. Interfaces 11 (10) (2019) 10373–10379.

[63] C.F. Yang, L.L. Li, J.X. Zhao, et al., Highly sensitive wearable pressure sensors based on three-scale nested wrinkling microstructures of polypyrrole films, ACS Appl. Mater. Interfaces 10 (30) (2018) 25811–25818.

[64] N. Hui, F.L. Chai, P.P. Lin, et al., Electrodeposited conducting polyaniline nanowire arrays aligned on carbon nanotubes network for high performance supercapacitors and sensors, Electrochim. Acta 199 (2016) 234–241.

[65] X. Zhou, G. Lu, X. Qi, et al., A method for fabrication of graphene oxide nanoribbons from graphene oxide wrinkles, J. Phys. Chem. C 113 (44) (2009) 19119–19122.

[66] Z.Y. Yuan, Z. Pei, M. Shahbaz, et al., Wrinkle structured network of silver-coated carbon nanotubes for wearable sensors, Nanoscale Res. Lett. 14 (1) (2019) 8.

[67] Z. Qin, Y.Y. Yin, W.Z. Zhang, et al., Wearable and stretchable triboelectric nanogenerator based on crumpled nanofibrous membranes, ACS Appl. Mater. Interfaces 11 (13) (2019) 12452–12459.

[68] P.T. Toi, T.Q. Trung, T.M.L. Dang, et al., Highly electrocatalytic, durable, and stretchable nanohybrid fiber for on-body sweat glucose detection, ACS Appl. Mater. Interfaces 11 (11) (2019) 10707–10717.

[69] Y.P. Yang, C.Z. Luo, J.J. Jia, et al., A wrinkled ag/CNTs-PDMS composite film for a high-performance flexible sensor and its applications in human-body single monitoring, Nanomaterials 9 (6) (2019) 17.

[70] L.M. Miao, J. Wan, Y. Song, et al., Skin-inspired humidity and pressure sensor with a wrinkle-on-sponge structure, ACS Appl. Mater. Interfaces 11 (42) (2019) 39219–39227.

[71] H. Kang, C.L. Zhao, J.R. Huang, et al., Fingerprint-inspired conducting hierarchical wrinkles for energy-harvesting E-skin, Adv. Funct. Mater. 29 (43) (2019) 10.

Tactile sensors based on ionic liquids

Yapei Wang*, Naiwei Gao, and Yonglin He
Department of Chemistry, Renmin University of China, Beijing, China
*Corresponding author: e-mail address: yapeiwang@ruc.edu.cn

11.1 Introduction of ionic liquids

Generally, the term "ionic liquids" was used as early as 1942, which was customarily defined as room-temperature molten salts [1–3]. Recently, this term was extended in some cases to salts with a melting point below 100°C. Generally, ionic compounds are composed of ions with a small radius. Regardless of solvation by particular solvents, their phase transition from solid to liquid can only occur at very high temperature due to the strong electrostatic forces between ions. In 1888, Gabriel and Weiner reported the first salt, ethanolammonium nitrate, with the melting point lower than 100°C, corresponding to a solid-liquid transition temperature at 52°C [4]. However, it was not until 1914 that the first room-temperature ionic liquid, ethlammonium nitrate ((C_2H_5)$NH_3^+ \cdot NO_3^-$), with a melting point of 12°C, was synthesized by Paul Walden [5]. Since then, a great number of ions have been successfully combined into room-temperature ionic liquids (Table 11.1). In principle, increasing the size of ions by means of uneven charge distributions and bulky substituents is able to weaken the intermolecular electrostatic forces, thus decreasing the melting point of the ionic complex. By virtue of the negligible saturated vapor pressure, ultralow elastic modulus, wide electrochemical window, good thermal stability, and conductive performance, ionic liquids have shown great potentials in chemical synthesis, catalysis, gas adsorption, processing biomass, and electrochemistry [7–11].

Some ion species that are commonly used to formulate ionic liquids are exemplified in Fig. 11.1. Cations mainly include imidazolium, pyridinium, pyrolidium, piperidium, sulfonium, ammonium, phosphonium, and tetramethylguanidinium. Among these cations, imidazole-based ionic liquids are one of the most popular choices in practical applications. This type of ionic liquid can be readily obtained via a two-step synthesis, including the functionalization of imidazolium cations and the anion exchange (Fig. 11.2) [12, 13]. As the other half of an ionic liquid, anions also have many choices. Anions such as chloride (Cl^-), bromide (Br^-), acetate (COO^-), hexafluorophosphate (PF_6^-), tetrafluoroborate (BF_4^-), and trifluoromethanesulfonylimide ($TFSI^-$) with a relatively large radius have enabled the preparation of a variety of ionic liquids with a low melting point. In theory, more than 1×10^{18} kinds of ionic liquids are predicted if many other ion species are taken into account.

Table 11.1 Physical properties of some well-established ionic liquids.

Ionic liquids	T_m (K)	ρ (g/cm^3)	η (cp)	σ (S/m)
[EMIM][NTf$_2$]	256.15	1.518	18	0.763
[EMIM][BETI]	272.15	–	61	0.34
[EMIM] [dca]	252.15	1.06	21	2.2
[EMIM][TSAC]	275.75	1.46	24	0.978
[EMIM] [(CH$_3$SO$_2$)$_2$N]	223.15	1.343	787	0.017
[EMIM][C1C2]	180.15	1.56	48	0.44
[EMIM][FSI]	260.1	–	18	1.54
[EMIM][PF$_6$]	193	–	–	0.52
[EMIm][n-C$_5$F$_{11}$BF$_3$]	289.15	1.065	88	0.27
[EMIm][CH$_2$CHBF$_3$]	167.15	1.161	41	1.05
[EMIm][TCM]	263.15	–	19.56	2
[PMIm][BF$_4$]	256.15	1.24	105	0.59
[PMIm][CF$_3$BF$_3$]	252.15	1.31	43	0.85
[PMIm][n-C$_4$F$_9$BF$_3$]	261.15	1.48	59	0.35
[BMIm]Cl	314.15	1.08	3950	–
[BMIM] [CF$_3$BF$_3$]	165.15	1.27	49	0.59
[BMIM][C$_2$F$_5$BF$_3$]	231.15	1.34	41	0.55
[C$_5$O$_2$MIm][C$_2$F$_5$BF$_3$]	183.15	1.37	51	0.34
[HMIm]Cl	198.15	1.03	7500	0.0425
[HMIm][BF$_4$]	190.75	1.16	310	0.12
[HMIm][CF$_3$BF$_3$]	173.15	1.22	77	0.28
[HMIm][C$_2$F$_5$BF$_3$]	263.15	1.28	59	0.27
[HMIm][NTf$_2$]	266	1.372	87.3	0.218
[HMIM][TfO]	302.5	1.2	160	–
[HMIm][PF$_6$]	212.15	1.302	560	0.0434
[OMIm]Cl	191.15	1.01	16,000	0.0338
[OMIm][BF$_4$]	193.15	1.11	440	0.0298
[VBIm]Br	356.15	–	–	–
[P11][TFSI]	378.15	–	–	–
[P11]M	346.85	–	–	0.353
[P13][dca]	238.15	0.92	45	–
[P14][CF$_3$BF$_3$]	255.15	1.2135	137	0.33
[P14][C$_2$F$_5$BF$_3$]	295.15	1.2877	71	0.35
[P14][TFSI]	251.5	1.379	85	0.22
[PP1.1O1][TFSI]	188.15	1.47	68	0.22
[PP1.1O2][BF$_4$]	196.15	1.2162	1240	0.06
[PP1.1O2][C$_3$F$_7$BF$_3$]	264.15	1.3884	131	0.15
[S111][TFSI]	317.65	1.58	44	0.82
[S222][TFSI]	259.15	1.49	33	3.66
[N1111][TFSI]	409.5	–	–	–
[N111C$_2$O][TFSI]	277.65	1.51	50	0.47
[N1113][TFSI]	290.15	1.44	72	0.327

Tactile sensors based on ionic liquids   221

Table 11.1 Continued

Ionic liquids	T_m (K)	ρ (g/cm³)	η (cp)	σ (S/m)
[N1113][C1C2]	292.15	1.48	174	0.12
[P666,14][PF₆]	312.65	–	–	–
[(C₄H₉)₄P][Lys]	208.01	0.973	755.71	0.0104
[P2224][TFSI]	328.15	–	–	–
[P2225][TFSI]	290.15	1.32	88	0.173

Note. All of these data are measured at the temperature of 298.15 K [6].

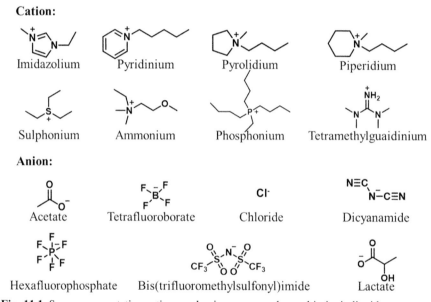

Fig. 11.1 Some representative cations and anions commonly used in ionic liquids.

Fig. 11.2 The typical preparation of imidazolium-based ionic liquids.

The conductivity of most ionic liquids lies at a level close to that of solid semiconductors. Acting like a particular class of "liquid semiconductors," some room-temperature ionic liquids are electrically sensitive to temperature change and external pressure. The sensing signal may result from the change of ion mobility or ion reorganization on electrodes (Fig. 11.3). In contrast to solid electrical materials, ionic liquids possess two attractive advantages. On the one hand, they have outstanding deformability compared to solids such that the appearance of a little pressure

Fig. 11.3 The mechanism of ion migration under a certain electric field.

enables them to significantly deform yet maintain their continuity. On the other hand, they own a unique self-healing ability. Isolated ionic liquids immediately merge together and recover conductive performance whenever they come in contact. Benefiting from these characters, ionic liquids are envisioned to open a new avenue to make tactile sensors with promises of high flexibility, self-healing performance, and easy recyclability [14–16]. This book chapter summarizes some successful tactile sensors based on ionic liquids and also explains their sensing mechanism accordingly.

11.2 Pressure sensors based on ionic liquids

As one of the most important tactile sensors, the pressure sensor is generally used to feel the change of pressure and convert it into an electrical signal. As stated above, ionic liquids can make a remarkable deformation under a little pressure owing to the liquid performance. Following the deformation of ionic liquids, two strategies including piezoresistive and capacitive sensors are proposed to correlate pressure with electrical responses.

11.2.1 Piezoresistive sensor

Piezoresistive sensors are sensors that are built on the basis of a piezoresistive effect, which refers to the resistance change of a conductor as a result of geometry change led by the mechanical force. The piezoresistive effect can be presented by the resistance Eq. (1) derived from Ohm's law.

$$R = \rho \frac{l}{A} \tag{1}$$

where R is the resistance of a piezoresistive sensor, ρ is the resistivity of the piezoresistive sensor, l is the length of the piezoresistive sensor, and A is the cross-sectional area of the current flow.

One simple way to formulate ionic liquids as piezoresistive sensors is to load them in elastic microchannels. For example, a highly stretchable and sensitive strain sensor was constructed by filling a straight Eco-Flex microchannel with the ionic liquid of [EMIm][TFSI] [17]. Such a sensor could detect a wide range of strains from 0.1% to 400% with a high gage factor of 7.9. However, a hysteresis problem was

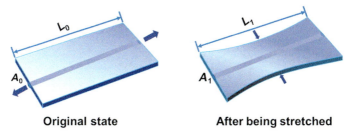

Fig. 11.4 Schematic illustration of the Poisson effect.

encountered in which the electrical response of ionic liquids was delayed after tensile stress was applied. Instead of a straight microchannel, a wavy-shaped microfluidic channel exhibited better affinity to ionic liquids so that ionic liquids could catch up with the deformation of the microchannel in time [18]. The strain sensor based on this topological design corresponded to a low-level hysteresis of 0.15% at a strain of 250% and a low overshoot of 1.7% at a strain of the 150% strain. A model was built to understand the mechanism of the piezoresistive effect in the cases of ionic liquid-based strain sensors. As illustrated in Fig. 11.4, once the microchannel is elongated along its axial direction, the cross-sectional area of the microchannel will decrease due to the Poisson effect [19]. If the ionic liquid filled in the microchannel flows with the microchannel deformation, its electrical resistance can be specifically correlated with the tensile strain according to Ohm's law in the following equation:

$$dR = \rho \frac{l_0}{A_0} \left(\frac{(1+2\nu)\varepsilon - \nu^2\varepsilon^2}{(1-\nu\varepsilon)^2} \right) \quad (2)$$

where dR is the resistance change of an ionic liquid-based piezoresistive sensor, ρ is the resistivity of the ionic liquid, L_0 is the original length of the microchannel, A_0 is the original cross-sectional area of the microchannel without tensile stretching, ν is Poisson's ratio of the elastomeric material, and ε is the applied strain.

A porous elastic polydimethylsiloxane (PDMS) sponge was also exploited as the supporting matrix to accommodate the ionic liquid of 1-ethyl-3-methylimidazolium acetate ([EMIm][Ac]) [20]. The PDMS sponge with an interconnected microporous structure was typically prepared from a cubic sugar template. Driven by capillary force, ionic liquids could be loaded into those interconnected micro-reservoirs without evidence of leakage. The piezoresistive effect was established based on the redistribution of ionic liquids under external pressure. In case the sponge was partially filled by ionic liquids, the air trapped in unoccupied micropores was extruded out when the PDMS sponge received mechanical compressing, leading to an increased number of conductive pathways. As a consequence, the sponge became more conductive if pressure was applied to it (Fig. 11.5). The sensitivity of the pressure sensors, as represented by the change of resistance, relied on two factors, including the cross-linking degree of PDMS and the volume fraction of ionic liquids relative to air in the PDMS sponge. The study revealed that the higher sensitivity of the pressure sensor was achieved with

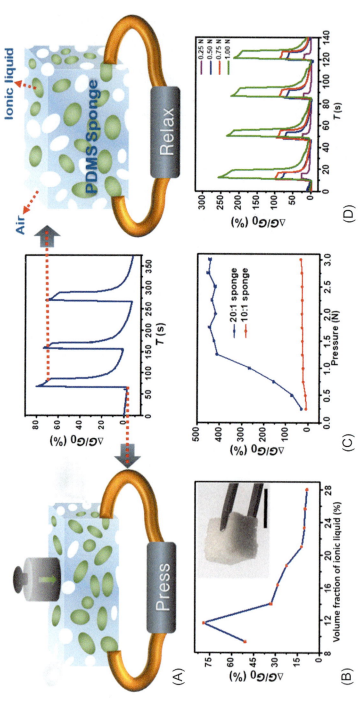

Fig. 11.5 The pressure sensor based on the ionic liquid-containing PDMS sponge. (A) Schematic illustration of the sensing principle of pressure sensors via partially filling [OMIm][Ac] into a PDMS sponge. (B) The relationship between the volume fraction of ionic liquid and the conductivity change at a given pressure of 0.5 N. Scale bar: 1 cm. (C) Pressure sensing performance of PDMS sponge with different cross-linking degrees. The pressure ranges from 0.25 to 3.00 N with an interval of 0.25 N. (D) Pressure sensing measurements of PDMS sponge with a cross-linking degree of 20:1 at different pressure forces.
Copyright 2015, John Wiley and Sons.

the use of a lower cross-linking degree of PDMS sponges. Meanwhile, the volume fraction of ionic liquid relative to air at 11.68% corresponded to the most sensitive pressure response at a given cross-linking degree of PDMS sponge.

Besides microchannels and porous sponges, ionic liquids could be also confined in gel networks [21, 22], forming elastic ion gels that were successfully exploited as piezoresistive sensors. Ion gels could be prepared by the cross-linking of curable ionic liquids or vinyl monomers in an ionic liquid. The cross-linked polymer network was able to prevent the flow of ionic liquids, while preserving the physical properties of ionic liquid, including high thermal stability, negligible saturated vapor pressure, and excellent conductivity. More intriguingly, the strategy of confining ionic liquids within ion gels did not require additional encapsulation which was commonly used in microchannel and porous sponge strategies. For example, a semitransparent ion gel with high stretchability and conductivity was prepared via the polymerization of 1-vinyl-3-ethylimidazolium dicynamide ([VMIm][DCA]) [23]. This ion gel functioned as a wonderful pressure sensor with high sensitivity of 15.4 kPa^{-1} and wide detection range from 5 to 5000 Pa. Practically, this pressure sensor could recognize the external stimulus of subtle signals, such as touching, torsion, and bending.

11.2.2 Capacitive sensors

The capacitive sensor is another choice to exploit ionic liquids as pressure sensors. In general, a capacitor is composed of two conductors separated by a layer of dielectric such as vacuum or electrical insulator materials. The capacitance is intrinsically relevant to the size of conductors and the permittivity of the dielectric as defined in the following equations:

$$C = \frac{Q}{V} \tag{3}$$

$$C = \frac{\varepsilon A}{d} \tag{4}$$

where C is the capacitance of the capacitor, Q is the charge capacity on each conductor, V is the applied voltage between two conductors, ε is the permittivity of the dielectric, A is the overlap area of two conductors, and d is the distance between two conductors. When an external pressure is applied to the capacitor, the distance will decrease, accounting for an increase of capacitance, which further causes the change of charge or voltage. Any form of capacitance, charge, or voltage is appropriate for monitoring applied pressure.

In order to formulate ionic liquids as a capacitive sensor, one way is to establish an electrical double layer (EDL) between one electrode and ionic liquids (Fig. 11.6). In the presence of a bias voltage between two electrodes, cations and anions of ionic liquids migrate in opposite directions, forming an EDL at the ionic liquid-electrode interface. The total capacitance includes the capacitance between two electrodes and the capacitance at the ionic liquid-electrode interface. When a pressure is applied to the sensor, the other electrode comes into contact with ionic liquids. This induces

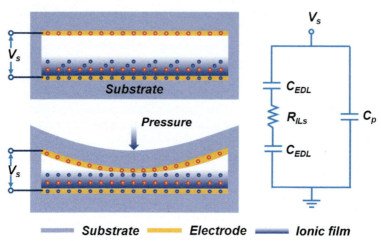

Fig. 11.6 The sensing mechanism of a capacitive sensor.

the formation of a new electrical double layer so that the capacitance is improved. This kind of capacitive pressure sensor does not need the redistribution of ionic liquid along with the deformation of encapsulating layers, significantly reducing the response time. Based on this mechanism, an example of a three-layer capacitive pressure sensor made of conductive fabric layer and electrospun nanofibrous containing ionic liquids exhibited a pressure sensitivity as high as 114 nF/kPa, a resolution of 2.4 Pa, and a millisecond response time of 4.2 ms [24].

An electromechanical effect is another way to fabricate the capacitive pressure sensor with the use of ionic liquids. The electromechanical effect is based on the different migration of cations and anions (Fig. 11.7). When a pressure is applied on the electromechanical sensor, the instantaneous pressure gradient from the deformed electrode drives the migration of ions along the pressure direction. Cations and anions have different migration rates due to their different sizes so that the originally random distribution of ions is imbalanced. This imbalance leads to a potential difference between two electrodes of the sensor without external power supply. For example, an ionic-polymer-metal composite (IMPC) was prepared with the use of [EMIm][TFSI] which acted as a pressure sensor based on electromechanical effect [25]. Such an IMPC could generate an output voltage of 1.3 mV under a bending-induced strain of 1.8%.

11.3 Temperature sensors based on ionic liquids

The conductivity of ionic liquids is attributed to the migration of cations and anions in an electrical field, and it generally increases upon raising the temperature. The temperature-dependent conductivity can be explained by the viscosity change of ionic liquids. Raising the temperature can reduce the viscosity of ionic liquids as a result of weakened intermolecular forces, which facilitates the ion migration, thus improving the conductivity.

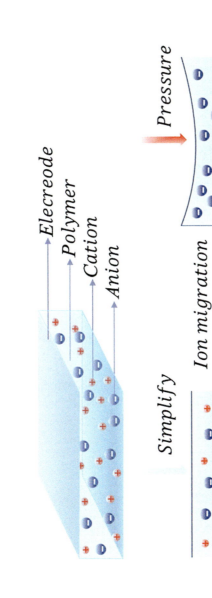

Fig. 11.7 The sensing mechanism of pressure sensors based on piezoelectric effect.

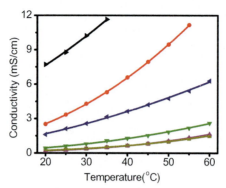

Fig. 11.8 The conductivity of ionic liquids at different temperatures. From *top* to *bottom*: ▶ [EMIm][TFSI], ● [EMIm][Ac], ◀ [HMIm][TFSI], ▼ [HMIm][PF$_6$], ▲ [OMIm][PF$_6$], and ■ [OMIm][Ac]. *Solid curves* are fitted by the VTF equation. Copyright 2015, John Wiley and Sons.

Previous studies revealed that the conductivity of imidazolium-based ionic liquids could be improved through shortening the alkyl chain length of imidazolium cations or increasing the size of anions (Fig. 11.8) [26]. The conductivity change of ionic liquids as a function of temperature can be described by the Volgel-Tamman-Fulcher (VTF) Eq. (5),

$$\sigma(T) = \sigma_\infty \exp\left(-\frac{B}{T-T_v}\right) \qquad (5)$$

where $\sigma(T)$ is the electrical conductivity of ionic liquids at a certain temperature T, σ_∞ is the theoretical conductivity at the limit of high temperature, T_v is the glass transition temperature of ionic liquids, and B is a constant only related to the species of ionic liquids. Through the VTF equation, the relationship between conductivity change and specific temperature can be described as the following equations:

$$\frac{\Delta G}{G_0} = \frac{\sigma(T)}{\sigma_0} - 1 \qquad (6)$$

$$\frac{\Delta G}{G_0} = \exp\left(\frac{B(T-T_0)}{(T-T_v)(T_0-T_v)}\right) - 1 \qquad (7)$$

where G_0 is the initial conductivity of ionic liquids, ΔG is the conductivity change or thermal response of ionic liquids, and T_0 is the initial temperature.

A representative example is the ionic liquid of 1-octyl-3-methylimidazolium hexafluorophosphate ([OMIm][PF$_6$]). The sensor made of this ionic liquid exhibited remarkable thermal sensitivity within a body temperature range of 35–45°C (Fig. 11.9A). The rate of change in conductivity can reach 7% at a temperature difference of only one degree, offering a potential for clinical diagnosis. Notably, the sensor also had outstanding sensing stability and reproducibility. The average electrical response retains almost 173% between 37°C and room temperature for nine cycles (Fig. 11.9B).

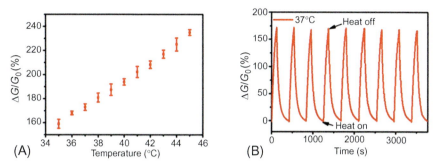

Fig. 11.9 Thermal response of ionic liquid-based sensors. (A) Thermal response of a single-channel-based ionic liquid sensor against the increase of temperature ranging from 35 to 45°C with an interval of 1 degree. Room temperature is 20°C. (B) On–off cycles of the thermal response of the self-healing sensor between 37°C and room temperature.
Copyright 2015, John Wiley and Sons.

Ionic liquids could be extended as self-healing thermometers by loading them into microchannels made of a supramolecular self-healing polymer (Fig. 11.10). The self-healing polymer was a product of the condensation reaction between fatty polybasic acid and diethylenediamine, followed by the conversion amine groups into amide bonds by urea. A great amount of hydrogen bonds between amide moieties were formed in such a polymer network, which enabled the reversible bonding of the polymer chains. The broken polymer could fully recover its mechanical performance if the fractured surfaces were touched in 5 min at 50°C. By sealing ionic liquids in this polymer, temperature sensors owned a unique self-healing ability. The broken sensor cut by a knife could fully recover its temperature sensing performance whenever the separated parts came into contact.

Temperature sensing array is a significant component for facilitating its spatially resolved applications in two-dimensional monitoring of the temperature field, medical imaging, and human skin. As a principal demonstration, an 8 × 8 temperature sensing array was fabricated by pen-writing or inkjet printing the ionic liquid of

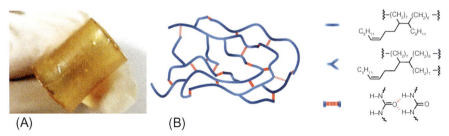

Fig. 11.10 The preparation of a supramolecular self-healing network. (A) The optical image of a self-healing network. (B) Schematic illustration of a self-healing network based on the hydrogen bond. *Blue lines* are the randomly branched network consisting of the linear and branched polymers; *Red lines* are primary hydrogen bonds between amide moieties.
Copyright 2015, John Wiley and Sons.

1-ethyl-3-methyl imidazolium bis(trifluoromethylsulfonyl)imide ([EMIm][Tf$_2$N]) on a regular 8 × 8 cm^2 A4 paper with each pixel size controlled as 25 mm^2 (Fig. 11.11A) [27]. When a heated metal rod was placed close to the sensing array at a specific position, the sensor was able to output a thermal mapping of temperature distribution, which well matched the vis-IR photos captured with a digital camera

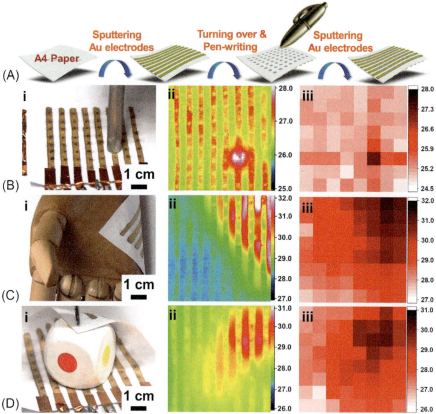

Fig. 11.11 Fabrication of a temperature sensing array based on ionic liquids. (A) Schematic illustration of a fabrication process of an ionic liquid paper chip with an 8 × 8 pixel array. (B) Data visualizationtest of a paper chip with an 8 × 8 pixel array contacted with a heated metal rod: (i) optical image; (ii) IR image recorded by an infrared camera exhibit surface temperature of paper chip arrays after contact with a heated metal rod; (iii) two-dimensional thermal imaging of a paper chip array after contact with a heated metal rod. (C) Data visualization test of a paper chip with an 8 × 8 pixel array contacted with a simulated human hand: (i) optical image; (ii) IR image recorded by an infrared camera exhibit surface temperature of paper chip arrays after contact with a simulated human hand; (iii) two-dimensional thermal imaging of a paper chip array after contact with a simulated human hand. (D) Data visualization test of a paper chip with an 8 × 8 pixel array contacted with a heated dice: (i) optical image; (ii) IR image recorded by an infrared camera exhibit surface temperature of paper chip arrays after contact with a heated dice; (iii) two-dimensional thermal imaging of a paper chip array after contact with a heated dice. Copyright 2017, American Chemical Society.

and an infrared camera (Fig. 11.11B). Superior to infrared cameras which can only read surface temperature in one direction, the paper thermometer with sensing arrays could simultaneously detect the temperature of curved surfaces. For example, the surface temperature of a handlike model and a dice with six faces was measured by using the infrared camera; however, only one specific side of these samples could be measured. The ease of bending and folding allowed the paper thermometer to surround the object, which was convenient to read temperatures from all directions. As shown in Fig. 11.11C, a temperature thermometer with a sensing array was bent and attached to the hand model. More than reading the temperature of the palm, the back of the hand was also spatially monitored in real time. Additionally, folding the thermometer to match the shape of the dice, several dice faces could be monitored simultaneously (Fig. 11.11D).

11.4 Signal separation and integration of tactile sensors based on ionic liquids

As stated above, the thermal-sensitive ionic liquids also change their conductivity in cases of deformation under external forces. In other words, the appearance of pressure or tensile stress will cause intervention in the thermal sensing test of the ionic liquid-based sensor. The reverse case may occur as well. This mutual intervention is a serious problem of impeding the multiple-signal sensing of ionic liquids, which even cripples the purpose of their single-signal sensing. Here, we discuss several strategies to bypass the problem of signal interference in the simultaneous temperature and pressure sensing tests.

11.4.1 Eliminating the interference of temperature on the pressure sensors

The temperature interference on pressure sensors can be eliminated by proper electrical circuits, namely, temperature compensating (TC). For example, the electrical signal led by temperature change can be avoided by introducing a Wheatstone bridge in the ionic liquid-based pressure sensor. The Wheatstone bridge is typically a bridge circuit consisting of four resistances (Fig. 11.12). By replacing one of the

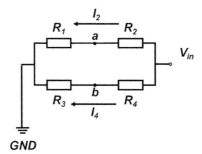

Fig. 11.12 Scheme of a Wheatstone bridge.

resistances by a pressure sensor, a potential difference occurs between the points of a and b when an external pressure is applied on the pressure sensor. The current from R_2 to R_1 can be specified according to the following equation:

$$I_2 = \frac{V_{in}}{R_1 + R_2} \tag{8}$$

Then the voltage across R_2 can be expressed by the following equation:

$$V_2 = R_2 \times \frac{V_{in}}{R_1 + R_2} \tag{9}$$

Similarly, the voltage across R_4 can be calculated by the following equation:

$$V_4 = R_4 \times \frac{V_{in}}{R_3 + R_4} \tag{10}$$

Therefore, the voltage between the points a and b can be calculated by the following equation:

$$V_{ab} = V_4 - V_2 = V \times \left(\frac{R_4}{R_3 + R_4} - \frac{R_2}{R_1 + R_2} \right) \tag{11}$$

If the Wheatstone bridge is composed of four identical resistances based on ionic liquids, the voltage of V_{ab} should be zero at the relaxed state. In the case of changing temperature, the conductivity change of four resistances is synchronous and proportional, so it theoretically has no influence on V_{ab}. Following this strategy, the interference of temperature on the pressure sensor can be eliminated completely.

A successful example of a temperature-independent pressure sensor was fabricated by loading the ionic liquid of 1-ethyl-3-methylimidazolium dycyanamide ([EMIm][DCA]) into a PDMS microfluidic channel with a topology of the Wheatstone Bridge. The pressure sensor could detect the minor pressure signal of 2.5 psi and owned a sensitivity of 8.45 mV/psi [28].

11.4.2 Eliminating the interference of strain on the temperature sensors

In practice, the deformation of ionic liquids is routinely inevitable when ionic liquids only serve as a temperature sensor. Hence, the temperature sensing signal may involve the undesired contribution led by the shape change of ionic liquids. In order to diminish this interference, a serpentine microchannel structure was introduced into a PDMS chip to retain the effective length unchanged upon mechanical strain (Fig. 11.13) [20]. Much better than the straight microchannel with a resistance change of 45.1%, the resistance change of the serpentine channel was no more than 2.2% under a given

Tactile sensors based on ionic liquids

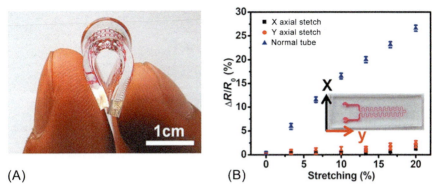

Fig. 11.13 The strategy of using a serpentine microchannel to eliminate the impact of applied strain. (A) Optical image of the serpentine microchannel. (B) The electrical response of a temperature sensor with a serpentine microchannel upon different stretching conditions. Copyright 2015, John Wiley and Sons.

strain of 20%, which ensured the limited influence of mechanical operation on the thermal sensing tests.

11.4.3 Separating the multiple signals in the integrated tactile sensors

Besides the single-signal tactile sensors, multiple signal tactile sensors derived from ionic liquids may be also possible if the combined signals, especially temperature and pressure signals, can be distinguished or separated accordingly. Fig. 11.14 demonstrates three different signals, including the pressure, the temperature, and the pressure-temperature mixing signals, that were recorded on a piezoresistive sensor

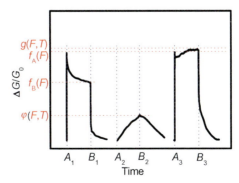

Fig. 11.14 The electrical responses of the PDMS sponge-based piezoresistive sensor filled with [OMIm][Ac] serving as the sensing material upon different external stimulations. The *left*, *middle*, and *right curves* refer to the pressure response, the temperature response, and the pressure-temperature mixing response, respectively.
Copyright 2015, John Wiley and Sons.

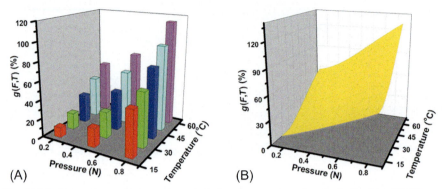

Fig. 11.15 A pressure and temperature dual-signal sensor based on ionic liquids. (A) The experimentally electrical responses of the sensor under the complex signals containing different temperature and pressure. (B) Theoretical prediction of temperature and pressure from a complex sensing signal.
Copyright 2015, John Wiley and Sons.

based on the PDMS sponge loaded with [EMIm][Ac] as mentioned in Fig. 11.5 [20]. Specifically, the pressure sensing curve contains two peaks that can be defined as two functions of pressure force $f_A(F)$ and $f_B(F)$. But the temperature curve of the sensor has only one peak that can be defined as the function of $\varphi(F,T)$. The mixing sensing curve also has two peaks that are assigned to the functions of $f_A(F)$ and $g(F,T)$. Because the electrical response caused by the pressure sensing is much faster than the response by temperature sensing, the function of $f_A(F)$ predominately depends on the pressure force (F), while the function of $g(F,T)$ is determined by both pressure force and temperature. Therefore, the mixing response of $g(F,T)$ is separated into two functions of $f_B(F)$ and $\varphi(F,T)$. The signal of $g(F,T)$ can be expressed as the following equation,

$$g(F,T) = f_B(F) + \varphi(F,T) \tag{12}$$

$$f_B(F) = d \times F^2 + h \times F \tag{13}$$

$$\varphi(F,T) = a \times \exp\left(\frac{b}{T+c}\right) \tag{14}$$

$$a = 715.78F^2 - 301.48F + 175.23 \tag{15}$$

$$b = -120.01F^2 + 38.33F - 59.00 \tag{16}$$

$$c = 21.10F^2 - 4.64F - 11.35 \tag{17}$$

where F and T are the force and temperature, respectively, applied on the piezoresistive sensor. Both d and h are constants. The parameters of a, b, and c rely on the pressure.

The relationship among $g(F,T)$, pressure force (F), and temperature (T) was correlated from the measured data in Fig. 11.15A, so the values of constant a–c, d, and f could be identified. Therefore, the relationship among $g(F,T)$, pressure force (F), and temperature (T) was successfully plotted as a 3D diagram (Fig. 11.15B). Consequently, the measured electrical response as a mixing signal was calculated and separated as two independent signals, including pressure force (F') and temperature (T'). For example, a sensing signal was separated into pressure and temperature signals corresponding to external force and temperature as 0.552 N and 46.8°C, which closely matched the actual force and temperature of 0.550 N and 45°C, respectively.

11.5 Preventing the leakage of ionic liquid-based sensors

Ionic liquids cannot be shaped because of the low storage modulus that is a common phenomenon to most liquid materials. Therefore, ionic liquids may encounter the leakage problem when the accidental breaking happens to those ionic liquid-based tactile sensors. This disadvantage may limit the repeatability and reliability of sensing tests. Several methods are proposed as below to solve this issue.

11.5.1 Confining ionic liquids by external capillary effect

In order to prevent the leakage of ionic liquids from supporting material of sensors, one strategy is to encapsulate the ionic liquids into the structures with a strong capillary effect. One simple model is loading ionic liquids in straight microchannels with small inner diameters. As presented in Fig. 11.16, the capillary effect in this model can be described by the following capillary equation:

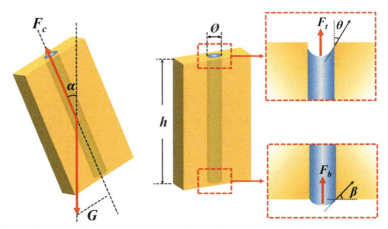

Fig. 11.16 Schematic illustration of the capillary effect of ionic liquids in a straight channel of supporting material.
Copyright 2015, John Wiley and Sons.

$$h = \frac{4\gamma\cos\theta}{\rho g \varnothing} \tag{12}$$

where h is the height of an ionic liquid column, γ is the ionic liquid-air surface tension, θ is the contact angel, ρ is the density of the ionic liquid, g is the local gravitational acceleration, and $\varnothing$ is the inner diameter of the straight tube. Notably, the theoretical supporting height of ionic liquids in straight microchannels opening at both ends is inversely proportional to the inner diameter. Referring to the work by loading [OMIm][PF$_6$] in a self-healing polymer channel, the theoretical threshold below which no leakage happened was estimated to be as high as 22.82 mm when the inner diameter of the channel was 0.709 mm [26].

In the work of preparing a temperature sensing array [27], cellulose paper provided substantial capillary force to confine ionic liquids within its interstitial voids so that no leakage of ionic liquids occurred during mechanical bending and folding of the paper sensors. To be specific, by means of pen-writing, a line of ionic liquid of [EMIm][Tf$_2$N] was drawn between two gold electrodes on a piece of A4 paper (Fig. 11.17A and B). The cellulose fibers in the region written with ionic liquid formed a denser network, strongly distinguished from the adjacent area without added ionic liquid (Fig. 11.17C). [EMIm][Tf$_2$N] was chosen as the ink with respect to its low viscosity and hydrophobic property in comparison with other ionic liquids (Fig. 11.17D), which ensured the high stability of paper thermometers during long-term use in air. This paper-based temperature sensor exhibited an ultrasensitive thermal response upon externally environmental temperature change because noncoverage of ionic liquids by other materials facilitated easier thermal conduction. Remarkably,

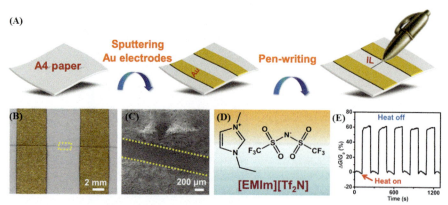

Fig. 11.17 (A) Schematic illustration of the fabrication process of a pen-written ionic liquid paper sensor. (B) Optical images of an ionic liquid-based paper sensor. The *yellow dotted box region* marked in (B) is magnified as an SEM image in (C). (D) Molecular structure of ionic liquid [EMIm][Tf$_2$N]. (E) On–off cycles of thermal response of paper chip operated between 45°C and room temperature (25°C).
Copyright 2017, American Chemical Society.

conductivity change of such a paper thermometer reached 60.5% by increasing the ambient temperature from room temperature (25°C) to 45°C and the result was reliable as it kept unchanged after multiple cycles of heating and cooling (Fig. 11.17E).

11.5.2 Confining ionic liquids by internal capillary effect

Instead of confining ionic liquids by an additional solid matrix with strong capillary effect, an ionic liquid binary system was exploited to effectively confine ionic liquids by internal capillary effect [29]. This kind of crystal-confined ionic liquids (CCILs) included one fluidic ionic liquid as the continuous phase and another crystallizable ionic liquid as the disperse phase. Specifically, nanocrystal rods formulated by 1-methyl-3-octyl-1H-imidazolium (E)-4-(phenyldiazenyl) phenolate ([OMIm]AzoO) as the disperse phase provided sufficient pinning capillary force to immobilize the continuous phase of fluidic ionic liquid of 1-octyl-3-methylimidazolium hexafluorophosphate ([OMIm]PF$_6$) (Fig. 11.18A). In spite of looking like a freestanding solid product, such a binary ionic liquid blend owned all the intrinsic advantages of high flexibility, self-healing, and reconfigurable properties that liquids generally have (Fig. 11.18B).

When adding [OMIm]AzoO to [OMIm]PF$_6$, four different states appeared stepwise with the continuous addition of [OMIm]AzoO, including solution state, suspended state, loose accumulation state, and close-packed state (Fig. 11.19). When kept in

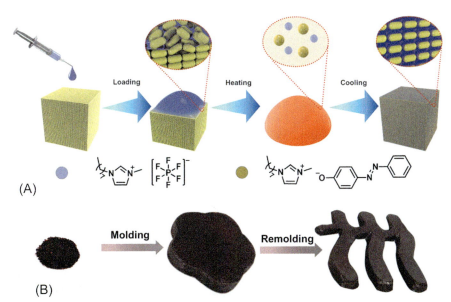

Fig. 11.18 (A) Scheme of the preparation of the crystal-confined ionic liquids. (B) The freestanding ionic liquid blend with different shapes, from left to right: powderlike product, plum blossom, and simplified logo of the Renmin University of China.

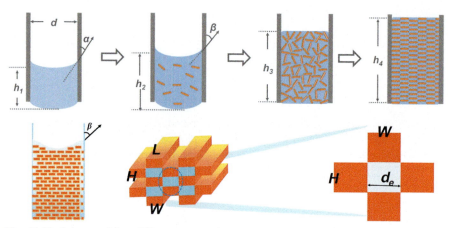

Fig. 11.19 Scheme of four different states of crystal-confined ionic liquids as the increased proportion of [OMIm][AzoO] in ionic liquid crystals.

glass tubes, pure [OMIm]PF$_6$ rapidly flowed out of the glass tube owing to the gravity effect. The ionic liquid blends with the addition of 20 wt% [OMIm]AzoO or more, nevertheless, were spatially confined in the glass tube without any leakage over a long term. This stop-flowing phenomenon convinced that the internal capillary force supplied by [OMIm]AzoO crystal fibers, rather than the external capillary force provided by the glass tube, was critical to confine fluidic [OMIm]PF$_6$ in an open space. In principle, the apparent capillary force should be the sum of capillary force that was provided by each crystal fiber. In this regard, the apparent capillary force might reach a maximum only when each [OMIm]AzoO crystal fiber was fully wetted by [OMIm]PF$_6$.

Several representative proportions of [OMIm][AzoO] were investigated to specify these four different states, including 2 wt% (solution state), 3 wt% (suspended state), 20 wt% (loose accumulation state), and close-stacked state (40, 60, and 80 wt%). Some crucial parameters were summarized as follows: ρ_l and ρ_s are the densities of [OMIm][PF$_6$] and [OMIm][AzoO], respectively. d is the inner diameter of the glass tube; γ_{gl} is the surface tension of the saturated solution of [OMIm][AzoO] in [OMIm][PF$_6$]; α is the contact angle between the saturated solution and glass tube; and β is the contact angle.

For the solution state, the theoretical confinement height can be calculated by the following classical capillary equation.

$$h = \frac{4\gamma_{gl}\cos\alpha}{\rho_L g d} \tag{13}$$

For the suspended state, the traditional capillary Eq. (14) is still valid. It is noteworthy that the density of the liquid needs to be replaced by the average density of crystal-confined ionic liquids, expressed by the form of $(\rho_L\varphi_L + \rho_S\varphi_S)$. Then the equation is turned into the following equation.

$$h = \frac{4\gamma_{gl}\cos\beta}{\rho_e g d} \tag{14}$$

$$h = \frac{4\gamma_{gl}\cos\beta}{(\rho_L\varphi_L + \rho_S\varphi_S)gd} \tag{15}$$

In terms of a loose accumulation state, the capillary effect of crystals will become the main impact to confine the ionic liquids. However, it is difficult to calculate the theoretical height of [OMIm][PF$_6$] absorbed in [OMIm][AzoO] through a traditional capillary equation. So the complex calculation of diameters is replaced by an equivalent diameter. It is supposed that the liquid level rises with a tiny height of dh. Correspondingly, the volume of liquid will change a tiny value dV_i and the area of liquid will change a tiny value dS_i. Therefore, the new balance force at a certain level can be expressed as

$$\sum\left(\Delta P dV_i + \left(\gamma_{ls} - \gamma_{gs}\right)dS_i\right) = 0 \tag{16}$$

Then it can be deduced as

$$\overline{P} = \frac{\left(\gamma_{gl} - \gamma_{gs}\right)\sum dS_i}{\sum dV_i} \tag{17}$$

Since the average Laplace pressure $\overline{P}$ can also be calculated by

$$\overline{P} = \frac{4\gamma_{gl}\cos\theta}{d_e} \tag{18}$$

Then the equivalent diameter is obtained according to the following equation:

$$d_e = \frac{4\sum dV_i}{\sum dS_i} \tag{19}$$

The following equation

$$\begin{cases} \sum dV_i = V_l \\ \sum dS_i = (m_s + m_l)\sigma \end{cases} \tag{20}$$

The equivalent diameter can be expressed by

$$d_e = \frac{4\varphi_L}{\rho_S\varphi_S\sigma} \tag{21}$$

Then the theoretical confinement height can be estimated according to the following equation:

$$h = \frac{\gamma_{gl}\cos\theta\rho_s}{g\rho_L} \times \frac{\varphi_S}{\varphi_L} \times \sigma \tag{22}$$

Here, σ can be calculated based on the completely noncontact crystals

$$\sigma \leq \frac{2(WH + WL + HL)}{WHL \times \rho_s} \tag{23}$$

The theoretical confinement height can be calculated by

$$h_{max} = \frac{2\gamma_{gl}\cos\theta}{g} \times \frac{\rho_S\varphi_S}{\rho_L\varphi_L} \times \frac{WL + WH + HL}{WHL\rho_s} \tag{24}$$

The theoretical confinement heights of the ionic liquid blend with 20, 40, 60, and 80 wt% of [OMIm][AzoO] were estimated as 0.3048, 0.8122, 1.8275, and 4.8732 m, respectively. Those so high confinement heights well explained the freestanding phenomenon of ionic liquid blends without the assistance of encapsulating protection. Intriguingly, this freestanding ionic liquid blend was successfully exploited as thermometers and strain sensors that could be molded as diverse shapes. As an easy-to-read demonstration, a self-repairing arm was made which could self-heal itself without additional human intervention (Fig. 11.20). The material and the demonstration of tactile sensing applications offered inspirations to build liquid robots as conceived in the famous science fiction movie *Terminator 2: Judgment Day*.

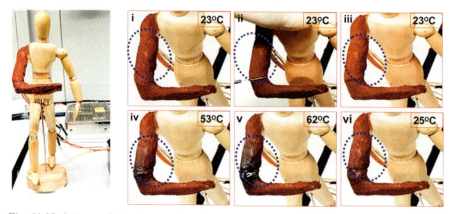

Fig. 11.20 Scheme of robotic arms based on crystal-confined ionic liquids.

11.6 Summary and outlook of the tactile sensor based on ionic liquid

This chapter summarized a few pioneering examples of pressure sensors and temperature sensors based on ionic liquids. The resistance change induced by mechanical deformation or varied ion mobility was explained to account for the mechanism of pressure sensing and temperature sensing. With proper signal separation, these two types of sensors could be integrated together to do synchronous detection of the changes of temperature and pressure. Regardless of the volatile issue according to the negligible saturated vapor pressure of ionic liquids at room temperature, the instability of sensors may be only caused by the leakage phenomenon of ionic liquids from encapsulated protection. This concern was addressed by the strategies of using a capillary effect to confine ionic liquids and impair their liquidity. From a thorough survey, the innovation of exploiting ionic liquids as tactile sensors is an extremely new application field of ionic liquids, though they have been well known for a long time, serving as green solvents in chemical synthesis, catalysis, gas adsorption, processing biomass, and electrochemistry. Ionic liquid-based tactile sensors offer great potentials in developing soft robotics, artificial skins, and medical implants where highly flexible and self-healing electrical components are essentially needed. Additionally, in consideration of easy recycling and pollution-free degradation, ionic liquid-based tactile sensors are also envisioned to generate less waste than traditional semiconductor-based devices, in accordance with the global goals to developing green electronics with less production of electronic waste.

In the end, some suggestions may be taken into account for who would like to expand the investigations of ionic liquid-based tactile sensors. (1) Specifically, for the purpose of health diagnosis, both implantable and adhesive sensors require the assurance of desired biocompatibility. It is necessary to exploit ionic liquids that are fully benign to the human body. (2) Beyond doubt, the ultimate aim of the fundamental researches on ionic liquid-based sensors is to promote their practical applications in the market. To do it, we have to be concerned about the necessity of replacing traditional semiconductor-based sensors upon evaluating the stability, reliability, repeatability, cost, and large-scale production of ionic liquid-based tactile sensors. (3) The category of ionic liquid sensors is still limited so far. Except temperature sensor and pressure sensor, some other underlying functions, e.g., light detection, chemical identification, and electromagnetic perception, may reinforce the special needs of tactile sensors through the choice of functional ionic liquids or the combination with other nanomaterials.

References

[1] R.D. Rogers, K.R. Seddon, Ionic liquids—solvents of the future? Science 302 (2003) 792–793.
[2] T. Welton, Room-temperature ionic liquids. Solvents for synthesis and catalysis, Chem. Rev. 99 (1999) 2071–2083.

[3] R.M. Barrer, The viscosity of pure liquids. 11. Polymerised ionic melts, Trans. Faraday Soc. 39 (1943) 59–67.

[4] S. Gabriel, J. Weiner, Ueber einige Abkömmlinge des Propylamins, Chem. Ber. 21 (1888) 2669–2679.

[5] V.N. Emel'yanenko, G. Boeck, S.P. Verevkin, R. Ludwig, Volatile times for the very first ionic liquid: understanding the vapor pressures and enthalpies of vaporization of ethylammonium nitrate, Chem. Eur. J. 20 (2014) 11640–11645.

[6] S. Zhang, X. Lu, Q. Zhou, X. Li, X. Zhang, S. Li, Ionic liquids, in: Physicochemical properties, Elsevier, Oxford, 2009.

[7] H. Jia, Z. Ju, X. Tao, X. Yao, Y. Wang, P-N conversion in a water-ionic liquid binary system for nonredox thermocapacitive converters, Langmuir 33 (2017) 7600–7605.

[8] H. Wang, Y. Zhao, Y. Wu, R. Li, H. Zhang, B. Yu, F. Zhang, J. Xiang, Z. Wang, Z. Liu, Hydrogenation of carbon dioxide to C2-C4 hydrocarbons catalyzed by Pd(PtBu3)2-FeCl2 with ionic liquid as cocatalyst, ChemSusChem 12 (2019) 4390–4394.

[9] Q. Yang, F. Mo, Z. Liu, L. Ma, X. Li, D. Fang, S. Chen, S. Zhang, C. Zhi, Activating C-coordinated iron of iron hexacyanoferrate for Zn hybrid-ion batteries with 10 000-cycle lifespan and superior rate capability, Adv. Mater. 31 (2019), 1901521.

[10] J. Wang, H. Wang, S. Zhang, H. Zhang, Y. Zhao, Conductivities, volumes, fluorescence, and aggregation behavior of ionic liquids [C4mim][BF4] and [Cnmim]Br (n = 4, 6, 8, 10, 12) in aqueous solutions, J. Phys. Chem. B 111 (2007) 6181–6188.

[11] F. Liu, Y. Lv, J. Liu, Z.-C. Yan, B. Zhang, J. Zhang, J. He, C.-Y. Liu, Crystallization and rheology of poly(ethylene oxide) in imidazolium ionic liquids, Macromolecules 49 (2016) 6106–6115.

[12] J. Gao, J. Cao, Z. Yin, X. Liu, H. Zhao, Y. Chu, X. Tan, X. Bian, F. Zhao, J. Zhang, A facilitate process to prepare hydrophilic ionic liquid monomers free of halide impurity and their electrochemical properties, Int. J. Electrochem. Sci. 8 (2013) 4914–4923.

[13] P. Naert, K. Rabaey, C.V. Stevens, Ionic liquid ion exchange: exclusion from strong interactions condemns cations to the most weakly interacting anions and dictates reaction equilibrium, Green Chem. 20 (2018) 4277–4286.

[14] Y. He, X.-Q. Xu, S. Lv, H. Liao, Y. Wang, Dark ionic liquid for flexible optoelectronics, Langmuir 35 (2018) 1192–1198.

[15] Q. Gui, Y. Zhou, S. Liao, Y. He, Y. Tang, Y. Wang, Inherently magnetic hydrogel for data storage based on the magneto-optical Kerr effect, Soft Matter 15 (2019) 393–398.

[16] Q. Gui, Y. He, N. Gao, X. Tao, Y. Wang, A skin-inspired integrated sensor for synchronous monitoring of multiparameter signals, Adv. Funct. Mater. 27 (2017), 1702050.

[17] S.-H. Zhang, F.-X. Wang, J.-J. Li, H.-D. Peng, J.-H. Yan, G.-B. Pan, Wearable wide-range strain sensors based on ionic liquids and monitoring of human activities, Sensors 17 (2017) 2621.

[18] D.Y. Choi, M.H. Kim, Y.S. Oh, S.H. Jung, J.H. Jung, H.J. Sung, H.W. Lee, H.M. Lee, Highly stretchable, hysteresis-free ionic liquid-based strain sensor for precise human motion monitoring, ACS Appl. Mater. Interfaces 9 (2017) 1770–1780.

[19] S.G. Yoon, H.J. Koo, S.T. Chang, Highly stretchable and transparent microfluidic strain sensors for monitoring human body motions, ACS Appl. Mater. Interfaces 7 (2015) 27562–27570.

[20] H. Jia, Y. He, X. Zhang, W. Du, Y. Wang, Integrating ultra-thermal-sensitive fluids into elastomers for multifunctional flexible sensors, Adv. Electron. Mater. 1 (2015), 1500029.

[21] T. Fukushima, A. Kosaka, Y. Ishimura, T. Yamamoto, T. Takigawa, N. Ishii, T. Aida, Molecular ordering of organic molten salts triggered by single-walled carbon nanotubes, Science 300 (2003) 2072–2074.

[22] Y. He, T.P. Lodge, A thermoreversible ion gel by triblock copolymer self-assembly in an ionic liquid, Chem. Commun. (2007) 2732–2734.
[23] S. Zhang, F. Wang, H. Peng, J. Yan, G. Pan, Flexible highly sensitive pressure sensor based on ionic liquid gel film, ACS Omega 3 (2018) 3014–3021.
[24] R. Li, Y. Si, Z. Zhu, Y. Guo, Y. Zhang, N. Pan, G. Sun, T. Pan, Supercapacitive iontronic nanofabric sensing, Adv. Mater. 29 (2017), 1700253.
[25] Y. Liu, Y. Hu, J. Zhao, G. Wu, X. Tao, W. Chen, Self-powered piezoionic strain sensor toward the monitoring of human activities, Small 12 (2016) 5074–5080.
[26] Y. He, S. Liao, H. Jia, Y. Cao, Z. Wang, Y. Wang, A self-healing electronic sensor based on thermal-sensitive fluids, Adv. Mater. 27 (2015) 4622–4627.
[27] X. Tao, H. Jia, Y. He, S. Liao, Y. Wang, Ultrafast paper thermometers based on a green sensing ink, ACS Sens. 2 (2017) 449–454.
[28] C.-Y. Wu, W.-H. Liao, Y.-C. Tung, A seamlessly integrated microfluidic pressure sensor based on an ionic liquid electrofluidic circuit, in: 2011 IEEE 24th international conference on micro electro mechanical systems, 2011, pp. 1087–1090.
[29] N. Gao, Y. He, X. Tao, X.-Q. Xu, X. Wu, Y. Wang, Crystal-confined freestanding ionic liquids for reconfigurable and repairable electronics, Nat. Commun. 10 (2019) 547.

Self-powered flexible tactile sensors

12

Xuan Zhang[a,b] and Bin Su[a,]*
[a]State Key Laboratory of Material Processing and Die & Mould Technology, School of Materials Science and Engineering, Huazhong University of Science and Technology, Wuhan, Hubei, PR China, [b]Department of Chemical Engineering, ARC Hub for Computational Particle Technology, Monash University, Clayton, VIC, Australia
[*]Corresponding author: e-mail address: xuan.zhang2@monash.edu, subin@hust.edu.cn

12.1 Introduction

The famous scientists Volt and Faraday pioneered the field of electricity and electromagnetism 200 years ago, which led to tremendous development in rigid electronics. In 1984, infrared e-skin was first reported, ushering in the age of flexible electronics [1]. Flexible sensors are a revolutionary and crucial technology with potential applications in human-machine interactivity, smart textiles, mobile communication, and biomedical healthcare [2–6]. Advances in flexible devices such as soft robots, wearable electronics, bendable displays, and bioinspired electronic skins have great promise for next-generation devices to be more portable, wearable, and even implantable [7–10]. This is of great significance and benefit for enhancing quality of life and national economic growth. Nevertheless, most flexible sensing systems require commercial batteries to constantly supply power, resulting in some inevitable constraints such as periodic replacement, troublesome assembly, and potential pollution to the environment [11, 12]. Thus, a green, self-powered system that captures ambient forms of energy and converts them into electricity is of prime importance.

There are various available energy sources that can be utilized in our surrounding environment, such as chemical energy, mechanical energy, thermal energy, solar energy, nuclear energy, and so on. Distinct from the others, mechanical energy mainly involves human-related movements and vibration, which are not only ubiquitously available but also easily ignored and largely wasted [13]. Thus, it is important to figure out how to harvest such mechanical energy and then convert it into usable electrical energy, instead of having it dissipate into the surroundings.

There are several energy-harvesting technologies to achieve self-powered flexible applications via mechanoelectrical conversion, including piezoelectricity, triboelectricity, and electromagnetic induction. Wang et al. [14–17] successively pioneered piezoelectric and triboelectric nanogenerators, both of which are capable of self-powered sensing under ambient stress stimuli. After the first ZnO nanowire (NW)-based piezoelectric nanogenerators (PENGs) were developed as flexible energy

Functional Tactile Sensors. https://doi.org/10.1016/B978-0-12-820633-1.00001-2
© 2021 Elsevier Ltd. All rights reserved.

harvesters in 2006 [18], diverse PENGs with various structures and materials such as PZT, BaTiO$_3$, polymeric polyvinylidine fluoride (PVDF), and P(VDF-TrFE) have been introduced by researchers [19–24]. More recently, the development of triboelectric nanogenerators (TENGs) has achieved high conversion efficiency and broad materials selections. However, the limited current output and high internal impedance of TENGs present some challenges for self-powered tactile sensors [25–27]. As such, novel self-powered magnetoelectrical elastomers based on electromagnetic induction have been proposed, paving a new way for flexible energy-capturing technology [28, 29]. In addition, the flexible application of electromagnetic induction has been successfully achieved.

This chapter focuses on the recent progress of PENGs and TENGs as well as magnetoelectric elastomers working as self-powered tactile sensors. We begin with a thorough review of the working mechanisms and the different approaches for PENGs based on inorganic, organic, and composite materials. Then, we elaborate on TENGs for self-powered sensors from operative principles to materials and structural designs. Next, we demonstrate magnetoelectric elastomers with self-powered capacity for tactile sensing applications. Finally, we present some challenges and perspectives for flexible tactile sensors.

12.2 Flexible piezoelectric nanogenerators

12.2.1 Piezoelectric mechanism

Piezoelectricity in inorganic materials (as in ZnO, PZT, BaTiO$_3$) is based on the positive/negative charge center separation in crystal structure. When applying external force (compressing/stretching) onto piezoelectric materials, the positive and negative charges inside crystal structure are separated due to the dislocation polarization of their charge centers, meanwhile forming an electric dipole to result in a piezopotential difference. If the deformed material is connected to an external load, the free electrons are propelled to flow thorough the external circuit, achieving a new equilibrium state [30, 31].

Piezoelectric phenomenon in organic materials, such as in PVDF and its copolymers, relies on the reorientation of molecular dipoles within the bulk polymer, which is induced and achieved by high electric poling or mechanical stretching. For PVDF containing five crystal phases (α, β, γ, δ, ε), β phases with all-trans confirmation show optimal piezoelectricity due to the highest dipole moment. In contrast to PZT, PVDF possesses negative piezoelectric effect [32].

12.2.2 Piezoelectric materials

12.2.2.1 Inorganic materials

In 2009, Yang et al. [33] developed a single ZnO NW nanogenerator, harvesting low-frequency energy from muscular movement to convert to a low electrical output (an output current of 400–750 pA). For improving the output performance, Xu et al. [34]

reported the integration of large numbers of ZnO NWs into a multiple NW-array, reaching to 29 nA. Considering the poor durability and stability for ZnO NW-based PENGs, Choi et al. [35] utilized single-walled carbon nanotube (SWCNT) sheets as the top contact electrodes. SWCNT films with numerous nanopores contributed to the ZnO nanorods contacting the electrode well, which resulted in increased current output and improved system stability. Lee et al. [36] developed an ultrathin flexible piezoelectric sensor employing a thin Al foil as the substrate and electrode. The surface of the Al foil had an insulating layer of anodic aluminum oxide, which favors the electron transfer between the Al electrode and the grown ZnO NW film. The highly sensitive and durable PENG was applied for monitoring tiny skin deformation from eyeball movement, generating 0.2 V and 2 nA.

PZT, as a kind of traditional piezoelectric ceramic material, has outstanding attributes such as high dielectric constants and piezoelectric voltage. However, the natural brittleness of PZT thin films (0.2% safe strain range) enable transferring technology for flexibility to be inevitable [37]. Qi et al. [38] reported a stretchable PENG with wavy-pattern PZT ribbons on PDMS rubber. The wavy PZT ribbons were obtained via a pre-stretched method, realizing more than 8% elastic strain without any cracks. Integrating ten PZT ribbons into a PENG yielded a current density of 2.5 $\mu A\ mm^{-2}$. Park et al. [39] fabricated a large-area flexible PENG via a laser lift-off process, transferring from the sapphire host substrates to flexible PET substrates. A 3.5 cm $\times$ 3.5 cm PZT-based PENG was able to generate 8 μA to light more than 100 commercial blue LEDs under slight bending of human fingers. Similarly, Lee et al. [40] utilized a laser lift-off technology to fabricate a flexible PZT-based piezoelectric acoustic nanosensor (PANS) on a silicon-based membrane (SM), aiming to biologically mimic artificial hair cells. The SM allowed PANS to convert tiny vibration displacement ($\sim$15 nm), driving incoming sound into a piezoelectric sensing signal of $\sim$55 μV.

Moreover, $BaTiO_3$ is an attractive, lead-free piezoelectric material due to its outstanding ferroelectric properties, high piezoelectric coefficient, and biocompatibility. There is great promise for $BaTiO_3$ in biomedical applications. A bio-inspired flexible PENG based on virus-directed $BaTiO_3$ nanostructures was described in 2013 [41], attaining an electrical output of $\sim$6 V and $\sim$ 300 nA. Chen et al. [42] developed a lead-free flexible PENG based on single-crystalline GaN thin film through wet chemical etching. The flexible GaN-based PENG attached to a human finger exhibited maximum output voltage of 28 V at bending testing.

12.2.2.2 Organic materials

Organic piezoelectric materials, such as PVDF and its copolymer P(VDF-TrFE), are widely used due to their relative flexibility, sufficient mechanical strength, outstanding biocompatibility, and conspicuous chemical resistance. Nevertheless, the disadvantage of polymeric PVDF is that it must experience electric poling to create more β phases for inducing piezoelectricity. Mechanical stretching is required to align the dipoles of β-phase PVDF to form oriented molecular chains [43]. Hansen et al. [44] utilized an electrospinning technique to obtain aligned PVDF nanofiber arrays, followed by in-plane poling. These PVDF nanofibers were encapsulated into PDMS to achieve a flexible thin-film PENG, yielding 20 mV and 0.3 nA at bending testing. In the same way, Persano et al. [45] synthesized a flexible pressure sensor consisting of highly aligned P(VDF-TrFE) electrospun fibers on a fast-rotating collector disk. The spinning process drove to overlap fibers to improve the mechanical robustness of the film without any damage undergoing bending or folding. The polymer-fiber PENG exhibited a superior sensitivity to monitor a minimum pressure of 0.1 Pa, showing potential in detecting vibration, acceleration, and orientation.

The aforementioned aligned fibers need to experience great bending to obtain an adequate fiber stretching for effective energy conversion. In order to enhance their output performance, PVDF and its copolymers have been designed and optimized with desired nanostructures. Mao et al. [46] proposed a sponge-like PVDF piezoelectric film with mesoporous network, employing ZnO nanoparticles as the template for porosity. The sponge-like porous PVDF nanogenerator can be directly attached on the surface of human skin or any electronic device, effectively scavenging mechanical oscillations to generate electricity. The 2 cm $\times$ 1 cm $\times$ 28 μm mesoporous PVDF device generated $\sim$11 V and $\sim$ 9.8 μA at oscillation frequency of 40 Hz. Cho et al. [47] presented a high-performance P(VDF-TrFE) thin film via a solvent annealing method, leading to the formation of β phases. This flexible PENG modified by surface morphology engineering exhibited almost eight times increased output voltage due to well-aligned electrical dipoles, displaying voltage output of 3 and 8 V via movement of the wrist and tapping of fingertips, respectively.

12.2.2.3 Composite materials

Composite-based piezoelectric nanogenerators disperse piezoelectric nanomaterials into an elastomeric matrix, which contributes to the large-scale applications of flexible energy harvesters. A hyper-stretchable, elastic-composite energy harvester was developed by Jeong et al. [48]. This piezo-elastic composite exhibited outstanding electrical output ($\sim$4 V and $\sim$ 500 nA) under 200% stretchability. The superior stretchability relied on the use of ultra-long Ag NW electrodes and flexible EcoFlex rubber elastomer. The stretchable nanogenerator consisted of PMN-PT particles and multi-walled CNTs that maintained mechanical and electrical robustness under twisting, crumpling, and folding deformation. Ma et al. [49] represented flexible, porous PDMS/PZT-based nanogenerators with the dual effect of piezoelectricity and ferroelectricity. The 2 $\times$ 2 $\times$ 0.3 cm^3 flexible porous nanogenerators (FPNGs) yielded a voltage output of 29 V and a current output of 116 nA under a pressure of 30 N, respectively. Due to their excellent adaptability in shape, FPNGs fixed onto bicycle tires were capable of monitoring bicycle moving states and transmitting signals.

Recently, a PZT/PDVF-composite piezoelectric sensor with baklava structure was developed for s wireless real-time monitoring system for table tennis training [50]. The unique structure contributed to enhance not only the piezoelectricity but also fast response through dispersing stress (Fig. 12.1A). The Voc and Isc were 2.51 V and 78.43 nA, respectively. More importantly, the sensor exhibited a high sensitivity of 6.38 mV/N. Fig. 12.1B and C show a smart table tennis racket integrated with

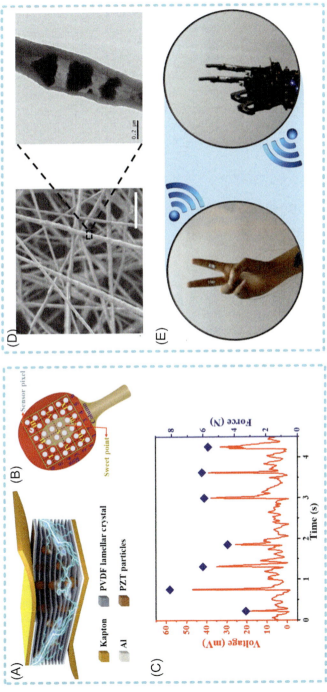

Fig. 12.1 Flexible PENGs with self-powered sensing capacity. (A–C) Baklava-structured piezoelectric sensors integrated into a smart table tennis racket. (D and E) Cowpea-structured ZnO/PVDF nanofibers for remote control of robot hand gestures. Panels (A–C) reproduced with permission from G. Tiana, W. Denga, Y. Gao, D. Xionga, C. Yan, X. He, et al. Rich lamellar crystal baklava-structured PZT/PVDF piezoelectric sensor toward individual table tennis training, Nano Energy 59 (2019) 574–581. Copyright 2019, Elsevier, Panels (D) and (E) reproduced with permission from W. Deng, T. Yang, L. Jin, C. Yan, H. Huang, X. Chu, et al., Cowpea-structured PVDF/ZnO nanofibers based flexible self-powered piezoelectric bending motion sensor towards remote control of gestures, Nano Energy 55 (2019) 516–525. Copyright 2019, Elsevier.

6 × 6 sensing units, gaining real-time information on hit location and contact force for individual training guidance. Deng et al. [51] introduced a piezoelectric bending motion sensor based on cowpea-structured ZnO/PVDF hybrid nanofibers for remote control of a robot hand (Fig. 12.1D and E). The synergistic piezoelectricity between ZnO and PVDF allowed for both bending and pressing, and their ratio was able to regulate sensitivity. The sensor was conformable for different curved surfaces, enabling recording and recognition of bending angles in human-machine interaction. In addition, PVDF-TrFE/ZnO composites reported by Han et al. [52] achieved better piezoelectric performance than pure PVDF-TrFE film, reaching 5.5 times with 7.5% ZnO nano-doping. These composite films also had high sensitivity.

12.3 Flexible triboelectric nanogenerators

12.3.1 Triboelectric mechanism

Flexible triboelectric nanogenerators for achieving mechanoelectrical conversion rely on the coupling effects of triboelectrification and electrostatic induction. Triboelectrification is an electrical charging process after two different materials come in contact with one other through friction [53]. The process gives rise to the formation of equal amounts but opposite charges on the attached surface in two different materials based on their different stability to capture electrons. Once the two materials are separated, a triboelectric potential at the interfacial region can drive free electrons to flow in the external load in order to balance the induced potential difference. Thus, the driving mechanical stress is able to periodically vary the induced potential difference through the contact and separation of two materials, resulting in a current pulse. Based on the mechanism, four kinds of TENGs with different operation modes have been elaborated. We discuss them in the following sections [54].

Vertical contact-separation mode. Two dielectric films with different electron-attracting ability are placed as friction layers on the inner surface of two electrodes. This mode efficiently harvests vertical pressure energy, which requires alternative switching between the fully contacted state and the separated state. Recently, various structure designs in TENGs, such as arch-shaped structures, planar structures with elastic spacers, stacking structures, and tangled fibers, have been developed for vibration and cyclic pressure.

Lateral sliding mode. This mode structure is similar to that of the vertical contact mode. It comprises dielectric-metal or dielectric-dielectric layers with two electrodes. Some vertical contact-based TENGs with a fully separated planar structure are also capable of working in the lateral sliding mode. The triboelectric charges on the two friction surfaces are generated by a relative parallel sliding, accompanied by a lateral polarization along the sliding direction. Such a sliding mode allows for harvesting new types of movements such as disk-like and cylindrical rotations. Additionally, the introduction of grated structures in TENGs increases the efficient charge transfer, leading to improved output current density.

Self-powered flexible tactile sensors 251

Single-electrode mode. This mode is suitable for some cases in which a mobile object can be one electrode of the TENG without electrically connecting to the load, such as our skin or clothes/shoes, which we touch and move every day.

Freestanding triboelectric-layer mode. In this mode, a pair of symmetric electrodes under a dielectric layer create an asymmetric electrical charge distribution as a moving object approaching to and departing from the electrodes. As a result, the free electrons can spontaneously flow from one electrode to another to maintain the local electrostatic equilibrium. The mode allows a non-contacted system between the moving object and the triboelectric layer deposited on the electrodes to be demonstrated.

12.3.2 Triboelectric materials

Triboelectric nanogenerators generally employ two types of materials with different electron-attracting ability as triboelectric layers. These materials are dependent on the triboelectrification phenomenon to realize mechanoelectrical conversion. In essence, almost all materials ranging from metals to polymers, both synthetic and natural, are options for triboelectric materials. Thus, the choice of two triboelectric materials with large difference in electron affinity, to some extent, obtains a highly efficient charge transfer.

More importantly, the modifying surface morphology of triboelectric films, such as various micro/nanopatterns, has a positive effect on the electrical output and other functionality of TENGs due to enhancement of the contact area. Fan et al. [55] designed flexible TENGs with three types of micropatterned PDMS films, including lines, cubes, and pyramids. The output performance of these different TENGs followed the order of: film (unstructured) < lines < cubes < pyramids, in light of air voids/bubbles presented in the microstructures. The micropatterned PDMS films with more sophisticated structures exhibited greater triboelectric effect due to more surface charges induced by the friction. The pyramid-patterned TENG had an optimal output voltage of 18 V at a current density of 0.13 $\mu A/cm^2$. Such triboelectric devices used as self-powered pressure sensors detected obvious responses for a droplet of water (8 mg, $\sim$3.6 Pa in pressure) as well as a piece of feather (20 mg, $\sim$0.4 Pa in touched pressure). Cheng et al. [56] introduced a wrinkle-feathered TENG via using fluorocarbon plasma to treat PDMS, fulfilling the stretchability due to the fluoropolymer deposited on the PDMS. The wrinkle microstructures increased the output performance of the triboelectric devices. Meanwhile, Chen et al. [57] proposed composite-based TENGs combining sponge-structured PDMS and high dielectric nanoparticles (SiO_2, TiO_2, $BaTiO_3$ and $SrTiO_3$), leading to the improvement of their capacitance and electrical output.

Furthermore, Zhu et al. [58] reported self-powered triboelectric sensors by employing a vertically aligned FEP NW film as the electrification layer. The triboelectric sensor showed ultrahigh sensitivity in the extremely low-pressure region ($<$0.15 kPa), which was promising for a wireless alarm system triggered by finger tapping, foot stepping, and hand grabbing. Song et al. [59] proposed a self-powered triboelectric sensor based on a patterned aluminum-plastic film (APLF) for sleep monitoring. The APLF film coated a polypropylene layer with nanopillar arrays played a key role in the sensitive sensing system for sleep movements. Considering the unique properties of carbon nanotubes (CNTs) including excellent conductivity, stretchability, and triboelectricity when touching the PDMS, CNT-based TENGs reported by Park et al. [60] electrically output up to tens of volts and several $\mu A/cm^2$. Natural fiber materials (silk, cotton) and synthetic fiber materials (nylon) are popular as triboelectric layers due to their mechanical compliance, wearability, and comfortability, which are all promising factors for self-powered smart clothing.

12.3.3 Structural designs

Air gap-introduced structures. The introduction of air gap contributes to the full separation between two triboelectric surfaces for high performance. The previous arch-shaped gap between a polymer film and a metal foil, as a typical example, enabled the TENG to obtain output voltage of 230 V and current density of 15.5 $\mu A/cm^2$ [61]. In order to enhance output performance, Zhang et al. developed alternatively stacked arch-shaped and anti-arch-shaped TENGs. [62]. Bai et al. [63] reported latex membrane-based triboelectric sensors with air-conducting channels for self-powered detection of pressure change. The sensor had extremely high sensitivity in response to air pressure increase and decrease of 0.34 and 0.16 Pa, respectively, allowing for practical applications in sensing footsteps, respirations, and heartbeat. Furthermore, the researchers applied this distinct structure to human-machine interfaces and in vivo cardiac monitoring. All-elastomer-based TENGs, as a smart keyboard cover to harvest typing energy, exhibited high flexibility and dual-mode operation due to optimized material and structure innovation [64]. For implantable TENGs with multi-layered structures, the nanostructured PTFE-Al film as the triboelectric core layer was double packaged by PTFE and PDMS-parylene for biocompatibility, leak-proof properties, and structure stability [65]. Driven by the heartbeat of adult swine, V_{oc} and corresponding I_{sc} reached up to 14 V and 5 μA, respectively (Fig. 12.2A and B).

Planar and zigzag structures. For lateral sliding mode in plane, the linear micrograting configuration of two triboelectric films is generally desired. A micrograting triboelectric film with complementary linear metal and PTFE grating arrays was displayed in 2014 [66]. Upon experiencing a 10 m/s relative sliding, the thin films (60 cm^2 in area) yielded an output current of up to 10 mA, with conversion efficiency of around 50%. Meanwhile, self-powered linear displacement and velocity sensors based on micrograting TENGs was reported by Wang et al. [67]. The TENG was constructed from two upper and lower microgratings with parylene and SiO_2 as triboelectrically dielectric material, respectively, and could periodically produce electrical output peak through the relative sliding between the gratings. Each output cycle corresponded to one period of two gratings from contact to separation, experiencing a 200 μm displacement in this device. Thus, the real-time displacement distance was derived from accumulating the peak number of output signals, and real-time motion speed was obtained by the time interval between two sensing signal peaks. The zigzag structure was a geometrically multilayered configuration by stacking more planar structures, achieving the improvement of more functional characteristics, especially

Self-powered flexible tactile sensors 253

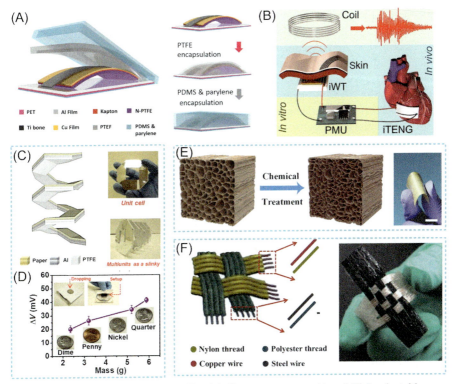

Fig. 12.2 Flexible self-powered TENGs with diverse structures. (A and B) Implantable triboelectric nanogenerator (iTENG) for wireless heartbeat monitoring of adult swine (iWT, implantable wireless transmitter; PMU, power management unit). (C and D) Paper-based TENG with slinky shape for self-powered sensing of the pressure of different coins. (E) Chemically treated, wood-based TENG. (F) 2D woven TENG with textile structure. Panels (A) and (B) reproduced with permission from Q. Zheng, H. Zhang, B. Shi, X. Xue, Z. Liu, Y. Jin, et al., In vivo self-powered wireless cardiac monitoring via implantable triboelectric nanogenerator, ACS Nano 10 (2016) 6510–651. Copyright 2016, American Chemical Society, Panels (C) and (D) reproduced with permission from P.-K.Yang, Z.-H. Lin, K.C. Pradel, L. Lin, X. Li, X. Wen, et al., Paper-based origami triboelectric nanogenerators and self-powered pressure sensors, ACS Nano 9 (2015) 901–907. Copyright 2015, American Chemical Society, Panel (E) reproduced with permission from J. Luo, Z. Wang, L. Xu, A.C. Wang, K. Han, T. Jiang, et al., Flexible and durable wood-based triboelectric nanogenerators for self-powered sensing in athletic big data analytics, Nat. Commun. 10 (2019) 5147. Copyright 2019, Springer Nature, Panel (F) reproduced with permission from J. Liu, L. Gu, N. Cui, S. Bai, S. Liu, Q. Xu, et al., Core-shell fiber-based 2D woven triboelectric nanogenerator for effective motion energy harvesting, Nanoscale Res. Lett. 14 (2019) 311. Copyright 2019, Springer.

for the pressure mode. Previously, a power-generating shoe insole based on zigzag-shaped, three-layer TENGs was proposed [68]. The contact-separation surfaces between aluminum foil and PTFE film coated with copper in each TENG induced charge transfer, yielding maximum electrical output of 220 V and 600 μA. Triggered by human walking, the insole could directly charge consumer electronics such as commercial LEDs and cellphones. Yang et al. [69] demonstrated a green, renewable paper-based TENG with slinky shape as another variation of zigzag structure (Fig. 12.2C). The new type of origami-slinky TENG enabled not only harvesting of mechanical energy from stretching, lifting, and twisting to light up four LEDs simultaneously, but also enabled a self-powered pressure sensor to distinguish standard coins with different masses (Fig. 12.2D).

3D porous structures. Chena et al. [70] developed a miniaturized, ultra-flexible, and three-dimensional triboelectric nanogenerator (3D-TENG) via utilizing hybrid UV 3D printing. The ultra-flexibility of this TENG stemmed from the distinct design of the 3D porous structure, where printed composite resin parts (composite resin I, II) and PAAm-LiCl ionic hydrogel served as the triboelectric layers and electrodes, respectively. The 3D-TENG working in a single-electrode mode was used as a self-powered source for not only LED flickering of an SOS distress signal system but also for portable temperature sensors. Luo et al. [71] proposed to transform natural wood into a high-performance and flexible triboelectric material. Motivated by the concepts of renewable and biodegradable energy, self-powered, flexible TENGs based on 3D porous wood were demonstrated by their group for monitoring and assisting table tennis training in the smart sports industry. The crumpled and porous structures of wood samples were obtained via chemical treatment involving partial removal of lignin/hemicellulose (Fig. 12.2E). The treated wood exhibited excellent bending durability (1000 cycles) and tensile strength (284.2 MPa). The wood-based TENGs showed maximum output power density of 57 mW/m^2 under an external load resistance of 40 MΩ.

Fiber-shaped and textile structures. Inspired by textile configuration, conductors have been constructed into fiber structures for stretchability and weaveability, rendering them applicable in wearable smart clothing. A coaxial structure is primarily adopted in fiber-based TENGs, involving insulating polymer fibers and operable conductive wires. Yu et al. [72] proposed a coaxial fiber-based TENG with the capacity of sensing under multidirectional deformation. Porous PDMS fiber and PMMA microspheres wrapped onto a CNT sheet as electrodes were the core and shell layer of contact electrification, respectively. The introduction of an air gap between two triboelectric layers plays a critical role in guaranteeing their effective contact and separation. The fiber-based TENG was capable of self-powered sensing triggered by diverse mechanical stimuli such as bending, pressing, stretching, twisting, and vibrating, yielding a maximum voltage output of 5 V and a current output of 240 nA. Xie et al. [73] demonstrated a fiber-shaped, tailorable TENG composed of a spiral steel wire as electrode and silicone elastomer as triboelectric layer, enabling an active gesture sensor through being knitted on the dorsum of a glove. The maximum voltage and current output for the single TENG were $\sim$ 59.7 V and $\sim$ 2.67 μA under 2.5 Hz, respectively, allowing a wearable fabric woven from several TENG to light up green LEDs through harvesting biomechanical energy from human movements. Meanwhile, Liu et al. [74] reported 2D textile-structured TENGs woven from two different core-shell composite fibers, with copper/steel wire as the central electrode and nylon/polyester thread as the outer triboelectric layer (Fig. 12.2F). This TENG was adaptable to covert motion energy in arbitrary sliding directions into electricity, making it potentially useful in fiber-based wearable generators.

12.4 Flexible, self-powered magnetoelectric elastomers

The electromagnetic induction effect is well known for power generation, inspiring the investigation of electromagnetic harvesters to convert biomechanical energy into electricity. Biomechanical energy harvesters reported by Donelan et al. [75] successfully produced enough electricity to charge prosthetic knee braces during walking. However, the critical component for power generation was completely rigid and presented some challenges for a miniaturized, self-powered, flexible system. In order to overcome this limitation, Zhang et al. [29] initiated an innovative binary material design strategy for flexible, self-powered sensors in 2019 (Fig. 12.3A). Integrating electrical and magnetic binary functional blocks into the flexible silicone that served as mechanical supporting, resulting in a binary cooperative magnetoelectric elastomer. The new magnetoelectric elastomers had self-powered sensing capacity in response to external mechanical

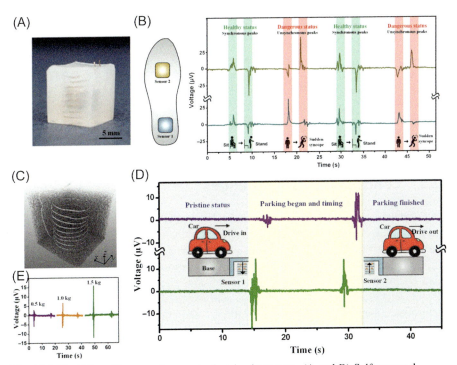

Fig. 12.3 Flexible, self-powered magnetoelectric elastomers. (A and B) Self-powered magnetoelectric sensor integrated into smart shoes for warning of sudden syncope in the elderly. (C and D) Flexible magnetoelectric composites for self-powered sensing of car parking. Panels (A) and (B) reproduced with permission from X. Zhang, J. Ai, Z. Ma, Z. Du, D. Chen, R. Zou, et al., Binary cooperative flexible magnetoelectric materials working as self-powered tactile sensors, J. Mater. Chem. C 7 (2019) 8527. Copyright 2019, Royal Society of Chemistry, Panels (C) and (D) reproduced with permission from X. Zhang, J. Ai, Z. Ma, Z. Du, D. Chen, R. Zou, et al., Magnetoelectric soft composites with a self-powered tactile sensing capacity, Nano Energy 69 (2020) 104391. Copyright 2020, Elsevier.

forces, verified via Maxwell numerical simulation. The change of magnetic flux before and after deformation stemmed from the controllable distance of electromagnetic interaction. The elastomer-based magnetoelectric sensor not only exhibited fast response (~20 ms) and minimum pressure detection of ~690 Pa, but also served as a smart shoe for real-time healthcare monitoring to avert sudden syncope in the elderly (Fig. 12.3B). Contrary to synchronous signals in healthy persons, asynchronous peaks appearing for two sensors embedded in a single insole suggested a potential risk. Afterwards, to unify Young's moduli of magnetic/polymeric building blocks, a magnetoelectric soft composite involving a powdery network technique was developed (Fig. 12.3C) [28]. Magnetic powders, instead of the magnetic cubes used previously, were uniformly dispersed in polymeric elastomers, allowing for magnetic/electrical/polymeric composites to deform/recover synchronously. Under optimal parameters, the open-circuit voltage for the assembly of magnetoelectric composites in a series reached up to 277 µV. Relative to piezoelectric/triboelectric counterparts, the magnetoelectric composite illustrated high current density (31.25 µA/cm^2) due to significantly small internal resistance (few ohms). The sensing cube array composed of the composites displayed the spatial distribution of pressure from different items onto it, indicating self-powered tactile capacity. Fig. 12.3D shows a smart parking system based on two composite-based sensors for precisely recording park time, regardless of the time spent time looking for a parking spot. The first signal of sensor 1 and the second signal of sensor 2 represented the beginning and finishing of the parking, respectively. Meanwhile, the intensity of output signals increased with the weight of the parked car, distinguishing the different loadings on the car (Fig. 12.3E).

12.5 Conclusion and challenges

In this chapter, we investigated the recent advances in self-powered, flexible tactile sensors based on piezoelectricity, triboelectricity, and electromagnetic induction. Since the first PENG and TENG developed in 2006 and 2012, respectively, flexible tactile sensors have advanced a new field in self-powered sensing. The studies presented elaborate on power-generating mechanisms, material selection, and structural designs, suggesting intimate correlation with excellent properties including output performance, sensitivity, durability, stability, and multifunctionality. In addition, magnetoelectric elastomers, as innovative self-powered flexible sensors, caused a paradigm shift of electromagnetic induction from rigid to flexible application. The three types of energy harvesters capable of mechanoelectrical conversion have tremendous potential in miniaturized, portable, environmentally friendly, multifunctional, and smart flexible sensors.

In terms of future development of mechanoelectrical-converted flexible systems, there are still tremendous constraints and challenges to address. First, given the inherent brittleness and high Young's modulus for piezoelectric materials, complicated fabrication techniques are unavoidable to improve output performance and flexibility. Compared to piezoelectricity, triboelectricity not only allows for broader material selection but also endows TENGs with more attractive properties such as high conversion efficiency, high output performance, versatility in structure design, low cost, and easy scalability. Unfortunately, triboelectrification requires TENGs to bear the risk of mechanical wear [76]. Furthermore, both triboelectric and piezoelectric devices have high output impedance, which negatively impacts their practical applications in real life. In order to reduce internal resistance, cost-effective magnetoelectric elastomers have been achieved for mechano-magneto-electrical conversion. Despite the current magnetoelectric sensors exhibiting superior current density, relatively low voltage output limits their sensitivity and multifunctionality, especially for powering flexible electronics and sensor networks. Thus, further investigation of operative mechanisms, novel structural designs, and advanced materials is needed.

Acknowledgements

This work was supported by National 1000 Young Talents Program of China, and initiatory financial support from HUST.

References

[1] M.L. Hammock, A. Chortos, B.C. Tee, J.B. Tok, Z. Bao, 25th anniversary article: the evolution of electronic skin (e-skin): a brief history, design considerations, and recent progress, Adv. Mater. 25 (2013) 5997–6038.

[2] Y. Wan, Y. Wang, C.F. Guo, Recent progresses on flexible tactile sensors, Mater. Today Phys. 1 (2017) 61e73.

[3] X. Wang, Z. Liu, T. Zhang, Flexible sensing electronics for wearable/attachable health monitoring, Small 13 (2017) 1602790.

[4] Z. Liang, J. Cheng, Q. Zhao, X. Zhao, Z. Han, Y. Chen, et al., High-performance flexible tactile sensor enabling intelligent haptic perception for a soft prosthetic hand, Adv. Mater. Technol. 4 (2019) 1900317.

[5] J. Shi, S. Liu, L. Zhang, B. Yang, L. Shu, Y. Yang, et al., Smart textile-integrated microelectronic systems for wearable applications, Adv. Mater. 32 (2019) 1901958.

[6] V. Amoli, S.Y. Kim, J.S. Kim, H. Choi, J. Koo, D.H. Kim, Biomimetics for high-performance flexible tactile sensors and advanced artificial sensory systems, J. Mater. Chem. C 7 (2019) 14816.

[7] A. Nathan, A. Ahnood, M.T. Cole, S. Lee, Y. Suzuki, P. Hiralal, et al., Flexible electronics: the next ubiquitous platform, Proc. IEEE 100 (2012) 1486–1517.

[8] Y. Wu, Y. Liu, Y. Zhou, Q. Man, C. Hu, W. Asghar, et al., A skin-inspired tactile sensor for smart prosthetics, Sci. Robot. 3 (2018) eaat0429.

[9] Y.H. Jung, B. Park, J.U. Kim, T.-i. Kim, Bioinspired electronics for artificial sensory systems, Adv. Mater. 31 (2019) 1803637.

[10] C. Wang, K. Xia, H. Wang, X. Liang, Z. Yin, Y. Zhang, Advanced carbon for flexible and wearable electronics, Adv. Mater. 31 (2019) 1801072.

[11] Z.L. Wang, Toward self-powered sensor networks, Nano Energy 5 (2010) 512–514.

[12] C. Dagdeviren, Z. Li, Z.L. Wang, Energy harvesting from the animal/human body for self-powered electronics, Annu. Rev. Biomed. Eng. 19 (2017) 85–108.

[13] L. Jin, B. Zhang, L. Zhang, W. Yang, Nanogenerator as new energy technology for self-powered intelligent transportation system, Nano Energy 66 (2019) 104086.

[14] X. Wang, J. Zhou, J. Song, J. Liu, N. Xu, Z.L. Wang, Piezoelectric field effect transistor and nanoforce sensor based on a single ZnO Nanowire, Nano Lett. 6 (2006) 2768–2772.

[15] Y. Hu, Z.L. Wang, Recent progress in piezoelectric nanogenerators as a sustainable power source in self-powered systems and active sensors, Nano Energy 14 (2015) 3–14.

[16] F.-R. Fan, Z.-Q. Tian, Z.L. Wang, Flexible triboelectric generator, Nano Energy 1 (2012) 328–334.

[17] Z.L. Wang, Triboelectric nanogenerators as new energy technology for self-powered systems and as active mechanical and chemical sensors, ACS Nano 7 (2013) 9533–9557.

[18] Z.L. Wang, J. Song, Piezoelectric nanogenerators based on zinc oxide nanowire arrays, Science 312 (2006) 242–246.

[19] C. Chang, V.H. Tran, J. Wang, Y.-K. Fuh, L. Lin, Direct-write piezoelectric polymeric nanogenerator with high energy conversion efficiency, Nano Lett. 10 (2010) 726–731.

[20] Y. Qi, N.T. Jafferis, K.L. Jr, C.M. Lee, H. Ahmad, M.C. McAlpine, Piezoelectric ribbons printed onto rubber for flexible energy conversion, Nano Lett. 10 (2010) 524–528.

[21] Y. Zhao, Q. Liao, G. Zhang, Z. Zhang, Q. Liang, X. Liao, High output piezoelectric nanocomposite generators composed of oriented $BaTiO_3$ NPs@PVDF, Nano Energy 11 (2015) 719–727.

[22] J. Yan, Y.G. Jeong, High performance flexible piezoelectric nanogenerators based on $BaTiO_3$ nanofibers in different alignment modes, ACS Appl. Mater. Interfaces 8 (2016) 15700–15709.

[23] W. Park, J.H. Yang, C.G. Kang, Y.G. Lee, H.J. Hwang, C. Cho, et al., Characteristics of a pressure sensitive touch sensor using a piezoelectric PVDF-TrFE/MoS2 stack, Nanotechnology 24 (2013) 475501.

[24] V. Bhavanasi, V. Kumar, K. Parida, J. Wang, P.S. Lee, Enhanced piezoelectric energy harvesting performance of flexible PVDF-TrFE bilayer films with graphene oxide, ACS Appl. Mater. Interfaces 8 (1) (2016) 521–529.

[25] L. Wang, F.G. Yuan, Energy harvesting by magnetostrictive material (MsM) for powering wireless sensors in SHM, Proc. SPIE 6529 (1) (2007) 652941–6529411.

[26] F. Narita, M. Fox, A review on piezoelectric, magnetostrictive, and magnetoelectric materials and device technologies for energy harvesting applications, Adv. Eng. Mater. 20 (2018) 1700743.

[27] C. Zhang, W. Tang, C. Han, F. Fan, Z.L. Wang, Theoretical comparison, equivalent transformation, and conjunction operations of electromagnetic induction generator and triboelectric nanogenerator for harvesting mechanical energy, Adv. Mater. 26 (2014) 3580–3591.

[28] X. Zhang, J. Ai, Z. Ma, Z. Du, D. Chen, R. Zou, et al., Magnetoelectric soft composites with a self-powered tactile sensing capacity, Nano Energy 69 (2020) 104391.

[29] X. Zhang, J. Ai, Z. Ma, Z. Du, D. Chen, R. Zou, et al., Binary cooperative flexible magnetoelectric materials working as self-powered tactile sensors, J. Mater. Chem. C 7 (2019) 8527.

[30] Z.L. Wang, X. Wang, J. Song, J. Liu, Y. Gao, Piezoelectric nanogenerators for self-powered nanodevices, IEEE Pervasive Comput. 7 (2008) 49–55.

[31] M.T. Chorsi, E.J. Curry, H.T. Chorsi, R. Das, J. Baroody, P.K. Purohit, et al., Piezoelectric biomaterials for sensors and actuators, Adv. Mater. 31 (2019) 1802084.

[32] X. Wang, F. Sun, G. Yin, Y. Wang, B. Liu, M. Dong, Tactile-sensing based on flexible PVDF nanofibers via electrospinning: a review, Sensors 18 (2018) 330.

[33] R. Yang, Y. Qin, L. Dai, Z.L. Wang, Power generation with laterally packaged piezoelectric fine wires, Nat. Nanotechnol. 4 (1) (2009) 34–39.

[34] S. Xu, Y. Qin, C. Xu, Y. Wei, R. Yang, Z.L. Wang, Self-powered nanowire devices, Nat. Nanotechnol. 5 (5) (2010) 366–373.

[35] D. Choi, M.-Y. Choi, H.-J. Shin, S.-M. Yoon, J.-S. Seo, J.-Y. Choi, et al., Nanoscale networked single-walled carbon-nanotube electrodes for transparent flexible nanogenerators, J. Phys. Chem. C 114 (2010) 1379–1384.

[36] S. Lee, R. Hinchet, Y. Lee, Y. Yang, Z.-H. Lin, G. Ardila, et al., Ultrathin nanogenerators as self-powered/active skin sensors for tracking eye ball motion, Adv. Funct. Mater. 24 (8) (2014) 1163–1168.

[37] F.R. Fan, W. Tang, Z.L. Wang, Flexible nanogenerators for energy harvesting and self-powered electronics, Adv. Mater. 28 (2016) 4283–4305.

[38] Y. Qi, J. Kim, T.D. Nguyen, B. Lisko, P.K. Purohit, M.C. McAlpine, et al., Enhanced piezoelectricity and stretchability in energy harvesting devices fabricated from buckled PZT ribbons, Nano Lett. 11 (3) (2011) 1331–1336.

[39] K.-I.I. Park, J.H. Son, G.-T. Hwang, C.K. Jeong, J. Ryu, M. Koo, et al., Highly-efficient, flexible piezoelectric PZT thin film nanogenerator on plastic substrates, Adv. Mater. 26 (16) (2014) 2514–2520.

[40] H.S. Lee, J. Chung, G.-T. Hwang, C.K. Jeong, Y. Jung, J.-H. Kwak, et al., Flexible inorganic piezoelectric acoustic nanosensors for biomimetic artificial hair cells, Adv. Funct. Mater. 24 (2014) 6914–6921.

[41] C.K. Jeong, I. Kim, K.-I. Park, M.H. Oh, H. Paik, G.-T. Hwang, et al., Virus-directed design of a flexible $BaTiO_3$ nanogenerator, ACS Nano 7 (12) (2013) 11016–11025.

[42] J. Chen, S.K. Oh, H. Zou, S. Shervin, W. Wang, S. Pouladi, et al., High-output lead-free flexible piezoelectric generator using single-crystalline GaN thin film, ACS Appl. Mater. Interfaces 10 (15) (2018) 12839–12846.

[43] Q. Zheng, B. Shi, Z. Li, Z.L. Wang, Recent progress on piezoelectric and triboelectric energy harvesters in biomedical systems, Adv. Sci. 4 (2017) 1700029.

[44] B.J. Hansen, Y. Liu, R. Yang, Z.L. Wang, Hybrid nanogenerator for concurrently harvesting biomechanical and biochemical energy, ACS Nano 4 (2011) 3647–3652.

[45] L. Persano, C. Dagdeviren, Y. Su, Y. Zhang, S. Girardo, D. isignano, et al., High performance piezoelectric devices based on aligned arrays of nanofibers of poly[(vinylidenefluoride-co-trifluoroethylene)], Nat. Commun. 4 (2013) 1633.

[46] Y. Mao, P. Zhao, G. McConohy, H. Yang, Y. Tong, X. Wang, et al., Sponge-like piezoelectric polymer films for scalable and integratable nanogenerators and self-powered electronic systems, Adv. Energy Mater. 4 (7) (2014) 1301624.

[47] Y. Cho, J.B. Park, B.-S. Kim, J. Lee, W.-K. Hong, I.-K. Park, et al., Enhanced energy harvesting based on surface morphology engineering of P(VDF-TrFE) film, Nano Energy 16 (2015) 524–532.

[48] C.K. Jeong, J. Lee, S. Han, J. Ryu, G.-T. Hwang, D.Y. Park, et al., A hyper-stretchable elastic-composite energy harvester, Adv. Mater. 27 (18) (2015) 2866–2875.

[49] S.W. Ma, Y.J. Fan, H.Y. Li, L. Su, Z.L. Wang, G. Zhu, Flexible porous polydimethylsiloxane/lead zirconate titanate-based nanogenerator enabled by the dual effect of ferroelectricity and piezoelectricity, ACS Appl. Mater. Interfaces 10 (39) (2018) 33105–33111.

[50] G. Tiana, W. Denga, Y. Gao, D. Xiong, C. Yan, X. He, et al., Rich lamellar crystal baklava-structured PZT/PVDF piezoelectric sensor toward individual table tennis training, Nano Energy 59 (2019) 574–581.

[51] W. Deng, T. Yang, L. Jin, C. Yan, H. Huang, X. Chu, et al., Cowpea-structured PVDF/ZnO nanofibers based flexible self-powered piezoelectric bending motion sensor towards remote control of gestures, Nano Energy 55 (2019) 516–525.

[52] J. Han, D. Li, C. Zhao, X. Wang, J. Li, X. Wu, Highly sensitive impact sensor based on PVDF-TrFE/nano-ZnO composite thin film, Sensors 19 (2019) 830.

[53] R. Hinchet, W. Seung, S.-W. Kim, Recent progress on flexible triboelectric nanogenerators for self-powered electronics, ChemSusChem 8 (2015) 2327–2344.

[54] S. Wanga, L. Lina, Z.L. Wang, Triboelectric nanogenerators as self-powered active sensors, Nano Energy 11 (2015) 436–462.

[55] F.R. Fan, L. Lin, G. Zhu, W. Wu, R. Zhang, Z.L. Wang, Transparent triboelectric nanogenerators and self-powered pressure sensors based on micropatterned plastic films, Nano Lett. 12 (2012) 3109.

[56] X. Cheng, B. Meng, X. Chen, M. Han, H. Chen, Z. Su, et al., Single-step fluorocarbon plasma treatment-induced wrinkle structure for high-performance triboelectric nanogenerator, Small 12 (2016) 229–236.

[57] J. Chen, H. Guo, X. He, G. Liu, Y. Xi, H. Shi, et al., Enhancing performance of triboelectric nanogenerator by filling high dielectric nanoparticles into sponge PDMS film, ACS Appl. Mater. Interfaces 8 (2016) 736–744.

[58] G. Zhu, W.Q. Yang, T. Zhang, Q. Jing, J. Chen, Y.S. Zhou, et al., Self-powered, ultrasensitive, flexible tactile sensors based on contact electrification, Nano Lett. 14 (2014) 3208–3213.

[59] W. Song, B. Gan, T. Jiang, Y. Zhang, A. Yu, H. Yuan, et al., Nanopillar arrayed triboelectric nanogenerator as a self-powered sensitive sensor for a sleep monitoring system, ACS Nano 10 (2016) 8097–8103.

[60] S. Park, H. Kim, M. Vosgueritchian, S. Cheon, H. Kim, J.H. Koo, et al., Stretchable energy-harvesting tactile electronic skin capable of differentiating multiple mechanical stimuli modes, Adv. Mater. 26 (2014) 7324–7332.

[61] S. Wang, L. Lin, Z.L. Wang, Nanoscale triboelectric-effect-enabled energy conversion for sustainably powering portable electronics, Nano Lett. 12 (2012) 6339–6346.

[62] W. Tang, B. Meng, H.X. Zhang, Investigation of power generation based on stacked triboelectric nanogenerator, Nano Energy 2 (2013) 1164–1171.

[63] P. Bai, G. Zhu, Q. Jing, J. Yang, J. Chen, Y. Su, et al., Membrane-based self-powered triboelectric sensors for pressure change detection and its uses in security surveillance and healthcare monitoring, Adv. Funct. Mater. 24 (2014) 5807–5813.

[64] S. Li, W. Peng, J. Wang, L. Lin, Y. Zi, G. Zhang, et al., All-elastomer-based triboelectric nanogenerator as a keyboard cover to harvest typing energy, ACS Nano 10 (2016) 7973–7981.

[65] Q. Zheng, H. Zhang, B. Shi, X. Xue, Z. Liu, Y. Jin, et al., In vivo self-powered wireless cardiac monitoring via implantable triboelectric nanogenerator, ACS Nano 10 (2016) 6510–6518.

[66] G. Zhu, Y.S. Zhou, P. Bai, X.S. Meng, Q. Jing, J. Chen, et al., A shape-adaptive thin-film-based approach for 50% high-efficiency energy generation through micro-grating sliding electrification, Adv. Mater. 26 (2014) 3788–3796.

[67] Y.S. Zhou, G. Zhu, S. Niu, Y. Liu, P. Bai, Q. Jing, et al., Nanometer resolution self-powered static and dynamic motion sensor based on micro-grated triboelectrification, Adv. Mater. 26 (2014) 1719–1724.

[68] G. Zhu, P. Bai, J. Chena, Z.L. Wang, Power-generating shoe insole based on triboelectric nanogenerators for self-powered consumer electronics, Nano Energy 2 (2013) 688–692.

[69] P.-K. Yang, Z.-H. Lin, K.C. Pradel, L. Lin, X. Li, X. Wen, et al., Paper-based origami triboelectric nanogenerators and self-powered pressure sensors, ACS Nano 9 (2015) 901–907.

[70] B. Chena, W. Tang, T. Jiang, L. Zhu, X. Chen, C. He, et al., Three-dimensional ultraflexible triboelectric nanogenerator made by 3D printing, Nano Energy 45 (2018) 380–389.

[71] J. Luo, Z. Wang, L. Xu, A.C. Wang, K. Han, T. Jiang, et al., Flexible and durable wood-based triboelectric nanogenerators for self-powered sensing in athletic big data analytics, Nat. Commun. 10 (2019) 5147.

[72] X. Yu, J. Pan, J. Zhang, H. Sun, S. He, L. Qiu, et al., A coaxial triboelectric nanogenerator fiber for energy harvesting and sensing under deformation, J. Mater. Chem. A 5 (2017) 6032.

[73] L. Xie, X. Chen, Z. Wen, Y. Yang, J. Shi, C. Chen, et al., Spiral steel wire-based fiber-shaped stretchable and tailorable triboelectric nanogenerator for wearable power source and active gesture sensor, Nanomicro Lett. 11 (2019) 39.

[74] J. Liu, L. Gu, N. Cui, S. Bai, S. Liu, Q. Xu, et al., Core-shell fiber-based 2D woven triboelectric nanogenerator for effective motion energy harvesting, Nanoscale Res. Lett. 14 (2019) 311.

[75] J.M. Donelan, Q. Li, V. Naing, J.A. Hoffer, D.J. Weber, A.D. Kuo, Biomechanical energy harvesting: generating electricity during walking with minimal user effort, Science 319 (2008) 807–810.

[76] G. Zhu1, P. Bai, J. Chen, Q. Jing, Z.L. Wang, Triboelectric nanogenerators as a new energy technology: from fundamentals, devices, to applications, Nano Energy 14 (2015) 126–138.

Self-healable tactile sensors

13

Jinqing Wang[a,b,], Xianzhang Wu[a,b], Zhangpeng Li[a,c], and Shengrong Yang[a,b]*

[a]State Key Laboratory of Solid Lubrication, Lanzhou Institute of Chemical Physics, Chinese Academy of Sciences, Lanzhou, China, [b]Center of Materials Science and Optoelectronics Engineering, University of Chinese Academy of Sciences, Beijing, China, [c]Qingdao Center of Resource Chemistry & New Materials, Qingdao, China
[*]Corresponding author: e-mail address: jqwang@licp.cas.cn

13.1 Introduction

Self-healing tactile sensors possessing remarkable properties of tactile sensation and autonomous self-healing properties have been of interest during the past decades. The design and development of self-healing tactile sensors for measuring electrical signals generated by external stimuli provide an important approach for artificial intelligence, soft robotics, and wearable devices. However, to enable the maintenance of high sensing and self-healing performances of tactile sensors is still a daunting challenge. To date, various functional nanomaterials have been applied as active layers to improve the sensing ability of tactile sensors. For instance, carbon nanotubes (CNTs) [1], graphene and graphene oxide (GO) [2], and silver nanowires (NWs) [3] were used as active materials to fabricate sensors. The as-assembled tactile sensors presented ideal responses to stretching, bending, and shearing. However, these tactile sensors still have some limitations, and they do not show sufficient sensitivity in some subtle detection because of their disordered and non-stereoscopic sensing units. On the other hand, the self-healing properties of tactile sensors that should self-repair after being damaged have fascinated many researchers. By virtue of the healable substrate materials, some researchers have succeeded in synthesizing self-healing tactile sensors. However, these devices still suffer from the following several critical issues: low sensing sensitivity, low stretchability, and low mechanical stability. Recently, the rapid improvements in new active and substrate materials, synthesis procedure, and signal acquisition techniques have led to remarkable progress in self-healing tactile sensors, which possess fascinating functions including automatic healing, high sensitivity, stretchability, and high flexibility. In particular, a wide range of interlocked microstructures have been introduced into the sensing components to enhance the sensitivity and distinguishing ability of tactile sensors. For example, micro-pyramid dielectric arrays have been applied for building pressure sensors based on their multidirectional force-sensing behaviors, and the pressure sensor could detect multiple mechanical stimuli and obtain an obviously elevated sensitivity [4]. Moreover, a micropillar interlocking structure was introduced into the sensing unit to enhance the sensitivity of tactile sensors, which presented the advantage of effective resistance change due to

Functional Tactile Sensors. https://doi.org/10.1016/B978-0-12-820633-1.00013-9
© 2021 Elsevier Ltd. All rights reserved.

flexible contact between the micropillars and electrodes [5]. In addition, the introduction of microwrinkle structures into the sensing unit could improve the sensing performance of tactile sensors [6]. Therefore, incorporating self-healable and highly sensitive tactile sensors capable of converting external stimuli into electrical signals would be very promising in various areas.

In this chapter, recent progress in the area of self-healing tactile sensors is highlighted. Furthermore, the self-healing behavior of substrate materials after mechanical damage is discussed, mainly focusing on the latest innovations in sensitive self-healing sensors that are able to detect multiaxial tactile stimuli including pressing and shearing. Finally, the key challenges in the field of self-healing tactile sensors are elaborated and potential developments for the device with enhanced performances are proposed.

13.2 The structure and functional materials of self-healing tactile sensors

Generally, a self-healing tactile sensor possesses a sandwich structure including substrates and active materials (Fig. 13.1A) [7], which starts from an elastic packaging substrate, then a sensing active material, and finally an elastic packaging substrate. The elastic packaging substrate protects the device from the harsh environments. Moreover, it contributes to the self-healing ability of tactile sensors after mechanical damage. The mechanical properties of the elastic packaging substrate should most match the characteristics of the human skin so as to provide adequate functions such as stretchability, bending, and self-healing. Therefore, researchers must consider flexibility to suit different application conditions when designing substrate materials with the intention of mimicking human skin properties. On the other hand, active materials play a key role in converting external stimuli into electrical signals. Active materials for self-healing tactile sensors have been developed mainly based on conductive materials, such as graphene, carbon nanotubes (CNTs), and conductive polymers. Nevertheless, the sensing performance of the tactile sensors based on these conductive materials suffers from poor sensitivity due to their planar structure (Fig. 13.1B) [8]. Therefore, new designs in active materials, such as converting conductive materials into stereoscopic nanostructure (Fig. 13.1C) [9], are desired. In a word, to acquire an optimal combination of sensing and self-healing properties, the design and selection of substrate and active materials are critical. The following section describes the most commonly studied substrate and active materials for the self-healing tactile sensors.

13.2.1 Substrates

Tactile sensors with self-healing and flexible features normally possess soft substrates and hence can enable self-repair and continuous use. To acquire a suitable substrate for tactile sensing application, flexibility and self-healing properties are pivotal factors that should be carefully considered. Polyurethane has been widely used as a

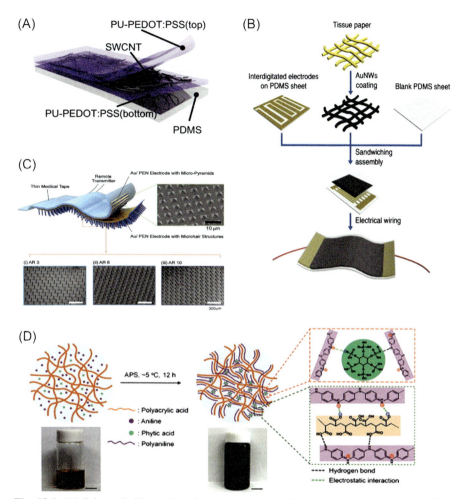

Fig. 13.1 (A) Schematic illustration of the cross section of the strain sensor consisting of the three-layer stacked nano-hybrid structure. (B) Pressure sensor based on the Au nanowires (NWs) coated tissue paper. (C) Cross-sectional diagram of the microhair-structured sensor and SEM images of dense microhair arrays. (D) Schematic showing the synthetic process for the ternary polymer composite and the corresponding interactions between polyaniline chains. Panel (A) reproduced with permission from E. Roh, B.U. Hwang, D. Kim, B.Y. Kim, N.E. Lee, Stretchable, transparent, ultrasensitive, and patchable strain sensor for human machine interfaces comprising a nanohybrid of carbon nanotubes and conductive elastomers, ACS Nano 9 (2015) 6252–6261. Copyright 2015, American Chemical Society. Panel (B) reproduced with permission from S. Gong, W. Schwalb, Y. Wang, Y. Chen, Y. Tang, J. Si, B. Shirinzadeh, W. Cheng, A wearable and highly sensitive pressure sensor with ultrathin gold nanowires, Nat. Commun. 5 (2014) 3132. Copyright 2014, Springer Nature Limited. Panel (C) reproduced with permission from C. Pang, J.H. Koo, A. Nguyen, J.M. Caves, M.G. Kim, A. Chortos, K. Kim, P.J. Wang, J.B. Tok, Z. Bao, Highly skin-conformal microhairy sensor for pulse signal amplification, Adv. Mater. 27 (2015) 634–640. Copyright 2015, Wiley-VCH. Panel (D) reproduced with permission from T. Wang, Y. Zhang, Q. Liu, W. Cheng, X. Wang, L. Pan, B. Xu, H. Xu, A self-healable, highly stretchable, and solution processable conductive polymer composite for ultrasensitive strain and pressure sensing, Adv. Funct. Mater. 28 (2017) 1705551. Copyright 2017, Wiley-VCH.

substrate to fabricate self-healing tactile sensors ascribing to its dynamic polymerization characteristics and high elasticity. It is well known that polyurethane has unique attributes including high strength, decent toughness, and excellent resilience. The optimal properties are beneficial to enhance the self-healing and damage resistance capabilities of the sensor device. For example, self-healing disulfide-cross-linked polyurethane and polyurethane/silver composites were developed specifically for tactile sensor substrates [10]. The cutting groove of the polyurethane composite could rapidly self-heal after 0.5 h at ambient conditions. Moreover, automatic healing of the polyurethane composite could be conducted in a wide temperature range from $-20°C$ to $40°C$, which is easy to swell, and the polyurethane chains become free to be mobile. In addition, recent research on the self-healing tactile sensor has demonstrated that the modified polydimethylsiloxane (PDMS) is a promising elastomer for the flexible substrate owing to its outstanding elasticity and self-healing capability. Besides, this elastomer is transparent, which can be employed for the fabrication of multifunction transparent devices. Except for the synthetic polymers, a lot of natural supramolecular materials have also been applied to fabricate a substrate for self-healing tactile sensors. For instance, gelatin not only displays excellent mechanical properties but also meets the self-healing requirement to ensure the self-repair of the device [11]. Furthermore, it exhibits obvious advantages including being odorless, nontoxic, and degradable, thus pushing forward the future development of tactile sensors toward a biologically integrated sensing system with biocompatibility and tactile sensing characteristics. Other supramolecular materials, such as polyaniline (Fig. 13.1D), polypyrrole, and poly(3,4-ethylenedioxythiophene)-poly (styrenesulfonate), have also presented great potentials as substrate materials for the self-healing tactile sensors [12].

13.2.2 Active materials

13.2.2.1 CNTs

As a carbon allotrope, CNT presents excellent mechanical and electrical properties. Moreover, it is easily embedded in the stretchable elastic substrate to impart the composite with electrical conductivity to the composite. Therefore, CNT is a promising candidate for developing tactile sensors. Highly aligned CNTs can be assembled into macroscopic aerogels, foams, or films, which show high sensitivity to external stimulation because of the variety in their band structures. Importantly, utilizing a self-healable polymer substrate for support, the CNT-based tactile sensors can autonomously self-heal at room temperature after being mechanically damaged. In addition, the flexible and healable polymer substrates can effectively disperse stress and thus avoid necking and stress concentration. For example, CNTs are sandwiched between polymer layers to form a tactile sensor. Higher CNT loading levels are inclined to result in lower sensing sensitivity because the contact resistance becomes very small here. Contact resistance plays a key role in the CNT-based tactile sensors, and too high or too low contact resistance will lead to the lower sensitivity of the tactile sensor.

Therefore, it is necessary to adjust the loading level of carbon nanotubes and the interfacial interaction between CNT and polymer reasonably.

The CNT-based tactile sensors can detect multiple stimuli such as pressuring, stretching, bending, and shearing, which can be used in monitoring various human motions in real time. Mu et al. [13] demonstrated a high-sensitive tactile sensor based on two-sublayered CNTs 3D conductive networks, which are anchored on a thin porous PDMS layer. At the low tensile strain, the bridging CNTs go across each other; thus, the conductive ability will steadily increase with the strain, resulting in a relaxed resistance increment with a lower sensitivity. Further strain leads to the detachment of the CNTs and destroys the initially conductive channels, thus resulting in marked increase of resistance and a much higher sensitivity. With these features, the assembled CNTs/PDMS-based tactile sensor exhibited high sensitivity and excellent reproducible response over 5000 loading cycles for stretching, bending, and shearing. In addition, CNTs were also dispersed on a polymer substrate to prepare a tactile sensor [14], which showed significant change in resistance upon stretching and maintained excellent conductivity at a high strain. After cycles of stimulation, however, the conductivity of this device decreased, which was caused by the severe structural collapse and local cracks. To overcome this issue, a variety of new strategies have been developed. For instance, a robust CNT network was constructed by chemical vapor deposition (CVD) of graphene in the nanotube voids (Fig. 13.2A) [15]. The resulting CNT networks could effectively resist structural collapse during repeated external stimuli of stretching, pressing and shearing, etc. The effect of structure collapse of CNTs networks on conductivity was further investigated by Lipomi et al. [16]. When the sensor was touched, the contact resistance reduced, while the resistance would dramatically increase as the stimulus was removed; meanwhile, the structure of the CNT networks remained intact. Moreover, the tactile sensor can accommodate to great deformation because CNTs networks can deform and recover via bending or twisting individual CNT to dispersive stress (Fig. 13.2B). A thin film constituted by aligned CNTs as tactile sensors is also developed. On stretching or bending, the CNT films demonstrated a structural deformation, which endows the tactile sensors with high sensitivity, excellent durability, and fast response. CNTs are also promising materials for tactile sensing mapping. A high sensitive tactile sensing array is assembled by using CNTs as the channel materials, where CNTs are attached to a thin polyimide (PI) substrate with an array of the hexagonal shape. The stretchability of the tactile sensing mapping is controllable through tuning the side length of the hexagons. Also, a tactile-sensitive sensing array on an active-matrix backplane is anticipated to enable a high-performance electronic skin.

13.2.2.2 Graphene

Graphene is another attractive candidate for the development of flexible tactile sensors because it possesses unique properties including ultrahigh conductivity, excellent mechanical property, and optical transparency. Graphene has a two-dimensional (2D) layer structure, and hence it can be easily assembled into a variety of macrostructures by various techniques. When stress is applied, the structure of graphene will partially

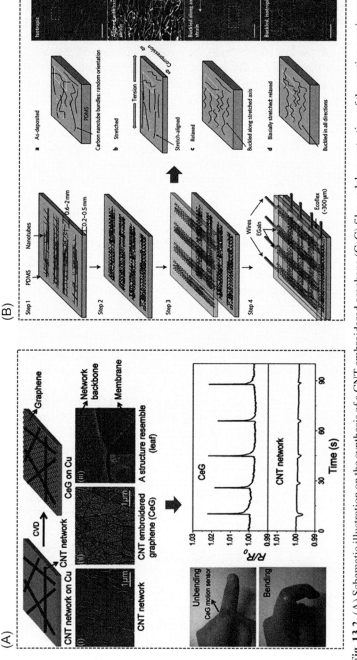

Fig. 13.2 (A) Schematic illustration on the synthesis of a CNT embroidered graphene (CeG) film and demonstration of the motion sensing application of CeG. (B) Summary of processes used to fabricate arrays of transparent, compressible, capacitive sensors and devolution of morphology of films of carbon nanotubes with stretching.

Panel (A) reproduced with permission from J. Shi, X. Li, H. Cheng, Z. Liu, L. Zhao, T. Yang, Z. Dai, Z. Cheng, E. Shi, L. Yang, Z. Zhang, A. Cao, H. Zhu, Y. Fang, Graphene reinforced carbon nanotube networks for wearable strain sensors, Adv. Funct. Mater. 26 (2016) 2078–2084. Copyright 2016, Wiley-VCH. Panel (B) reproduced with permission from D.J. Lipomi, M. Vosgueritchian, B.C. Tee, S.L. Hellstrom, J.A. Lee, C.H. Fox, Z. Bao, Skin-like pressure and strain sensors based on transparent elastic films of carbon nanotubes, Nat. Nanotechnol. 6 (2011) 788–792. Copyright 2011, Springer Nature Limited.

collapse, thus resulting in the change of resistance. Based on this, many studies have been carried out to explore the applications of graphene for tactile sensors. For example, a large-area graphene synthesized by CVD has been used to fabricate a tactile sensor [17], which showed a high sensitivity. The contact resistance of the tactile sensor decreased slightly under low strain due to the relaxation of wrinkles of the graphene layer, and then it significantly increased under high strain. The result can be ascribed to the structure defects, surface functional groups, and interfacial interaction between graphene and the substrate.

To enhance the sensitivity of the graphene-based tactile sensors, the characteristic microstructures, such as defect and crack, are often required. Therefore, high-performance tactile sensors can be fabricated by coating graphene nanosheets on a flexible plastic substrate. The sensitivity of these devices can be easily adjusted by changing the size, the number of the defects, and other deposition parameters of graphene nanosheets. The synthesis strategy of this device exhibits simple, time-saving, and cost-effective advantages. Structural optimization of the graphene using CVD or self-assembly strategies can further improve the sensitivity of the graphene-based tactile sensors. In these devices, segregated graphene flakes are used instead of overlapped flakes, at an available regulating distance, resulting in remarkable change of contact resistance under external stimuli.

Reduced graphene oxide (rGO) is another interesting material for the tactile sensing layer. The conductive properties of rGO are mainly determined by the following two characteristics: (i) the intra-nanosheet resistance, which is governed by structural defects of rGO nanosheets; (ii) the inter-nanosheet resistance, which is controlled by the junctions of rGO nanosheets. Under the external stimulus, the resistance of rGO is expected to be strongly influenced because of the change in the contact point of the nanosheets. Attributing to Attributing to the resistance variation with the external stimulus, rGO can be utilized to assemble tactile sensors. Wu et al. [18] reported a novel 3D cellular GO material with high elasticity and low density for application in highly sensitive flexible tactile sensors. The sensing device based on 3D cellular GO material exhibits a particularly high sensitivity of 0.28 kPa^{-1}, fast response of 40 ms (rising time), and excellent cycling stability. Moreover, practical application of the tactile sensor has also been realized through monitoring a variety of human motions in real time, such as human pulse, finger folding, and swallowing actions. Huang et al. [19] fabricated 3DG/PI composite sponges by introducing a PI layer into the as-prepared 3DG (3D graphene) sponge. The resulting 3DG/PI composite sponges with a robust 3D network structure exhibited high electrical conductivity (3.7 S/cm), compression strength (175 kPa), elasticity, and flexibility, as well as outstanding compression sensitivity to resistance and stable tactile sensing effect. Moreover, the sponges possess a large change in resistance in response to the applications of small strain and low density, and the resistance change remains favorably stable after performing 300 compress-release cycles, which means that the prepared composite sponges can find wide applications in a tactile-sensing or stimulus-responsive graphene system. In addition, a structurally ordered and less defective GO/CNT composite aerogel (abbreviated as GCNT) featuring both high conductivity and superelasticity has been developed through alkali induced self-assembly of GO liquid crystals (GO LCs) and

CNTs [20]. The methodology of a GCNT aerogel relies on the double roles of KOH solution as a dispersant for CNTs and an inducer for the self-assembly of GO LCs nanosheets. The less-defective CNTs acting as reinforcement material contribute to the robust network structures, leading to the significantly improved conductivity (2.4 S m^{-1}) and elasticity (14.3 kPa) of GCNT. Benefiting from these outstanding properties of GCNT aerogels, the assembled tactile sensor exhibits an ultrahigh sensitivity of 1.22 kPa^{-1}, a rapid response time of 28 ms, and excellent cycling stability, which enables it as a high-performance sensing platform to monitor various human motions in real time (Fig. 13.3).

With excellent mechanical property and high conductivity, graphene can also be utilized as the active material for the fabrication of some tactile sensors. According to the tactile sensing principle, upon external stimulus, resistance will change due to the variation in contact sites between two electrode sheets or electrode-active layer interfaces. In addition, graphene can be applied as a channel material in a flexible sensor array to achieve tactile sensing and data addressing. Moreover, numerous designs of sensing materials based on graphene and its derivatives for tactile sensing devices have also been reported, which enable the easy fabrication of a flexible electronic skin, opening a new avenue for next-generation electrical devices.

13.2.2.3 Conductive elastomers

Incorporating conductive elastomers into specific microscopic topographies has been widely utilized to fabricate flexible tactile sensors for years. The tactile sensors integrated with different conductive microstructures are more sensitive to tactile stimuli because various microstructures exhibit different force-deformation behaviors specified by their geometries. For instance, the tactile sensor integrated with micropillar-wrinkle hybrid structures exhibits multidirection-perception capability, high sensitivity, wide detection range, and tunable sensing performance (Fig. 13.4A) [21]. The hybrid tactile sensor is highly compressible, flexible, and stretchable. Moreover, it can be used to detect multiaxial tactile stimuli including normal pressure, bending, and stretching over a wide range of applied stresses in a highly sensitive, fast, and durable manner. Micropillars are also fabricated to improve the sensing performance of tactile sensors. Micropillar-based tactile sensors exhibit better sensing performance over external stimuli compared to planar tactile sensors. Ha et al. [22] developed a highly sensitive and fast responsive tactile sensor based on the interlocked geometry of hierarchical microstructures of PDMS micropillars decorated with ZnO NWs arrays (Fig. 13.4B). The device can be utilized to detect static and dynamic tactile signals. The effective contact area between NWs of the tactile sensor can be significantly changed in response to the external stimuli, thus resulting in the highly sensitive sensing performance. These results indicate that the sensing ability of the tactile sensor can be enhanced by integrating conductive microstructures with proper microscopic architectures. However, the fabrication process of these conductive microstructures usually involves photolithography, which is typically tedious and expensive, seriously restricting the large-scale application of the tactile sensor.

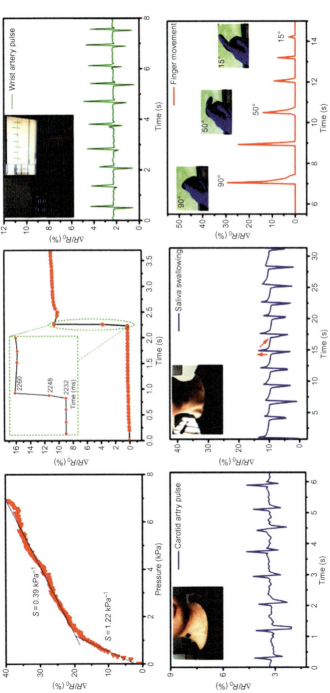

Fig. 13.3 Tactile sensor as a high-performance sensing platform to monitor various human motions in real time. Reproduced with permission from X. Wu, X. Liu, J. Wang, J. Huang, S. Yang, Reducing structural defects and oxygen-containing functional groups in GO-hybridized CNTs aerogels: simultaneously improve the electrical and mechanical properties to enhance pressure sensitivity, ACS Appl. Mater. Interfaces 10 (2018) 39009–39017. Copyright 2018, American Chemical Society.

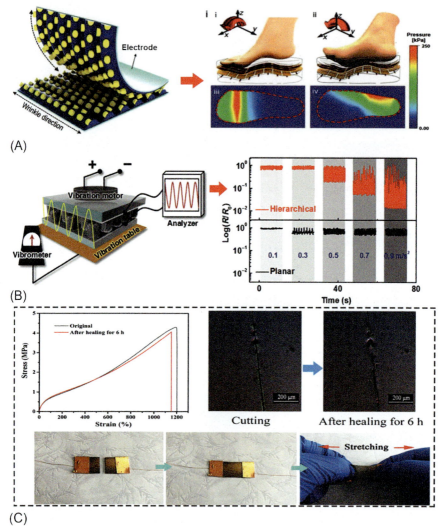

Fig. 13.4 (A) Conceptual illustration showing the e-skin and different foot postures on the hybrid e-skin-integrated insole. (B) Schematic illustration of vibration measurements based on hierarchical ZnO NWs and change in relative resistance of e-skins in response to various vibration intensities. (C) Optical microscopy images of the damaged and healed toothbrush-hair-based electronic sensor.

Panel (A) reproduced with permission from K. Sun, H. Ko, H.H. Park, M. Seong, S.H. Lee, H. Yi, H.W. Park, T.I. Kim, C. Pang, H.E. Jeong, Hybrid architectures of heterogeneous carbon nanotube composite microstructures enable multiaxial strain perception with high sensitivity and ultrabroad sensing range, Small 14 (2018) e1803411. Copyright 2018, Wiley-VCH, Panel (B) reproduced with permission from M. Ha, S. Lim, J. Park, D.S. Um, Y. Lee, H. Ko, Bioinspired interlocked and hierarchical design of ZnO nanowire arrays for static and dynamic pressure-sensitive electronic skins, Adv. Funct. Mater. 25 (2015) 2841–2849. Copyright 2015, Wiley-VCH, Panel (C) reproduced with permission from X. Wu, Z. Li, H. Wang, J. Huang, J. Wang, S. Yang, Stretchable and self-healable electrical sensors with fingertip-like perception capability for surface texture discerning and biosignal monitoring, J. Mater. Chem. C 7 (2019) 9008–9017. Copyright 2019, Royal Society of Chemistry.

To solve this issue, a flexible, multifunctional, and self-healable tactile sensor was fabricated by Wu et al. [23], which consisted of Au-deposited toothbrush-hairs micro-prick arrays as the sensing component and the composite elastomer (P-TDI-IP) obtained via the block polymerization of polytetramethylene glycol (P), 2,4′-tolylene diisocyanate (TDI), and isophorone diisocyanate (IP) as the substrate component. In the device, the sensing micro-prick arrays are derived from the commercial toothbrush-hair; thus, its fabrication process is simple, time-saving, and cost-effective. Moreover, the self-healable P-TDI-IP elastomer acting as a substrate endows the electronic sensor with an excellent self-healing capability (96% with 6 h) (Fig. 13.4C) and an outstanding stretchability (tensile strain: 1200%). Importantly, due to the presence of unique micro-prick arrays, the tactile sensor could detect multiaxial tactile stimuli including pressing and shearing. Furthermore, this tactile sensor exhibited high sensitivity (3.32 kPa^{-1} for pressing, GF = 2.82 for shearing), excellent cycling stability (3000 pressing and shearing cycles), a rapid response time of 25 ms, and a low detection limit of 5 Pa. Owing to these prominent properties, the tactile sensor presented anticipated abilities for distinguishing ultrafine surfaces with different roughnesses and monitoring various human motions. In a word, this work provides a simple, cost-effective, and feasible approach for the fabrication of multifunctional electronic sensors, which can find promising applications in smart robotics, health-care monitoring devices, and other soft electronics.

13.3 Self-healing mechanisms of tactile sensors

Like the human skin, which can directly translate external stimulus information into electrical signals and realize rapid self-healing after being mechanically damaged, self-healing tactile sensors can not only be applied as a source of information regarding external contact with the physical world, but also can be used to compensate incidental scratches and/or mechanical cuts. In general, self-healing tactile sensors can be classified into two major categories according to their healing mechanisms: (1) extrinsic and (2) intrinsic healing behaviors. In the extrinsic healing behavior, catalysts and external healing agents are normally used to form small capsules or vascular networks. When these capsules or vascular networks are broken, entrap monomers and initiators are mixed, resulting in the formation of new polymer chains that mend crack faces (Fig. 13.5A–C) [24, 25]. For instance, Toohey et al. [24] reported that a bioinspired coating-substrate design delivers the healing agents to cracks in polymer coating via a three-dimensional (3D) microvascular network embedded in the substrate. Crack damage in the epoxy coating is healed repeatedly. This method is reliable, but it suffers from the incapability of achieving rapid healing since the polymeric reaction between entrap monomers and initiators is time-consuming. In contrast, intrinsic healing systems enable tactile sensors to heal repeatedly, in some cases at ambient temperatures, through incorporating dynamic covalent bonds or noncovalent interactions into the polymer matrix (Fig. 13.5D and E) [26, 27]. Designing new tactile sensors with an intrinsic self-healing ability requires a profound understanding of the dynamic covalent bonds which can be broken and reformed repeatedly. For example,

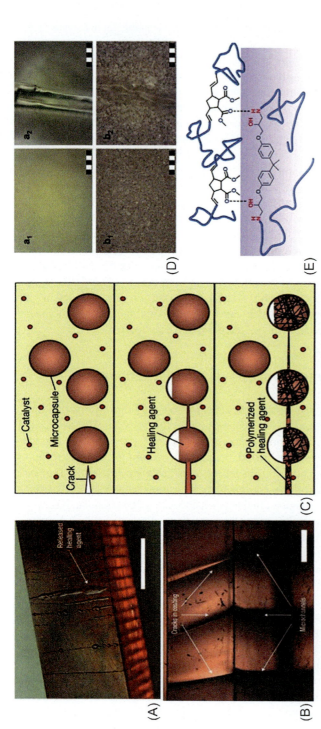

Fig. 13.5 Self-healing mechanism of the tactile sensor. (A) High-magnification cross-sectional image of the coating shows the cracks. (B) Optical image of self-healing structure after cracks are formed in the coating. (C) A microencapsulated healing agent is embedded in a structural composite matrix containing a catalyst capable of polymerizing the healing agent. (D) Illustration characterizing the self-healing components of the chemiresistor. (E) Schematic concept of noncovalent adhesion promotion in a self-healing system. Panel (A and B) reproduced with permission from K.S. Toohey, N.R. Sottos, J.A. Lewis, J.S. Moore, S.R. White, Self-healing materials with microvascular networks, Nat. Mater. 6 (2007) 581–585. Copyright 2007, Springer Nature Limited. Panel (C) reproduced with permission from S.R. White, N.R. Sottos, P.H. Geubelle, J.S. Moore, M.R. Kessler, R. Sriram, E.M. Brown, S. Viswanathan, Autonomic healing of polymer composites Nature 409 (2001) 794–797. Copyright 2001, Springer Nature Limited. Panel (D) reproduced with permission from T. P. Huynh, H. Haick, Self-healing, fully functional, and multiparametric flexible sensing platform, Adv. Mater. 28 (2016) 138–143. Copyright 2016, Wiley-VCH. Panel (E) reproduced with permission from G.O. Wilson, M.M. Caruso, S.R. Schelkopf, N.R. Sottos, S.R. White, J.S. Moore, Adhesion promotion via noncovalent interactions in self-healing polymers, ACS Appl. Mater. Interfaces 3 (2011) 3072–3077. Copyright 2011, American Chemical Society.

Tee et al. [28] developed a tactile sensor featuring a repeatedly self-healing ability at ambient conditions. The self-healing capabilities of tactile sensors were driven by the reassociation of hydrogen bonds between the cut surfaces. The obtained self-healing tactile sensor is pressure- and flexion-sensitive, and thus can be used to integrate electronic skin systems. Once fractured, the mechanical properties of the device were fully restored within only 10 min and the initial conductivity restored nearly 90% within 15 s. Therefore, the excellent self-healing and sensing abilities expand the application scopes of tactile sensors.

13.3.1 Interdiffusion of molecules

The healing of polymers through interdiffusion cracks has been widely studied since the 1980s. The studies covered amorphous polymers, semicrystalline polymers, block copolymers, and fiber-reinforced composites. The study found that when two identical polymers were in contact above their glass transition temperature (T_g), the interface between two polymers would gradually disappear, much like the healing of molecular diffusion cracks at the interface, which increased the mechanical strength of the polymer interface. The self-healing process is carried out under vacuum or atmospheric pressure, and the healing time varies from minutes to years. The self-healing temperature is usually higher than T_g. Jud and Kausch [29] studied the molecular weight and copolymerization degree of polymethyl methacrylate (PMMA) on crack self-healing behavior. After a period of clamping at the fracture interface of the polymer and heating the sample, many parameters were tested, including the time of compression and entry into the fracture, the time and the temperature of self-healing, as well as the pressure of compression. The results show that the self-healing ability of polymers was better than that of those individual polymers, while T_g of the former was higher than that of the latter. They also found that the longer the time that was required for self-healing from the initial fracture to the fracture surface, the lower the self-healing efficiency would be; and the healing rate decreased from 120% to 80% during the period. Visual self-healing of fractured surfaces has emerged before significant recovery of strength, and many segments have been found to be the most likely self-healing mechanisms. Using minimum thermal stress to investigate the fluidity of thermoplastic polymer chains, Lin et al. investigated the crack healing of PMMA treated with methanol at 40–60°C [30]. They found that the tensile strength of PMMA treated with methanol could be completely restored to the original state of the material. The degree and process of this healing can be determined by the recovery of tensile strength depending on wetting and diffusion, as the presence of methanol facilitates the reduction of T_g and the diffusion of polymer chains throughout the interface. Subsequent studies examined the crack healing of PMMA, similarly induced by ethanol, and compared the differences in the use of methanol. The results showed that the crack healing process of ethanol was similar to that of methanol in the plasticizing effect and decreasing T_g. However, ethanol caused excessive plasticization and the expansion of PMMA, which resulted in incomplete mechanical strength recovery. Yang and Pitchumani [31] studied the interfacial healing of polyetheretherketone (PEEK) and polyetherketone (PEK) after carbon fiber enhancement under non-isothermal conditions. After different treatment times, comparing the strength of this thermal adhesive plate to its final shear strength, the efficiency of all systems is 100%. For this purpose, they proposed a healing model of the surface of non-isothermal plastic polymer materials, but this model seems to be more suitable for polymer processing.

13.3.2   Light-induced healing

The first example of light-induced healing of polymer was proposed by Chung et al. [32]. The cycloaddition reaction of light was used to produce the cyclobutane structure, so the cycloaddition reaction of light chemistry was considered as a healing mechanism. In the solid state, the formation and diffusion of surface cracks make cyclobutane easy to revert to its original cinnamon structure. The feasibility of this mechanism was tested by mixing light-cross-linked cinnamic acid monomers, 1,1,1-1-tri (cinnamyl-methoxy) ethane with polyurethane dimethylacrylate, homotriethylene glycol dimethylacrylate monomers, and visible-light-sensitive initiator camphor brain quinone. After 10 min of irradiation with 280 nm light, the mixture was polymerized into a very tough transparent film. The healing results indicate that the reaction occurs only when the wavelength is consistent, which proves that the healing is caused by light.

13.3.3   Reversible bond formation

The fracture of the polymer can be cured by the introduction of reversible bonds in the polymer base at ambient temperature, which uses hydrogen or ionic bonds to treat damaged polymer networks, providing an alternative to UV or catalyst-triggered covalent bond healing. In 2004, Harreld et al. [33] cooperated systematically and patented the self-healing of molecules that form reversible bonds, where intermediate-strength cross-linked polymers were formed through hydrogen and ionic bonds. As a result, medium-strength cross-linking provides a material with good overall toughness while making reversible cross-linking possible for self-healing. When the fracture surface is physically closed or initiated through a solvent-induced chain, healing begins. This self-healing method uses specific chemical cross-linking to achieve self-healing. However, the system proposed by Harreld et al. is not based on covalent bonds, so healing can still occur when energy is lacking. Moreover, the self-healing number of polymers can be adjusted by changing the structure of the polymer, the degree of cross-linking, or the strength of cross-linking.

13.3.4   Active polymerization

To provide protection for space applications from specific damage factors, such as ionizing radiation damage, it has been proposed to develop self-healing materials using active polymer materials such as resin-based materials [34]. These initiators suggested the preparation of active polymers using some macromolecular groups (free radical polymer chains). Theoretically, only active polymers can be synthesized by ionic polymerization or radical polymerization in the case of chain growth, no chain

Self-healable tactile sensors   277

transfer, and chain termination [35–38]. Therefore, if other monomers are added to the system, a self-healing polymer can obtain active groups with the ability to reunite. In this approach, the material may be degraded after exposure to ionizing or UV radiation, which is due to the possible recombination reaction between the new radical generation and the macromolecular groups on the end chain. This scale molecular healing process is influenced by the diffusion rate of macromolecular groups and correspondingly by T_g of polymers. When the temperature is lower than T_g, the diffusion rate of the macromolecular groups in the condensed state is very slow, resulting in the slow healing process. However, this polymer shows excellent self-healing ability at temperatures above 127°C.

13.4   Applications of self-healing tactile sensors

Generally, self-healing tactile sensors are incorporated with actuators or stimulators for monitoring human activities and personal health care. The sensor devices can perform the analysis of the body condition to gain insights into various aspects of physiological health. Meanwhile, a self-healing tactile sensor integrated on the gripper of a robot can monitor the command execution, and then the resulting feedback stimulation is applied to the operator through an electro-tactile stimulator from the bottom of the user's wrist. The self-healing tactile sensors can be utilized in various complex motions, such as shaking hands, tapping a keyboard, and grasping an object. These results indicate the unique opportunities that can be provided by self-healing tactile sensors for emerging classes of prostheses and peripheral nervous-system-interface technologies.

13.4.1   Medicine and health

The flexible and stretchable properties of the self-healing tactile sensor make it compatible with the skin surface of the human body, which is beneficial to the application in the medical health field (Fig. 13.6) [39]. Traditional medical testing equipments are mainly rigid, with many of them needing to be connected by trained medical staff in the designated clinic, and the psychological state of the examination will affect the test results to a certain extent. If a flexible tactile sensor is used, the discomfort will be greatly reduced, and information collection of the human body will be more accurate and humanized. The tactile sensor can be used as a tool not only to detect body signals but also help to treat diseases [40, 41]. However, the application of tactile sensors in the medical and health field also faces many challenges [42]. For example, low efficiency sensing and undirectional delivery of therapeutic drugs are major disadvantages of the tactile sensor. The long-term and stable electrical connection between flexible medical equipment and biological tissue can be realized by utilizing the structure design and application of conjugated polymer materials [43, 44]. Human signals (such as human sweat or other body fluids) often contain a lot of information. The health or physiological condition of the human body can be inferred by detecting and analyzing these signals [45, 46]. The tactile sensor can measure the physiological

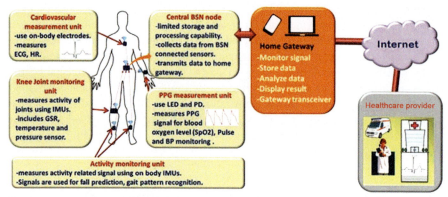

Fig. 13.6 Application of tactile sensor in medical health.
Reproduced with permission from S. Majumder, T. Mondal, M.J. Deen, Wearable sensors for remote health monitoring, Sensors 17 (2017) 130. Copyright 2017, Multidisciplinary Digital Publishing Institute.

indexes of tissue pressure, movement, temperature, and humidity. Moreover, the device can fit well with the human body and collect the active signals of the heart, skeletal muscle, and brain efficiently [47–50]. The reticular structure can integrate well with the human body or tissue, and ensure that the detected parts are breathable, without irritation and restraint [51–54]. For clinical applications, researchers have developed bionic tactile sensors for different conditions. Lee et al. [55] reported that an Au-doped graphene-based mimic-sensual tactile sensor could be used for sweat-based diabetes detection and feedback therapy. The patch ensures that there is enough sweat on the skin for glucose, pH detection by measuring relative humidity, while tremor sensing determines whether the patient is in hypoglycemia, thereby determining the amount of medication used. The design enhances the integration of the human skin and equipment, and also improves the efficiency of sensor work and drug delivery. Dagdeviren et al. [56] developed a bioelectronic bandage for dermatosis, which consists of sensors and actuators made of zirconia titanate nanomaterials. The device can be used in contact with the surface conformation of organs and tissues, and to detect skin stiffness, local changes, which can help to understand and study the age, skin properties, and skin health, thus helping to treat skin canceration. Yokota et al. [57] have developed a flexible conformal reflective pulse-blood meter that laminates the device on the finger and emits light using a high-efficiency polymer light-emitting diode. An organic photodetector detects reflected light as a measure of oxygen concentration in the pulse and blood, which can be used to monitor oxygen levels before and after surgery. The tactile sensors not only act on the surface of the skin for detection and adjuvant treatment but also can be attached to the surface of biological tissue for monitoring or treatment. Flexible electrons have good biocompatibility and do not cause infection problems in organisms. Minev et al. [58] designed and fabricated soft protective films of the brain and spinal cord, which can monitor the nervous system in real time for a long time.

13.4.2 Wearable equipments

In today's society, the standard of living and the level of science and technology are further improved, the demands for the quality and diversity of daily life is increasing, and fitness and health have become a topic of increasing concern. Flexible sensors can be attached to clothing such as stockings [59] or can be directly integrated into the skin surface to monitor human activity signals. In Fig. 13.7, flexible tactile sensors acting as wearable devices can seamlessly integrate the man-machine interface and fit the human body at the measured site [60]. The tactile sensor equipment with a small size, portable and comfortable to wear, biocompatible, is expected to occupy the future market of wearable electronic equipment. Dong et al. [61] thus designed a human-machine interface with a self-similar snakelike structure, and the capacitive electrodes measured their corresponding actions and reduced motion artifacts caused by skin movements. Dagdeviren et al. [62] designed a sensor by coupling the sensor on an elastic substrate to enhance its piezoelectric effect in order to monitor tiny movements of the skin surface (such as changes in arterial pressure, language-related throat vibrations).

For self-healing tactile sensors, one of the factors to be considered is the equipment energy supply aspect. Xu et al. [63] developed rechargeable lithium batteries with a low modulus silicone elastomer as basal. The device can be used through the way of wireless charging without external physical connections. In addition to the uses of charging and power supply, researchers are also exploring self-powered supply for self-healing tactile sensors. For example, Bu et al. [64] have presented a stretchable triboelectric photonic smart skin composed of a soft AIE-active substrate with a tightly bonded microcracked copper film and a middle stretchable conducting layer of an Ag NWs network (Fig. 13.8A). By developing a microcrack structure-like skin stripe, the strain-dependent tactile sensor is used for lateral tensile sensing in a large range. At the same time, the tactile sensor is found to be pressure-sensitive when served as the nanogenerators, and possesses the ability for vertical pressure sensing with high sensitivity (Fig. 13.8B and C). The tactile sensors can be further integrated on a robotic hand, which demonstrates multidimensional mechanical sensing for external stimuli and different gestures. The application of tactile sensors in wearable electronic devices is promising and can meet the increasing demand of contemporary consumers. Some companies, including large multinational corporations, have entered the wearable sensory device market [65–67]. Therefore, self-healing tactile sensors will offer many opportunities and there is much room for development in the future.

13.4.3 Legs with tactile function

The traditional prosthesis is relatively rigid and lacks touch, which makes it difficult for owners to use freely without a sense of isolation. The self-healing tactile sensors with the same characteristics as skin have the potential to replace damaged skin [68]. Flexible electronics should be able to replace the skin, which can let the user control the flexibility of movements and feel the stimulation around. The human skin is very

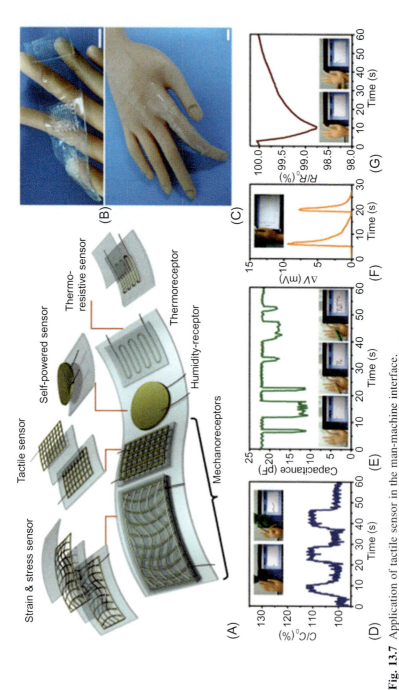

Fig. 13.7 Application of tactile sensor in the man-machine interface.
Reproduced with permission from Z. Lei, P. Wu, Adaptable polyionic elastomers with multiple sensations and entropy-driven actuations for prosthetic skins and neuromuscular systems, Mater. Horiz. 6 (2019) 538–545. Copyright 2019, Royal Society of Chemistry.

Self-healable tactile sensors

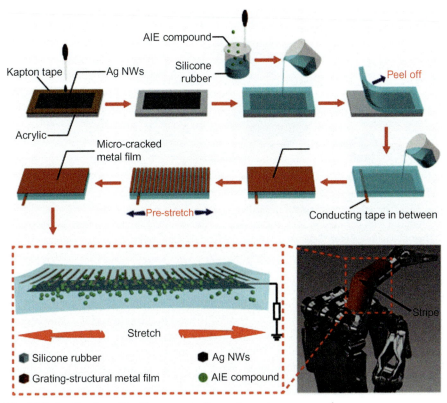

Fig. 13.8 Application of tactile sensor in the equipment energy supply.
Reproduced with permission from T. Bu, T. Xiao, Z. Yang, G. Liu, X. Fu, J. Nie, T. Guo, Y. Pang, J. Zhao, F. Xi, C. Zhang, Z.L. Wang, Stretchable triboelectric–photonic smart skin for tactile and gesture sensing, Adv. Mater. 30 (2018) 1800066. Copyright 2018, Wiley-VCH.

fine, and our fingers can sense the texture, soft and hard properties of the object by touch [69]. Therefore, the self-healing tactile sensor for prosthesis needs to have high precision and sensitivity to sense different tactile factors such as roughness, hardness, and force size. The prosthesis is fully in contact with and interacts with the human nervous system before it can play an alternative role. Organic semiconductors and biological structures have similar carbon frameworks, and their seamless integration is highly likely [70–72]. The neural interface can act as an artificial extension of the lost nervous system function, capable of recording and stimulating neural activity, linking the external prosthesis to the internal nerve [73]. Organic materials, conductive polymers, and composites are all important materials for neural interface applications. Jensen et al. [74] developed a large area of epidermal electronic physiological sensors that could be covered in a large area. They successfully covered the electrophysiological device in the entire scalp and forearm position, and demonstrated the multifunctional control of the patient's prosthesis with such an electronic interface.

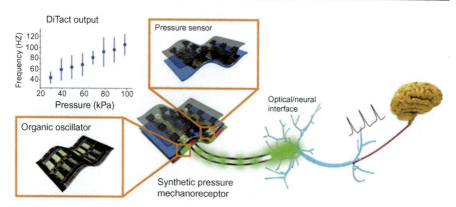

Fig. 13.9 The schematic diagram of DiTact.
Reproduced with permission from B.C.K. Tee, A. Chortos, A. Berndt, A skin-inspired organic digital mechanoreceptor, Science 350 (2015) 313–316. Copyright 2016, the American Association for the Advancement of Science.

The successful use of prosthesis should not only have the ability of perception, but also have the function of generating and processing signals [75]. When the skin is stimulated by the outside world, the nerve cells deep in the skin feel the compression and produce a signal that will be transmitted to the brain with the nerve fibers, and then identify and judge the response. Therefore, the application of tactile sensors to the field of prosthesis, signal transmission, and processing are also aspects to be considered. There are different ways of sensory feedback, such as targeting nerve innervation and sensory substitution. Tee et al. [76] reported a digital tactile system (digital tactile system, DiTact). As shown in Fig. 13.9, DiTact consists of a pressure-sensitive tactile element and an organic annular oscillator, which uses an organic transistor circuit to collect pressure and convert it into a digital signal to simulate.

13.4.4 Software robots

At first, the concept of the tactile sensor was proposed for robot technology. Nowadays, artificial intelligence is an enormous field. Traditional robots are rigid, difficult to handle lightweight and fragile objects in complex environments. Therefore, it is difficult for rigid robots to adapt and perform tasks because irregular or changing environments require complex algorithms to create scenarios in which robots cannot get tactile contact information from visual graphics [77]. In the development of software robots, scholars are inspired by many soft organisms (such as octopus and elephant nose) [78, 79]. It is hoped that software robots will be developed by means of structure or movement similar to biological tissues. The material builds a robot hand to achieve a good grasping ability. Shin et al. [80] implemented a flexible actuator consisting of elastomeric materials that simulated the mechanical properties of soft muscle tissue, which could be used to match physiological and pathological movements. They also created modeling methods based on this material and mimicked the relationship of biological morphology to function in the left ventricle of the heart. Inspired by plant deformable systems, Gladman et al. [81] demonstrated a 4D printing method with hydrogel composites as ink using the anisotropy of cellulose protofibers. By programming control materials and geometry structure, it provided ideas for the development of software robots, and also was beneficial for robots to conduct directional detection of a special environment of underwater.

With the further development of intelligence, human-computer interaction deserves more and more attention. The tactile sensor manufactured by Wang et al. [82] can visualize pressure distribution, which can serve as a medium for good human-computer interaction and promote the development of intelligence. Many organic soft materials possess excellent photoelectric properties and can be selected as flexible display devices [83]. Robots with self-healing tactile sensors can more accurately sense the surroundings to adapt to the complex and changeable environments. For instance, Su et al. [84] developed a humidity sensor based on Au nanoparticles and GO. Humidity sensing is very helpful for software robots to perceive the environment. The complexity and flexibility of the environment will make the machine more vulnerable, and the self-healing function described above is very meaningful. The realization of software robots requires self-healing tactile sensors to cover the surface of robots, which has a high requirement for the size of the self-healing tactile sensors. Zou et al. [85] developed a biomimetic tactile sensor with a dynamic covalent thermosetting. The sensor is self-healing and permanently adapts to complex surfaces without introducing too much interfacial stress. Lee et al. [86] used carbon nanotubes and graphene composite nanofibers to reduce the bending sensitivity of the sensor. It can achieve a large area of real-time conventional pressure detection under different complex bending conditions. That is, the bending of the device itself does not affect the results of pressure detection. The material can be successfully applied to the artificial heart sensing system. The experimental device for measuring the pressure of the artificial blood vessels is shown in Fig. 13.10.

13.5 Outlook and future challenges

In this chapter, we describe the main progress in self-healable tactile sensors and the efforts made to develop self-healable electronic systems used in peripheral nervous-system-interface technologies. In spite of the great success, their current performance is still not comparable to that of human skin. However, combining flexible sensors with self-healing capability should perform potential applications of the tactile sensing systems in the artificial intelligence field.

To implement these functions of the device, we should identify the key challenges of the current tactile sensor. First, tactile sensors should be able to self-heal at room temperature or even at low temperatures, even after damage. In particular, issues such as tensile fatigue, toughness, and self-healing efficiency of the tactile sensors should be considered. We also need to gain insight into synergies in terms of rational design, toughness, self-healing, and the dynamic properties of the stretchable polymer. In order to strengthen this assumption, we refer to the two previous reports of Parida et al. [87] and Hou et al. [88] On one hand, they focused on improving the intrinsic dynamic properties of polymers to promote the self-healing ability of tactile sensors.

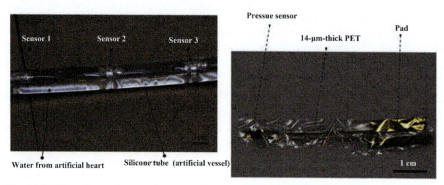

Fig. 13.10 Experimental setup for measuring pressure propagation from the artificial blood vessel.
Reproduced with permission from S. Lee, A. Reuveny, J. Reeder, A transparent bending-insensitive pressure sensor, Nat. Nanotechnol. 11 (2016) 472–478. Copyright 2016, Springer Nature Limited.

On the other hand, they explained that tactile sensors, which are intrinsically self-healable tactile sensors, should maintain the rapid movement of polymer molecular chains after damage. Second, novel integrated strategies are needed to achieve robust and self-healing tactile sensing systems. Although great advances have been made over the past few years, most tactile sensors are still in the prototype stage, and the related integrated approaches are facing significant challenges. It is anticipated that tactile sensors made from self-healing materials and liquid metal will be a promising direction to easily fabricate highly integrated structures without additional complexity or glue. In addition, for self-healable tactile sensing systems, similar multimodal methods are used to process tactile information, usually using different sensors. Furthermore, low power consumption and appropriate energy collection and storage are ideal features of the tactile sensing platform to reduce battery replacement, suggesting that practical applications require further optimization of material and device configurations. In conclusion, the application of self-healing, self-function, and effective integration strategies to tactile sensors has led to a wide range of applications for soft and robust tactile sensing systems.

Acknowledgment

We gratefully acknowledge the funding support from the National Natural Science Foundation of China (Grant Nos. 51975562 and 51675514) and the National Defense Basic Research Program of China (JCKY2019130C105).

References

[1] I. Kang, M.J. Schulz, J.H. Kim, V. Shanov, D. Shi, A carbon nanotube strain sensor for structural health monitoring, Smart Mater. Struct. 15 (2006) 737.

[2] S. Basu, P. Bhattacharyya, Recent developments on graphene and graphene oxide based solid state gas sensors, Sensors Actuators B Chem. 173 (2012) 1–21.

[3] M. Amjadi, A. Pichitpajongkit, S. Lee, S. Ryu, I. Park, Highly stretchable and sensitive strain sensor based on silver nanowire-elastomer nanocomposite, ACS Nano 8 (2014) 5154–5163.

[4] S.H. Cho, S.W. Lee, S. Yu, H. Kim, S. Chang, D. Kang, I. Hwang, H.S. Kang, B. Jeong, E.H. Kim, S.M. Cho, K.L. Kim, H. Lee, W. Shim, C. Park, Micropatterned pyramidal ionic gels for sensing broad-range pressures with high sensitivity, ACS Appl. Mater. Interfaces 9 (2017) 10128–10135.

[5] H. Park, Y.R. Jeong, J. Yun, S.Y. Hong, S. Jin, S.J. Lee, G. Zi, J.S. Ha, Stretchable array of highly sensitive pressure sensors consisting of polyaniline nanofibers and Au-coated polydimethylsiloxane micropillars, ACS Nano 9 (2015) 9974–9985.

[6] N. Gao, X. Zhang, S. Liao, H. Jia, Y. Wang, Polymer swelling induced conductive wrinkles for an ultrasensitive pressure sensor, ACS Macro Lett. 5 (2016) 823–827.

[7] E. Roh, B.U. Hwang, D. Kim, B.Y. Kim, N.E. Lee, Stretchable, transparent, ultrasensitive, and patchable strain sensor for human machine interfaces comprising a nanohybrid of carbon nanotubes and conductive elastomers, ACS Nano 9 (2015) 6252–6261.

[8] S. Gong, W. Schwalb, Y. Wang, Y. Chen, Y. Tang, J. Si, B. Shirinzadeh, W. Cheng, A wearable and highly sensitive pressure sensor with ultrathin gold nanowires, Nat. Commun. 5 (2014) 3132.

[9] C. Pang, J.H. Koo, A. Nguyen, J.M. Caves, M.G. Kim, A. Chortos, K. Kim, P.J. Wang, J.B. Tok, Z. Bao, Highly skin-conformal microhairy sensor for pulse signal amplification, Adv. Mater. 27 (2015) 634–640.

[10] T.P. Huynh, H. Haick, Self-healing, fully functional, and multiparametric flexible sensing platform, Adv. Mater. 28 (2016) 138–143.

[11] A. Pettignano, M. Häring, L. Bernardi, N. Tanchoux, F. Quignard, Self-healing alginate-gelatin biohydrogels based on dynamic covalent chemistry: elucidation of key parameters, Mater. Chem. Front. 1 (2017) 73–79.

[12] T. Wang, Y. Zhang, Q. Liu, W. Cheng, X. Wang, L. Pan, B. Xu, H. Xu, A self-healable, highly stretchable, and solution processable conductive polymer composite for ultrasensitive strain and pressure sensing, Adv. Funct. Mater. 28 (2017), 1705551.

[13] C. Mu, Y. Song, W. Huang, A. Ran, R. Sun, W. Xie, H. Zhang, Flexible normal-tangential force sensor with opposite resistance responding for highly sensitive artificial skin, Adv. Funct. Mater. 28 (2018), 1707503.

[14] L. Hu, W. Yuan, P. Brochu, G. Gruner, Q. Pei, Highly stretchable, conductive, and transparent nanotube thin films, Appl. Phys. Lett. 94 (2009) 161108.

[15] J. Shi, X. Li, H. Cheng, Z. Liu, L. Zhao, T. Yang, Z. Dai, Z. Cheng, E. Shi, L. Yang, Z. Zhang, A. Cao, H. Zhu, Y. Fang, Graphene reinforced carbon nanotube networks for wearable strain sensors, Adv. Funct. Mater. 26 (2016) 2078–2084.

[16] D.J. Lipomi, M. Vosgueritchian, B.C. Tee, S.L. Hellstrom, J.A. Lee, C.H. Fox, Z. Bao, Skin-like pressure and strain sensors based on transparent elastic films of carbon nanotubes, Nat. Nanotechnol. 6 (2011) 788–792.

[17] P. Nicholas, Large-area graphene synthesized by chemical vapor deposition for high-performance, flexible electronics, ProQuest LLC 1346 (2014) 48106. –1346.

[18] X. Wu, K. Hou, J. Huang, J. Wang, S. Yang, Graphene-based cellular materials with extremely low density and high pressure sensitivity based on self-assembled graphene oxide liquid crystals, J. Mater. Chem. C 6 (2018) 8717–8725.

[19] J. Huang, J. Wang, Z. Yang, S. Yang, High-performance graphene sponges reinforced with polyimide for room-temperature piezoresistive sensing, ACS Appl. Mater. Interfaces 10 (2018) 8180–8189.

[20] X. Wu, X. Liu, J. Wang, J. Huang, S. Yang, Reducing structural defects and oxygen-containing functional groups in GO-hybridized CNTs aerogels: simultaneously improve the electrical and mechanical properties to enhance pressure sensitivity, ACS Appl. Mater. Interfaces 10 (2018) 39009–39017.

[21] K. Sun, H. Ko, H.H. Park, M. Seong, S.H. Lee, H. Yi, H.W. Park, T.I. Kim, C. Pang, H.E. Jeong, Hybrid architectures of heterogeneous carbon nanotube composite microstructures enable multiaxial strain perception with high sensitivity and ultrabroad sensing range, Small 14 (2018), e1803411.

[22] M. Ha, S. Lim, J. Park, D.S. Um, Y. Lee, H. Ko, Bioinspired interlocked and hierarchical design of ZnO nanowire arrays for static and dynamic pressure-sensitive electronic skins, Adv. Funct. Mater. 25 (2015) 2841–2849.

[23] X. Wu, Z. Li, H. Wang, J. Huang, J. Wang, S. Yang, Stretchable and self-healable electrical sensors with fingertip-like perception capability for surface texture discerning and biosignal monitoring, J. Mater. Chem. C 7 (2019) 9008–9017.

[24] K.S. Toohey, N.R. Sottos, J.A. Lewis, J.S. Moore, S.R. White, Self-healing materials with microvascular networks, Nat. Mater. 6 (2007) 581–585.

[25] S.R. White, N.R. Sottos, P.H. Geubelle, J.S. Moore, M.R. Kessler, R. Sriram, E.N. Brown, S. Viswanathan, Autonomic healing of polymer composites, Nature 409 (2001) 794–797.

[26] T.P. Huynh, H. Haick, Self-healing, fully functional, and multiparametric flexible sensing platform, Adv. Mater. 28 (2016) 138–143.

[27] G.O. Wilson, M.M. Caruso, S.R. Schelkopf, N.R. Sottos, S.R. White, J.S. Moore, Adhesion promotion via noncovalent interactions in self-healing polymers, ACS Appl. Mater. Interfaces 3 (2011) 3072–3077.

[28] B.C. Tee, C. Wang, R. Allen, Z. Bao, Pressure and flexion-sensitive electronic skin with repeatable ambient self-healing capability, Nat. Nanotechnol. 7 (2012) 825–832.

[29] K. Jud, H.H. Kausch, Load transfer through chain molecules after interpenetration at interfaces, Polym. Bull. 1 (1979) 697–707.

[30] C.B. Lin, S.B. Lee, K.S. Liu, Methanol-induced crack healing in poly(methyl methacrylate), Polym. Eng. Sci. 30 (1990) 1399–1406.

[31] F. Yang, R. Pitchumani, Healing of thermoplastic polymers at an interface under non-isothermal conditions, Macromolecules 35 (2002) 3213–3224.

[32] C.M. Chung, Y.S. Roh, S.Y. Cho, J.G. Kim, Crack healing in polymeric materials via photochemical [2+2] cycloaddition, Chem. Mater. 16 (2004) 3982–3984.

[33] J.H. Harreld, M.S. Wong, P.K. Hansma, Self-Healing Organosiloxane Materials Containing Reversible and Energy-Dispersive Crosslinking Domains, US Patent, 2004007792-A1[P], 2004.

[34] F.T. Moutos, F. Guilak, Functional properties of cell-seeded three-dimensionally woven poly(ε-caprolactone) scaffolds for cartilage tissue engineering, Tissue Eng. Part A 6 (2010) 1291–1301.

[35] H.K. Choi, M.H. Kim, S.H. Im, O.O. Park, Fabrication of ordered nanostructured arrays using poly(dimethylsiloxane) replica molds based on three-dimensional colloidal crystals, Adv. Funct. Mater. 19 (2009) 1594–1600.

[36] F.T. Mouyos, L.E. Fred, F. Guilak, A biomimetic three-dimensional woven composite scaffold for functional tissue engineering of cartilage, Nat. Mater. 6 (2007) 162–167.

[37] G. Ra Yi, J.H. Moon, S.M. Yang, Ordered macroporous particles by colloidal templating, Chem. Mater. 13 (2001) 2613–2618.

[38] P. Lin, S. Ma, X. Wang, F. Zhou, Molecularly engineered dual-crosslinked hydrogel with ultrahigh mechanical strength, toughness, and good self-recovery, Adv. Mater. 27 (2015) 2054–2059.

[39] S. Majumder, T. Mondal, M.J. Deen, Wearable sensors for remote health monitoring, Sensors 17 (2017) 130.

[40] Y. Liu, R. Bao, J. Tao, J. Li, M. Dong, C. Pan, Recent progress in tactile sensors and their applications in intelligent systems, Sci. Bull. 65 (2020) 70–88.

[41] D. Chen, Q. Pei, Electronic muscles and skins: a review of soft sensors and actuators, Chem. Rev. 117 (2017) 11239–11268.

[42] Z.W.K. Low, Z.B. Li, C. Owh, P.L. Chee, E. Ye, D. Kai, D.P. Yang, X.J. Loh, Using artificial skin devices as skin replacements: insights into superficial treatment, Small 15 (2019), 1805453.

[43] Z. Lei, Q. Wang, S. Sun, W. Zhu, P. Wu, A bioinspired mineral hydrogel as a self-healable, mechanically adaptable ionic skin for highly sensitive pressure sensing, Adv. Mater. 29 (2017), 1700321.

[44] E. D'Elia, S. Barg, N. Ni, V.G. Rocha, E. Saiz, Self-healing graphene-based composites with sensing capabilities, Adv. Mater. 27 (2015) 4788–4794.

[45] N. Maity, R. Ghosh, A.K. Nandi, Optoelectronic properties of self-assembled nanostructures of polymer functionalized polythiophene and graphene, Langmuir 34 (2018) 7585–7597.

[46] M. Xie, K. Hisano, M. Zhu, T. Toyoshi, M. Pan, S. Okada, O. Tsutsumi, S. Kawamura, C. Bowen, Flexible multifunctional sensors for wearable and robotic applications, Adv. Mater. Technol. 4 (2019), 1800626.

[47] H. Zhuo, Y. Hu, Z. Chen, X. Peng, L. Liu, Q. Luo, J. Yi, C. Liu, L. Zhong, A carbon aerogel with super mechanical and sensing performances for wearable piezoresistive sensors, J. Mater. Chem. A 7 (2019) 8092–8100.

[48] Q. Shi, T. He, C. Lee, More than energy harvesting–combining triboelectric nanogenerator and flexible electronics technology for enabling novel micro-/nano-systems, Nano Energy 57 (2019) 851–871.

[49] Y. Lu, Z. Liu, H. Yan, Q. Peng, R. Wang, M.E. Barkey, J.W. Jeon, E.K. Wujcik, Ultrastretchable conductive polymer complex as a strain sensor with a repeatable autonomous self-healing ability, ACS Appl. Mater. Interfaces 11 (2019) 20453–20464.

[50] T.H. Chang, K. Li, H. Yang, P.Y. Chen, Multifunctionality and mechanical actuation of 2D materials for skin-mimicking capabilities, Adv. Mater. 30 (2018) 1802418.

[51] V. Giurgiutiu, A. Zagrai, J.J. Bao, Piezoelectric wafer embedded active sensors for aging aircraft structural health monitoring, Struct. Health. Monit. 1 (2002) 41–61.

[52] H. Guo, G. Xiao, N. Mrad, J. Yao, Fiber optic sensors for structural health monitoring of air platforms, Sensors 11 (2011) 3687–3705.

[53] J. Leng, A. Asundia, Structural health monitoring of smart composite materials by using EFPI and FBG sensors, Sensors Actuators A Phys. 103 (2003) 330–340.

[54] H. Banaee, M.U. Ahmed, A. Loutfi, Data mining for wearable sensors in health monitoring systems: a review of recent trends and challenges, Sensors 13 (2013) 17472–17500.

[55] H. Lee, T.K. Choi, Y.B. Lee, H.R. Cho, R. Ghaffari, L. Wang, T.D. Chung, N. Lu, A graphene-based electrochemical device with thermoresponsive microneedles for diabetes monitoring and therapy, Nat. Nanotechnol. 11 (2016) 566–572.

[56] C. Dagdeviren, Y. Shi, P. Joe, R. Ghaffari, G.B.K. Usgaonkar, O. Gur, P.L. Tran, J.R. Crosby, M. Meyer, Y. Su, R.C. Webb, A.S. Tedesco, M.J. Slepian, Y. Huang, J.A. Rogers, Conformal piezoelectric systems for clinical and experimental characterization of soft tissue biomechanics, Nat. Mater. 14 (2015) 728–736.

[57] T. Yokota, P. Zalar, M. Kaltenbrunner, H. Jinno, N. Matsuhisa, Ultraflexible organic photonic skin, Sci. Adv. 2 (2016), e1501856.

[58] I.R. Minev, P. Musienko, A. Hirsch, Q. Barraud, N. Wenger, E. Martin Moraud, J. Gandar, M. Capogrosso, T. Milekovic, L. Asboth, R.F. Torres, N.S. Vachicouras, Q. Liu, N. Pavlova, S. Duis, A. Larmagnac, J. Vörös, S. Micera, Z. Suo, G. Courtine, S.P. Lacour, Electronic dura mater for long-term multimodal neural interfaces, Science 347 (2015) 159–163.

[59] T. Yamada, Y. Hayamizu, Y. Yamamoto, Y. Yomogida, A.I. Najafabadi, D.N. Futaba, K. Hata, A stretchable carbon nanotube strain sensor for human-motion detection, Nat. Nanotechnol. 6 (2011) 296–301.

[60] Z. Lei, P. Wu, Adaptable polyionic elastomers with multiple sensations and entropy-driven actuations for prosthetic skins and neuromuscular systems, Mater. Horiz. 6 (2019) 538–545.

[61] W. Dong, C. Zhu, W. Hu, L. Xiao, Y. Huang, Stretchable human-machine interface based on skin-conformal sEMG electrodes with self-similar geometry, J. Semicond. 39 (2018), 014001.

[62] C. Dagdeviren, Y. Su, P. Joe, R. Yona, Y. Liu, Conformable amplified lead zirconate titanate sensors with enhanced piezoelectric response for cutaneous pressure monitoring, Nat. Commun. 5 (2014) 4496.

[63] S. Xu, Y. Zhang, J. Cho, J. Lee, X. Huang, L. Jia, J.A. Fan, Y. Su, J. Su, H. Zhang, H. Cheng, B. Lu, C. Yu, C. Chuang, T. Kim, T. Song, K. Shigeta, S. Kang, C. Dagdeviren, I. Petrov, P.V. Braun, Y. Huang, U. Paik, J.A. Rogers, Stretchable batteries with self-similar serpentine interconnects and integrated wireless recharging systems, Nat. Commun. 4 (2013) 1543.

[64] T. Bu, T. Xiao, Z. Yang, G. Liu, X. Fu, J. Nie, T. Guo, Y. Pang, J. Zhao, F. Xi, C. Zhang, Z.L. Wang, Stretchable triboelectric–photonic smart skin for tactile and gesture sensing, Adv. Mater. 30 (2018) 1800066.

[65] R. Ma, V.V. Tsukruk, Seriography guided reduction of graphene oxide biopapers for wearable sensory electronics, Adv. Funct. Mater. 27 (2017) 1604802.

[66] F. Sorgini, A. Mazzoni, L. Massari, C.M. Oddo, M.C. Carrozza, Encapsulation of piezoelectric transducers for sensory augmentation and substitution with wearable haptic devices, Micromachines 8 (2017) 270.

[67] N. Liu, L.Q. Zhu, P. Feng, C.J. Wan, Y.H. Liu, Y. Shi, Q. Wan, Flexible sensory platform based on oxide-based neuromorphic transistors, Sci. Rep. 5 (2015) 18082.

[68] J.J. Cabibihan, D. Joshi, Y.M. Srinivasa, M.A. Chan, Illusory sense of human touch from a warm and soft artificial hand, IEEE Trans. Neural Syst. Rehabil. Eng. 23 (2014) 517–527.

[69] L.B. Dahlin, S. Thrainsdottir, R. Cederlund, N.O.B. Thomsen, K.F. Eriksson, I. Rosén, T. Speidel, G. Sundqvist, Vibrotactile sense in median and ulnar nerve innervated fingers of men with Type 2 diabetes, normal or impaired glucose tolerance, Diabet. Med. 25 (2008) 543–549.

[70] V. Coropceanu, J. Cornil, D.A. Filho, Y. Olivier, R. Silbey, J.L. Brédas, Charge transport in organic semiconductors, Chem. Rev. 107 (2007) 926–952.

[71] L.L. Chua, J. Zaumseil, J.F. Chang, E.C.W. Ou, P.K.H. Ho, H. Sirringhaus, R.H. Friend, General observation of n-type field-effect behaviour in organic semiconductors, Nature 434 (2005) 194–199.

[72] G.H. Kim, L. Shao, K. Zhang, K.P. Pipe, Engineered doping of organic semiconductors for enhanced thermoelectric efficiency, Nat. Mater. 12 (2013) 719–723.

[73] M.C.F. Castro, A. Cliquet Jr., Artificial grasping system for the paralyzed hand, Artif. Organs 24 (2000) 185–188.

[74] T. Jensen, Integral Patch Type Electronic Physiological Sensor, US Patent App. 10/325, 449, 2003.

[75] M.A.L. Nicolelis, Actions from thoughts, Nature 409 (2001) 403–407.

[76] B.C.K. Tee, A. Chortos, A. Berndt, A skin-inspired organic digital mechanoreceptor, Science 350 (2015) 313–316.

[77] J. Tegin, J. Wikander, Tactile sensing in intelligent robotic manipulation—a review, Ind. Robot. 32 (2005) 64–70.

[78] W. McMahan, I.D. Walker, Octopus-inspired grasp-synergies for continuum manipulators, IEEE ROBIO 20 (2009) 303–314.

[79] C. Laschi, B. Mazzolai, V. Mattoli, M. Cianchetti, P. Dario, Design and development of a soft actuator for a robot inspired by the octopus arm, Exp. Robot. 20 (2009) 25–33.

[80] S.R. Shin, C. Shin, A. Memic, S. Shadmehr, M. Miscuglio, H.Y. Jung, S.M. Jung, H. Bae, A. Khademhosseini, X. Tang, M.R. Dokmeci, Aligned carbon nanotube-based flexible gel substrates for engineering biohybrid tissue actuators, Adv. Funct. Mater. 25 (2015) 4486–4495.

[81] A.S. Gladman, E.A. Matsumoto, R.G. Nuzzo, Biomimetic 4D printing, Nat. Mater. 15 (2016) 413–418.

[82] D. Wang, Y. Mei, G. Huang, Printable inorganic nanomaterials for flexible transparent electrodes: from synthesis to application, J. Semicond. 39 (2018), 011002.

[83] B. Geffroy, P. Roy, C. Prat, Organic light-emitting diode (OLED) technology: materials, devices and display technologies, Polym. Int. 55 (2006) 572–582.

[84] P.G. Su, W.L. Shiu, M.S. Tsai, Flexible humidity sensor based on au nanoparticles/graphene oxide/thiolated silica sol-gel film, Sens. Actuators B 216 (2015) 467–475.

[85] Z. Zou, C. Zhu, Y. Li, Rehealable, fully recyclable, and malleable electronic skin enabled by dynamic covalent thermoset nanocomposite, Sci. Adv. 4 (2018), eaaq0508. https://www.nature.xilesou.top/articles/nnano.2016.38 - citeas.

[86] S. Lee, A. Reuveny, J. Reeder, A transparent bending-insensitive pressure sensor, Nat. Nanotechnol. 11 (2016) 472–478.

[87] K. Parida, V. Kumar, W. Jiangxin, V. Bhavanasi, R. Bendi, P.S. Lee, Highly transparent, stretchable, and self-healing ionic-skin triboelectric nanogenerators for energy harvesting and touch applications, Adv. Mater. 29 (2017) 1702181.

[88] C. Hou, T. Huang, H. Wang, H. Yu, Q. Zhang, Y. Li, A strong and stretchable self-healing film with self-activated pressure sensitivity for potential artificial skin applications, Sci. Rep. 3 (2013) 3138.

Index

Note: Page numbers followed by *f* indicate figures and *t* indicate tables.

A

Active materials
  for self-healing tactile sensors, 264
    CNT, 266–267, 268*f*
    conductive elastomers, 270–273, 272*f*
    graphene, 267–270, 271*f*
Active-matrix tactile sensory array, 55–57, 58*f*
Active polymerization, self-healable tactile sensors, mechanisms of, 276–277
AFM. *See* Atomic force microscopy (AFM)
AgNW-PDMS-based strain sensor, fabrication flow of, 81–82, 83*f*
Ag NWs network, microcracked copper film and middle stretchable conducting layer of, 279
Air gap-introduced structures, flexible triboelectric nanogenerators, 252
Aligned electrospun PVDF nanofibers, 178–179, 180*f*
Aluminum-plastic film (APLF), 251–252
Anions, ionic liquids, 219, 221*f*
Anisotropic CNT strain sensor, 81
Anomalous photovoltaic effects, 143–144
Apparent capillary force, 237–238
Arrhenius equation, 115
Artificial electronic skin (e-skin), 22–23
Artificial intelligence (AI), 159
Atomic force microscopy (AFM), 116, 117*f*
  experiments, 118–119
Au-deposited toothbrush-hairs micro-prick arrays, 273
Au nanowires (NWs), pressure sensor based on, 264, 265*f*

B

BaTiO$_3$, 247
Bell's model, 115, 126
Bending instability, 160–161
Biomonitoring, CNT-based tactile sensors, 267

Bond dissociation energy (BDE), 128–130
Buckled tactile sensor, 197
  conductive material for, 206–214
    hybrid composite materials, 212–214, 213*f*
    one-dimensionnanomaterials, 208–210, 209*f*
    two-dimensional materials, 210–212, 211*f*
    zero-dimensionnanomaterials, 206–208, 207*f*
  methods of, 198–205
    composite, 204–205
    mold, 201–202, 202*f*
    stretch–release, 200–201, 201*f*
    swelling, 203–204, 204*f*
    thermal, 202–203, 203*f*
  in tactile sensors, 197–198, 199*f*

C

Capacitance, 40–42
  type, 1–2
Capacitive pressure sensors, 32, 40–49, 55
  electronic textiles with tactile sensing ability, 47–49, 49–50*f*
  high capacitive dielectric materials for tactile sensors, 45–47, 46*f*, 48*f*
  microstructures of dielectric films, 42–45, 43–45*f*
  principles of, 40–42
Capacitive sensor, ionic liquids, pressure sensors based on, 225–226, 226–227*f*
Capacitive tactile sensor, 171–175, 174–175*f*
Capacitive transduction, 166
Capillary effect
  confining ionic liquids by
    external, 235–237
    internal, 237–240, 237–238*f*
Carbon-based 0D materials, 207

Carbon nanotubes (CNTs), 72, 76–78, 173–175
  active materials, for self-healing tactile sensors, 266–267, 268*f*
  fabrication process of, 81, 82*f*
  resistive composite strain sensors, conductive filler for, 81
  thermal method, 202–203
  triboelectric materials, 251–252
Cations, ionic liquids, 219
Chemical vapor deposition (CVD), of graphene, 267
Classical force distribution analysis, 121
CNT embroidered graphene (CeG) film, 267, 268*f*
CNT-PDMS-based active sensor, 78*f*
  for sliding detection, 72
CNT-PDMS composite, 83
CNT-PDMS sponge-based conductive composite, 72, 74*f*
Coaxial electrospinning method, 162
Composite methods,buckle structure, 204–205
Composites, 279–281
Conducting polymers, 15
Conductive composite-based tactile sensor, 67
  strategy, 67
Conductive elastomers, self-healable tactile sensors, structure and functional materials of, 270–273, 272*f*
Conductive material, 204–205, 205*f*
  for buckled tactile sensor, 206–214
    hybrid composite materials, 212–214, 213*f*
    one-dimensionnanomaterials, 208–210, 209*f*
    two-dimensional materials, 210–212, 211*f*
    zero-dimensionnanomaterials, 206–208, 207*f*
Conductive polymers, 279–281
Constrained geometries simulate external force (CoGEF), 119–120, 120*f*
  method, 127
  potential, 119–120
Contact resistance, 32–34, 50–51, 266–269
Conventional spinning techniques, 162
Covalent mechanophores, 123–128

Cowpea-structured ZnO/PVDF hybrid nanofibers, 248–250
Crosslinkable prepolymer, metal salt in, 80
Cross-linked polymer network, 68
Crystal-confined ionic liquids (CCILs)
  four different states of, 237–238, 238*f*
  preparation of, 237, 237*f*
  robotic arms based on, 240, 240*f*
Cusp model, 115
Cycloaddition reaction of light, 276
Cyclobutane, 276
Cyclobutene-based mechanophores, 126–127
Cyclopropane-based mechanophores, 123–126, 124*f*

D

Deformation luminescence (DML), 95–96
Density functional theory (DFT), 118
Dermatosis, bioelectronic bandage for, 277–278
Detection range, 19
Dielectric films, microstructures of, 42–45, 43–45*f*
Digital camera, 229–231
Digital tactile system (DiTact), 282, 282*f*
Direct-current-driven electroluminescent tactile sensors, 91, 92–93*f*
Drain-source current (I_{DS}), 53–54
Drain-source voltage (V_{DS}), 53–54
Dual-signal sensor, pressure and temperature, ionic liquids, 234*f*, 235
Dudko-Hummer model, 115, 126, 131–132

E

Elastic-composite energy harvester, 248
Elastomer
  mixing conductive filler into, 70–72, 72–73*f*
  synthesizing conductive fillers within, 78–80, 81*f*
Elastomeric dielectrics, 166
Elastomer surface, conductive film deposited on, 73–78, 75*f*
Electrical conductivity, 67
Electrical percolation threshold, 68–69
Electric field-driven electroluminescent tactile sensors, 92–93, 94*f*

Index 293

Electrochemical double layers (EDLs), 45–46
Electromagnetic induction effect, 255–256
Electronic skin (e-skin), 23, 40*f*, 55, 62–63, 91
Electronic textiles, with tactile sensing ability, 47–49, 49–50*f*
Electrospinning, 214
 method, 5–6
 technique, 160–164, 161*f*
Electrospun composite/doped PVDF nanofibers, 179–182, 181*f*
Electrospun nanofibers, for tactile sensors
 capacitive tactile sensor, 171–175, 174–175*f*
 electrospinning technique and, 160–164, 161*f*
 features/advantages of, 161–162
 form electrospun nanofibrous materials, 167–183
 functionalization methods of, 162–164, 163*f*
 overview of, 159–160
 of piezoelectric polymers, 182
 piezoelectric tactile sensor, 176–182
 piezoresistive tactile sensor, 168–171, 169–170*f*, 172–173*f*
 transduction mechanisms, 164–167, 165*f*
 triboelectric tactile sensor, 182–183
Electrostatic force, 160–161
Endoscopic surgery, 7
Energy harvesting electronic skin (EHES), 37, 37*f*
Energy-harvesting technologies, 245–246
Epidermal electronic physiological sensors, 279–281
Epipremnumaureum, 201–202
Epoxy coating, crack damage in, 273–275
Ethanol, 275–276
Ethanolammonium nitrate, 219
1-ethyl-3-methyl imidazoliumbis (trifluoromethylsulfonyl)imide ([EMIm][Tf2N]), 229–231
External capillary effect, confining ionic liquids, ionic liquids, 235–237, 237–238*f*
External force explicitly included (EFEI), 120–121, 126

External stimuli, multidimensional mechanical sensing for, 279
Extrinsic healing behavior, 273–275

F

Fabricated 3DG/PI composite sponges, 269–270
Face-to-face package, 82–83
Fiber-based TENG, 254
Fiber-shaped structures, flexible triboelectric nanogenerators, 254
Field-effect transistors, operation principle of, 53–55, 54*f*
Field emission scanning electron microscopy (FESEM), 148, 148*f*
Flexible electronics, 15
Flexible electrons, 277–278
Flexible PENGs, 248–250, 249*f*
Flexible piezoelectric nanogenerators, 246–250
 self-powered flexible tactile sensors, 246–250
  composite materials, 248–250
  inorganic materials, 246–247
  organic materials, 247–248
  piezoelectric mechanism, 246
Flexible porous nanogenerators (FPNGs), 248
Flexible pressure sensors, 20–21*t*
Flexible self-powered TENGs, 252, 253*f*
Flexible tactile sensors, 245
Flexible triboelectricnanogenerators
 self-powered flexible tactile sensors
  structural designs, 252–254
  triboelectric materials, 251–252
  triboelectric mechanism, 250–251
Force-clamp experiments, 248–250
Force-clamp molecular dynamics (FCMD), 118–119
Force distribution analysis (FDA), 118
 approaches, 121–123
Force-matching force distribution analysis (FM-FDA), 121–122
Force-probe experiments, 116–117
Force-probe molecular dynamics (FPMD), 118–119
Force-torque sensors, 13
Force-transformed potential energy surface, 120

294 Index

Freestanding triboelectric-layer mode, flexible triboelectricnanogenerators, 251

Functional nanomaterials, 263–264

G

Gate voltage (V_G), 53–54
Gauge factor (GF), 5–6, 168, 207–208
GCNT aerogel, 269–270
gem-dichlorocyclopropane (gDCC), 125
gem-difluorocyclopropane (gDFC), 123–125
Gradient CNTs, 81
Graphene, 78, 79*f*
    active materials, for self-healing tactile sensors, 267–270, 271*f*
    chemicalvapor deposition of, 267
Graphenenanocrystalline carbon (GNC) film, 207–208
Graphene oxide (GO), 82–83, 210, 213–214
GR-based piezoresistive pressure sensor, 204–205
GR film sensor, 200–201

H

Handlike model, surface temperature of, 229–231
Helmholtz equation, 45–46
High-speed rotating disc collector, 162
Humanoids, 141
Hybrid composite materials, conductive material, for buckled tactile sensor, 212–214, 213*f*
Hydrazine hydrate, 80
Hydrogels, 173–175
Hydrothermal reaction, 164
Hydroxyl-functionalized multiwalled carbon nanotubes (OH-fMWCNTs), 213
Hyper-stretchable energy harvester, 248
Hysteresis, 222–223

I

IDTechEx electronics company, 9–10
Imidazole-based ionic liquids, 219, 221*f*
Infrared camera, 229–231
Inorganic hybrid sensor, 91
Inorganic materials, flexible piezoelectric nanogenerators, 246–247
In situ polymerization, 78–80

Integrated tactile sensors, separating multiple signals, 233–235, 233–234*f*
Intense-pulsed-light irradiation, 76
Interdiffusion of molecules, self-healable tactile sensors, 275–276
Interdigital electrodes (IDE), 171
Internal capillary effect, confining ionic liquids by, 237–240, 237–238*f*
Internet of Things (IoT), 159
Intra-nanosheet resistance, 269–270
Intrinsic healing behaviors, 273–275
Ionic liquids (ILs), 45–46, 173–175, 219–222
    anions, 219, 221*f*
    cations, 219
    conductivity of, 221–222
    imidazolium-based ionic liquids, 219, 221*f*
    investigations, 241
    mechanism of, 222*f*
    physical properties of, 220–221*t*
    pressure sensors based on, 222–226
        capacitive sensor, 225–226, 226–227*f*
        piezoresistive sensor, 222–225, 223–224*f*
    preventing leakage of, 235–240
        external capillary effect, confining ionic liquids by, 235–237
        internal capillary effect, confining ionic liquids by, 237–240, 237–238*f*
    tactile sensors, signal separation and integration of, 231–235
        eliminating strain interference, temperature sensors, 231–232, 231*f*
        eliminating temperature interference, pressure sensors, 231–232, 231*f*
        integrated tactile sensors, separating multiple signals in, 233–235, 233–234*f*
    temperature sensors based on, 226–231, 228–230*f*
Isolated ionic liquids, 1
Isotropic CNTs, 81

J

Javey's group, 62
Judgment of energy distribution (JEDI), 122–123

K

Kramers theory, 115

Index

L

Laplace pressure, 239
Lateral sliding mode, flexible triboelectricnanogenerators, 250
Lead (Pb), 154
LED tactile sensor, 91
Light-induced healing, self-healable tactile sensors, mechanisms of, 276
Linear regime, 53–54
Linear response, 143, 150–153
Liquid semiconductors, 221–222

M

Magnetic/polymeric building blocks, young's moduli of, 255–256
Magnetoelectric elastomers, self-powered flexible tactile sensors, 255–256, 255f
Man-machine interface, application, tactile sensor in, 279, 281f
Mechanical compliance, 67
Mechanical energy, 245
Mechanical stretching, 247–248
Mechanochemistry, 113
  theory of, 115
Mechanoelectrical conversion, 245–246, 250, 256
Mechanoelectrical-converted flexible systems, 256–257
Mechanoluminescence (ML) materials
  application scenarios of, 98f
  categories of, 97f
  for tactile sensors, 93–106, 94–96f, 99–100f, 103–107f
Mechanophores, in polymer mechanochemistry
  computational approaches, 118–123
    constrained geometries simulate external force (CoGEF), 119–120, 120f
    external force explicitly included (EFEI), 120–121
    force-clamp molecular dynamics (FCMD), 118–119
    force distribution analysis approaches, 121–123
    force-probe molecular dynamics (FPMD), 118–119
  covalentmechanophores, 123–128

cyclobutene-basedmechanophores, 126–127
cyclopropane-basedmechanophores, 123–126, 124f
ring-openingmechanophores, 127–128
effect of polymer chain, 131–132
organometallicmechanophores, 128–131, 130f
metallocenemechanophores, 128–130, 129f
single-molecule approaches, 115–118
molecular force probe, 117–118, 118f
single-molecule force spectroscopy, 116–117, 117f
theory of mechanochemistry, 115
Medical health, self-healable tactile sensors, 277–278, 278f
Medium-strength cross-linking, 276
Metal-based materials, 208
Metal-based nanoparticles, 207
Metal-insulator-metal (MIM) structure, 40–42
Metallocenemechanophores, 128–130, 129f
Methylammonium lead iodide (MAPbI$_3$), 141–142
  device, 144
  ion migration effect in, 144, 146f
  perovskite materials, 143
  polarization effect in, 143–148
  polystyrene (PS) chains, 148
Micropillar-based tactile sensors, 270
Micropillar interlocking structure, 263–264
Micro-pyramid dielectric arrays, 263–264
Microsphere, 71–72
Microstructure-enhanced piezo-resistivity, 17–19, 18f
Minimally invasive surgery (MIS), 7
Mold, buckle structure, 201–202, 202f
Molecular dynamics (MD) simulation, 113–114
Molecular force probe, 117–118, 118f
Molecules, interdiffusion of, self-healable tactile sensors, 275–276
Multidimensional mechanical sensing, for external stimuli, 279
Multi-functional tactile sensors, 22–23, 24–25f
Multifunctional transistor-based pressure sensors, 57–62

296

Index

Multiscale coarse-graining (MS-CG), 121–122
Multiwalled carbon nanotubes (MWCNTs) networks, 212
MXene, 210–211

N

Nanocrystal rods, 237
Nanofibers, 162
Nanomaterials, 160
National Institute of Advanced Industrial Science and Technology (AIST), 95
Natural fiber materials, 251–252
Negative temperature coefficient (NTC), 84–88

O

Ohm's law, 222
1-octyl-3-methylimidazolium hexafluorophosphate ([OMIm][PF6]), 228
One-dimensional gold nanometer pressure sensor, 35f
One-dimension nanomaterials, conductive material, for buckled tactile sensor, 208–210, 209f
ON/OFF current ratio, 55
Organic field-effect transistors (OFETs), 53, 60f, 62
  parameter of, 55
Organic hybrid sensor, 91
Organic materials, 279–281
  flexible piezoelectric nanogenerators, 247–248
Organic photodetector, 277–278
Organic polymer materials, 102
Organometallic mechanophores, 128–131, 130f

P

Palladium, 78, 79f
Parallel-plate collector, 162
PDMS sponge-based piezoresistive sensor, electrical responses of, 233–234, 233f
PEDOT:PSS, 78, 79f
Percolation theory, stretchable conductive composite -based tactile sensor, 68–69

Perovskites for tactile sensors
  based light-powered tactile sensors, 150–153, 151–152f
  and properties, polarization effects in, 143–150, 145f, 147f, 149f
  self-powered sensors, 142–143
Photo-curable fluorinated elastomeric material (PVDF-HFP), 47
Physical vapor deposition (PVD), 162–164
Piezoelectric acoustic nanosensor (PANS), 247
Piezoelectricity type, 5–6
Piezoelectric nanogenerators (PENGs), 245–246
Piezoelectric polymers, electrospun nanofibers of, 182
Piezoelectric sensors, 31–32
Piezoelectric tactile sensor, 166–167, 176–182
  aligned electrospun PVDF nanofibers, 178–179, 180f
  electrospun composite/doped PVDF nanofibers, 179–182, 181f
  electrospunnanofibers of PVDF and PVDF copolymers, 176–178, 177–178f
  piezoelectric polymers, 182
Piezoelectric transduction, 166–167
Piezoluminescence, 95–96
Piezo-resistive effect, of semiconductors, 14–15, 14f, 16f
Piezo-resistive resistance, 32–34
Piezoresistive sensor, 55–57
  ionic liquids, pressure sensors based on, 222–225, 223–224f
  tactile sensor, 159–160, 168–171, 169–170f, 172–173f
Piezoresistive transduction, 165
Piezo-resistivity
  of conducting materials, 15–17, 17f
  type, 2–5, 3–4f
Planar structures, flexible triboelectric nanogenerators, 252–254
Poisson effect, 222–223, 223f
Poly(3,4-ethylenedioxythiophene) (PEDOT), 82–83
Polyacrylonitrile (PAN), 182
Polyaniline, 264–266, 265f
Polydimethylsiloxane (PDMS), 57, 59f, 72, 264–266

Index

gold microstructures on, 75*f*
pyramid structure of, 17–19
Polyetheretherketone (PEEK), 275–276
Polyetherketone (PEK), 275–276
Polyethylene (PE), 202–203
Polyethylene terephthalate (PET) packaging, 101, 102*f*
Polymer
chain, effect of, 131–132
characteristics of, 68
nanofibers, 168–170, 173
swelling method, 203–204
Polymer-fiber PENG, 247–248
Polymeric polyvinylidine fluoride (PVDF), 246–248
Polymer polyaniline (PANI)
electrodeposition of, 208–210
multiwalled carbon nanotubes, 212
Poly(methyl methacrylate) (PMMA), 78, 275–276
Poly (methacrylamide-co-acrylic acid) (PMAAc), 211–212
Poly (vinylidene fluoride) (PVDF),
copolymers, electrospun nanofibers of, 176–178, 177–178*f*
Polypyrrole (PPy), 35–37, 36*f*, 212
Polystyrene (PS), 202–203
chains, 148
Polystyrene sulfonate (PEDOT), 211–212
Polytetramethylene glycol, 2,4'-tolylene diisocyanate, and isophoronediisocyanate (P-TDI-IP), 273
Polyurethane, 264–266
cutting groove of, 264–266
Polyvinyl chloride (PVC), 202–203
Polyvinylidenefluoridehexafluoropropylene (PVDF-HFP), 213–214
Porous elastic polydimethylsiloxane (PDMS) sponge, 223–225, 224*f*
Porous pyramid dielectric layer (PPDL), 42–44
Positive temperature coefficient (PTC) effect, 84
Potential energy surface (PES), 119–120
Pressure sensitivity, 32
Pressure sensors, 13, 142, 203–204
stretchable conductive composite-based tactile sensor, 82–83, 84–85*f*

Pyramidal polydimethylsiloxane (PDMS), 34, 42
Pyramid-patterned TENG, 251
PZT, 247

Q

Quantum mechanical methods, 123

R

Reduced graphene oxide (rGO), 210, 269–270
Resistive composite strain sensors, conductive filler for, 81
Resistive pressure sensors, 32–40, 33–34*f*, 41*f*
Resistive signal-sensing mechanisms, 32
Resistive tactile sensors
multi-functional tactile sensors, 22–23, 24–25*f*
overview, 13
pressure detection of, 19–22, 22*f*
principle of, 14–19
microstructure-enhancedpiezo-resistivity, 17–19, 18*f*
piezo-resistive effect of semiconductors, 14–15, 14*f*, 16*f*
piezo-resistivity of conducting materials, 15–17, 17*f*
Reversible bond formation, self-healable tactile sensors, mechanisms of, 276
Ring-opening mechanophores, 127–128
Robotic arms, on crystal-confined ionic liquids, 240*f*
Robots, 141

S

Sandwich structured strain sensor, 76–78
SAO luminescence, 97
Saturation regime, 53–54
Self-healable tactile sensors, 263–264
applications of, 277–283
legs with tactile function, 279–282, 282*f*
medicine and health, 277–278, 278*f*
software robots, 282–283, 284*f*
wearableequipments, 279, 280–281*f*
functions, implementation, 283–284
mechanisms of, 273–277, 274*f*
active polymerization, 276–277
interdiffusion of molecules, 275–276
light-induced healing, 276

Self-healable tactile sensors *(Continued)*
  reversible bond formation, 276
  multimodal methods, 283–284
  polymers, intrinsic dynamic properties of, 283–284
  structure and functional materials of, 264–273
    CNT, 266–267, 268*f*
    conductive elastomers, 270–273, 272*f*
    graphene, 267–270, 271*f*
    substrates, 264–266
Self-healing polymer, 229
Self-powered flexible tactile sensors, 245
  flexible piezoelectric nanogenerators, 246–250
    composite materials, 248–250
    inorganic materials, 246–247
    organic materials, 247–248
    piezoelectric mechanism, 246
  flexibletriboelectricnanogenerators
    structural designs, 252–254
    triboelectric materials, 251–252
    triboelectric mechanism, 250–251
  magnetoelectric elastomers, 255–256, 255*f*
Self-powered magnetoelectrical elastomers, 245–246
Self-powered sensors, 142–143
Semiconductors, piezo-resistive effect of, 14–15, 14*f*, 16*f*
Semitransparent ion gel, 225
Sensing cube array, 255–256
Sensitivity *(S)*, 1–3, 5–6, 19
Sensory feedback, 282
Silicon-based membrane (SM), 247
Silver nanoparticles, conductive composite mat of, 80
Silver nanowire (AgNW), 73–76, 76–77*f* network
    nanocomposite of, 81–82
    stretch method, 200–201
Single-electrode mode, 182–183, 184*f*
  flexible triboelectric nanogenerators, 251
Single-molecule approaches, 115–118
Single-molecule force spectroscopy (SMFS), 113, 116–117, 117*f*
Singlewall carbon nanotubes (SWNTs), 38, 39*f*
Single-walled carbon nanotube (SWCNT) sheets, 246–247
Slip sensors, 13

Sodium dodecylbenzenesulfonate (SDBS), 72
Software robots, self-healable tactile sensors, 282–283, 284*f*
Solution-based fabrication method, 73–75
Spin coating method, 75–76
Sponge-like porous PVDF nanogenerator, 248
Strain sensor, stretchable conductive composite (SCC)-based tactile sensor, 80–82, 82*f*
Stress luminescence, 94–96
Stretchable conductive composite (SCC)-based tactile sensor, 68
  applications, 80–88
    pressure sensor, 82–83, 84–85*f*
    strain sensor, 80–82, 82*f*
    temperature sensor, 83–88, 86–87*f*
  preparation methods, 70–80
    elastomer, mixing conductive filler into, 70–72, 72–73*f*
    elastomer surface, conductive film deposited on, 73–78, 75*f*
    synthesizing conductive fillers within elastomer, 78–80, 81*f*
  working mechanism, 68–70
    percolation theory, 68–69
    tunnel current theory, 69–70, 69*f*, 71*f*
Stretching ratio, 34, 200–201
Stretch–release method, buckle structure, 200–201, 201*f*
Styrene-block-butadieneblock- styrene (SBS) fibers, 80
Substrates, self-healable tactile sensors, structure and functional materials of, 264–266
Sugar microparticles, 83
Supramolecular self-healing network, 229, 229*f*
Sweat-based diabetes, Au-doped graphene-based mimic-sensual tactile sensor, 277–278
Swelling methods,buckle structure, 203–204, 204*f*
Switchable photovoltaic effect, 143–144
Synthetic fiber materials, 251–252

T

Tactile field-effect transistors, 55–57, 56*f*, 58*f*
Tactile sensors, 13, 31, 141–142, 197
  advancements in 2010s, 8–9, 9*t*

Index

array, 55–57

buckled tactile sensor, 197–214, 199f
applications in, 214–215
composite, 204–205
hybrid composite materials, 212–214, 213f
mold, 201–202, 202f
one-dimensionnanomaterials, 208–210, 209f
stretch–release, 200–201, 201f
swelling, 203–204, 204f
thermal, 202–203, 203f
two-dimensional materials, 210–212, 211f
zero-dimension nanomaterials, 206–208, 207f
from electrospun nanofibrous materials, 167–183
evolution in 1990s, 7
high capacitive dielectric materials for, 45–47, 46f, 48f
improvement in 2000s, 8
on ionic liquids, 219–222
anions, 219, 221f
capacitive sensor, 225–226, 226–227f
cations, 219
conductivity of, 221–222
imidazolium-based ionic liquids, 219, 221f
investigations, 241
mechanism of, 222f
physical properties of, 220–221t
piezoresistive sensor, 222–225, 223–224f
pressure sensors based on, 222–226
preventing leakage of, 235–240
tactile sensors, signal separation and integration of, 231–235
temperature sensors based on, 226–231, 228–230f
organic field-effect transistors (OFETs), 53, 55
principles, 1–6, 2t
capacitance type, 1–2
piezoelectricity type, 5–6
piezoresistivity type, 2–5, 3–4f
seed in 1970s, 6
seedling in 1980s, 6
transduction mechanisms of, 164–167, 165f
capacitive transduction, 166
piezoelectric transduction, 166–167
piezoresistive transduction, 165
triboelectric transduction, 167
in 2020s, 9–10

Tactile transistor memory, 57–60, 61f

Tandem compound pattern, 76

Temperature compensating (TC), 231–232

Temperature-dependent conductivity, 226, 228–230f

Temperature-independent pressure sensor, 232

Temperature sensing array, 229–231
based on ionic liquids, 229–231, 230f

Temperature sensors, 13
based on ionic liquids, 226–231, 228–230f
eliminating strain interference, 232–233
stretchable conductive composite-based tactile sensor, 83–88, 86–87f

Temperature thermometer, 229–231

Tetrabutylammonium perchlorate (TBAP), 47

Textile structures, flexible triboelectric nanogenerators, 254

Thermal expansion polymer matrix, 127

Thermal methods,buckle structure, 202–203, 203f

3D porous structures, flexible triboelectric nanogenerators, 254

Threshold voltages, 55

Trade-off effect, 214–215

trans-3,4-dimethylcyclobutene (tDCB), 118, 126–127, 132

Transesterification reactions, catalysts for, 131

Triboelectric materials, flexible triboelectricnanogenerators, 251–252

Triboelectricnanogenerators (TENGs), 167, 182, 245–246, 250

Triboelectric tactile sensor, 182–183

Triboelectric transduction, 167

Triboelectrification, 250, 256–257

Tunnel current theory, stretchable conductive composite-based tactile sensor, 69–70, 69f, 71f

Two-dimensional materials, conductive material, for buckled tactile sensor, 210–212, 211f

U

Ultrasound, 116

Ultraviolet (UV) ray, 96
Uniaxial stretching, 200–201
User-interactive electronic skins, 62, 63*f*
UV recoverable ML materials, 96

V

Vertical contact-based TENGs, 250
Vertical contact mode, 183, 185*f*
Vertical contact-separation mode, flexible
triboelectric nanogenerators, 250
Vibration sensors, 13
Visual self-healing, of fractured surfaces,
275–276
Volgel-Tamman-Fulcher (VTF), 228

W

Wearable equipments, self-healable tactile
sensors, 279, 280–281*f*
Wearable flexible hybrid electronics
(WFHE), 8–9
Wearable strain sensor, Gr/Pd/PEDOT:PSS
for, 78, 79*f*

Wheatstone bridge, 231–232, 231*f*
Woodward-Hoffmann (WH) rules, 114

Y

Young's moduli, of magnetic/polymeric
building blocks, 255–256
Young's modulus, 210, 256–257

Z

Zero-dimension nanomaterials, conductive
material, buckled tactile sensor,
206–208, 207*f*
Zigzag structures, flexible triboelectric
nanogenerators, 252–254
Zinc sulfide (ZnS), 101
electroluminescent
phosphor, 92
ZnO nanowire- based piezoelectric
nanogenerators (PENGs), 245–246
ZnS:Mn-based ML material, 99

Printed in the United States
By Bookmasters